Gas Chromatographic Techniques
and Applications

Sheffield Analytical Chemistry

Series Editors: John M. Chalmers and Alan J. Handley

A series which presents the current state of the art in chosen sectors of analytical chemistry. Written at professional and reference level, it is directed at analytical chemists, environmental scientists, food scientists, pharmaceutical scientists, earth scientists, petrochemists and polymer chemists. Each volume in the series provides an accessible source of information on the essential principles, instrumentation, methodology and applications of a particular analytical technique.

Titles in the Series:

Inductively Coupled Plasma Spectrometry and its Applications
Edited by S.J. Hill

Extraction Methods in Organic Analysis
Edited by A.J. Handley

Design and Analysis in Chemical Research
Edited by R.L. Tranter

Spectroscopy in Process Analysis
Edited by J.M. Chalmers

Gas Chromatographic Techniques and Applications
Edited by A.J. Handley and E.R. Adlard

Gas Chromatographic Techniques and Applications

Edited by

ALAN J. HANDLEY
Separations Science and Spectroscopy
LGC, Runcorn, UK

and

EDWARD R. ADLARD
Scientific Editorial Consultant
Wirral, UK

CRC Press

First published 2001
Copyright © 2001 Sheffield Academic Press

Published by
Sheffield Academic Press Ltd
Mansion House, 19 Kingfield Road
Sheffield S11 9AS, England

ISBN 1-84127-118-7

Published in the U.S.A. and Canada (only) by
CRC Press LLC
2000 Corporate Blvd., N.W.
Boca Raton, FL 33431, U.S.A.
Orders from the U.S.A. and Canada (only) to CRC Press LLC

U.S.A. and Canada only:
ISBN 0-8493-0514-4

Printed on acid-free paper in Great Britain by
Antony Rowe Ltd., Chippenham, Wiltshire

British Library Cataloguing-in-Publication Data:
A catalogue record for this book is available from the British Library

Library of Congress Cataloging-in-Publication Data:
Gas chromatographic techniques and applications / edited by Alan J. Handley, Edward R. Adlard.
     p. cm. -- (Sheffield analytical chemistry)
   Includes bibliographical references and index.
   ISBN 0-8493-0514-4 (alk. paper)
   1. Gas chromatography. I. Handley, Alan J. (Alan John) II. Adlard, Edward R. III. Series.

   QD79.C45 G354 2000
   543'.0896--dc21 α C

                                                                    00-029733

Preface

Gas chromatography (GC) has been in existence now for over 50 years. Martin and his co-workers first disclosed their work at a meeting of the Biochemical Society as early as October 1950, but it was the paper by James and Martin at the Society for Analytical Chemistry meeting in Oxford in September 1952 that first drew the attention of analytical chemists to their technique. From that moment there was an explosion of developments which lasted until the mid-1960s. Gradually the pace of research slackened until, by the early 1980s and especially with the introduction of silica capillary columns and the use of microprocessors for control and data handling, GC had become a 'mature technique'—a technique known to all and the practice of which, allegedly, could be left safely in the hands of a technician.

By this time, development work was concentrated on high performance liquid chromatography (HPLC). This itself has now reached the mature technique stage, with research moving on to capillary electrophoresis and capillary electrochromatography. However, no matter how good any technique has become, it would be foolhardy to declare that no further advances were possible.

In the last few years, there have been significant advances in many aspects of GC and this book sets out to give an account of some of them. Injection systems, for many years the weak link in quantitative capillary column work, have now been rationalised and made more reliable. Column coating technology has made considerable progress. The wall-coated open tubular column is now much more robust than it was in its early days and, with bonded stationary phases, there is little danger of washing the phase off the column wall. It is also possible to design column coatings for specific applications and to optimise their performance in these areas. Another important aspect of this work is the improved stability of the silicone compounds, which are now synthesised specifically for use as stationary phases, in contrast to the earlier adaptation and use of materials synthesised for other purposes.

However, no matter how good a single column is, there will always be many complex mixtures, such as those encountered in the petroleum industry and essential oil analysis, that are incapable of complete separation. Column switching, carried out by early pioneers such as Boer and Deans with packed columns, has now reached new heights, as described in a chapter on this topic. Not only is resolution vastly improved by column switching, but significant gains can also be achieved in lower limits of detection by variations of peak focussing, as is used in column switching.

Chromatography in all its manifestations is, first and foremost, a separation method and, although peaks may be assigned a tentative identification by means of retention data, it is often essential to use other, independent methods of identification. GC-MS, which once only existed in the most well equipped research laboratories, is now commonplace. Furthermore, the performance of the various types of mass spectrometer available for this type of work has advanced almost beyond recognition. Electron impact ionisation, once the only method of ionising a sample in the mass spectrometer source, is now just one of half a dozen ionisation methods, albeit still the most important. There are also different types of mass spectrometer available, so that the analyst not only has a choice of the modern versions of magnetic sector, focussing instruments for high resolution MS but also ion trap and quadrupole designs. MS-MS is now readily available and may offer an alternative to high resolution MS.

In the early days of GC-MS, the time-of-flight mass spectrometer found some popularity because of its simplicity but fell out of favour because of the poor mass discrimination of these early instruments. Modern improvements in the design of time-of-flight mass spectrometers has caused a renewal of interest in their application in GC-MS, and one area in which TOF instruments are showing promise is high speed GC. The theoretical aspects of high speed GC have been known for a number of years but it is only relatively recently that manufacturers have taken tentative steps towards providing suitable equipment. It remains to be seen whether there exists a general need for high speed GC or whether it will end up as something of a niche technique. Alternative methods, such as far infra-red spectroscopy, now exist for plant analysis and are described, together with the many developments that are taking place in the application of GC and IR to at-plant and on-line analysis. In addition to the mass spectrometer as a detector for GC, there is still a need for selective detectors for specific applications and some of these have been covered in this book.

Writers on HPLC are prone to point out that only about 20% of all known organic compounds are amenable to GC, but this can be increased considerably by means of derivatization, which is dealt with in a chapter on this topic. The trials of trying to establish new approaches to analytical problems in the face of conservative legislators are highlighted in this chapter.

The subject of legislation leads on to the topics of quality control, method validation and method standardisation. The petroleum industry has an excellent record of establishing standard, validated, analytical methods over a period of more than 50 years, but it would seem that some other parts of the chemical industry have realised the needs and benefits of this type of work only relatively recently. This has stimulated a number of national and international bodies to initiate method validation over a much wider field of application.

What of the future? It seems clear that there is still plenty of room for developments in specific fields. Particular topics acquire popularity for a time and, to be in the forefront of research and development, it is almost obligatory to become involved in whatever topic is currently fashionable. Some of these fashionable topics disappear altogether in a few years. Others settle down and become established techniques with advantages and limitations—that is the path that has been followed by supercritical fluid chromatography.

The most fashionable topic at the time of writing is undoubtedly miniaturisation and it is clear that there are many advantages to making GC equipment very much smaller that the present bench-top instruments. It is patently absurd to use about 2 kilowatts of electricity to heat a filament, to heat air, to heat a few grams of silica tubing, and finally to heat a few milligrams of stationary liquid on the walls of the tubing. However, it is easier to pose the problem than to suggest a solution. A number of manufacturers have tried without success to coat fused silica capillaries externally with various conducting layers such as aluminium, in order to heat the column by passing a current directly through the conducting layer. Such an approach has not so far proved successful but it would appear that it is only a matter of time before someone solves the problem. Until then, the miniaturisation of temperature programmed GC will be held in abeyance. A number of manufacturers have fallen into the trap of miniaturising the column and detector, but have nullified their efforts by failing to miniaturise the whole system from carrier gas supply to data capture. As far as carrier gas is concerned, the best solution at present appears to be the use of hydrogen generated by the electrolysis of water. A simple calculation shows that the electrolysis of 10 ml of water would allow about 24 hours of operation at a column flow rate of 1 ml per minute—and considerably longer for narrow bore columns requiring a lower volumetric flow rate. Such an approach is attractive, even for laboratory installations, since it provides a high purity carrier gas, suitable for high speed separations and with greater safety and considerably less space requirements than bottled gas supplies. There should be little difficulty in the miniaturisation of the electronics required for instrument control, signal amplification and data capture. The computer industry has ensured the availability of the required equipment, and advances in this field are still taking place, as PCs become ever more powerful and smaller.

It is said that there is a Chinese curse to the effect that 'May you live in interesting times'. For better or worse, we live in such times.

A.J. Handley
E.R. Adlard

Contributors

J. Beens Laboratory of Instrumental Analysis, Department of Chemical Engineering, Eindhoven University of Technology, PO Box 513 (STO 3.33), 5600 MB Eindhoven, The Netherlands

Professor Carel A. Cramers Laboratory of Instrumental Analysis, Department of Chemical Engineering, Eindhoven University of Technology, PO Box 513 (STO 3.33), 5600 MB Eindhoven, The Netherlands

Dr Frank David Research Institute for Chromatography, Kennedypark 20, B-8500 Kortrijk, Belgium

J.G.M. Janssen Unilever Research Laboratorium, Department of Compositional Analysis, PO Box 114, 3130 AC Vlaardingen, The Netherlands

Russell M. Kinghorn Baseline Separation Technologies P/L, PO Box 4198, Burwood East, 3151 Victoria, Australia

Dr Paul Larson Agilent Technologies, Wilmington, Delaware, USA / Hewlett Packard, Little Falls Analytical Division, 2850 Centreville Road, Wilmington, DE 19808-1610, USA

Tom Lynch BP Amoco Chemicals, Analytical Service Laboratory, Saltend, Hull HU12 8DS, UK

Professor Philip J. Marriott Department of Applied Chemistry, Royal Melbourne Institute of Technology, GPO Box 2476V, Melbourne, 3001 Victoria, Australia

Dr R.D. McDowall McDowall Consulting, 73 Murray Avenue, Bromley, Kent BR1 3DJ, UK

Dr Mark Powell School of Pharmacy and Chemistry, Liverpool John Moores University, Byrom Street, Liverpool L3 3AF, UK

Dean Rood

Production Manager, Custom GC Columns, Agilent Technologies / J&W Scientific, 91 Blue Ravine Road, Folsom, CA 95630-4714, USA

Professor Pat Sandra

Department of Organic Chemistry, University of Ghent, Krijgslaan 281-S4, B-9000 Ghent, Belgium

Tony Taylor

Crawford Scientific, Holm St, Strathaven, Lanarkshire, ML10 6NB, UK

M. M. van Deursen

Laboratory of Instrumental Analysis, Department of Chemical Engineering, Eindhoven University of Technology, PO Box 513 (STO 3.33), 5600 MB Eindhoven, The Netherlands

Allen K. Vickers

Applications Engineer Manager, Agilent Technologies / J&W Scientific, 91 Blue Ravine Road, Folsom, CA 95630-4714, USA

Contents

1 Developments in sample preparation for capillary GC analysis   **1**
PAT SANDRA and FRANK DAVID

1.1   Introduction   1
1.2   Enrichment methods   2
    1.2.1   Introduction   2
    1.2.2   Gaseous (air) samples   3
    1.2.3   Liquid (aqueous) samples   8
    1.2.4   Solid samples   17
1.3   Fractionation and clean-up   26
    1.3.1   Introduction   26
    1.3.2   Chemical methods   26
    1.3.3   Chromatographic methods   26
1.4   Evaporation   30
1.5   Derivatisation   31
    1.5.1   Introduction   31
    1.5.2   Analysis of phenols in water samples   31
    1.5.3   Analysis of racemic amino acids in biological fluids   33
1.6   Some case studies   35
    1.6.1   Fractionation of nitro-PAHs from air particulates   35
    1.6.2   Versatile robotic system for static (SPME) and dynamic (GPE) sampling   37
    1.6.3   The 1999 Belgian dioxin crisis   41
References   50

2 Sample injection systems   **52**
TONY TAYLOR

2.1   Introduction   52
    2.2.1   Important concepts   52
2.2   Split/splitless injectors   53
    2.2.1   Split/splitless inlet design   53
    2.2.2   Septa   54
    2.2.3   Split injection mode   57
    2.2.4   Splitless injection mode   64
2.3   Packed-column inlets   70
    2.3.1   Inlet design   70
    2.3.2   Inlet parameters and sample considerations   71
2.4   Capillary direct interfaces   72
    2.4.1   Inlet design   73
2.5   Cool on-column inlets   74
    2.5.1   Inlet design   74
    2.5.2   Inlet parameters and sample considerations   77
    2.5.3   Retention gap   77

2.5.4   Temperature   78
2.5.5   Flow rates   78
2.6   Programmed temperature vaporising (PTV) inlets   78
2.6.1   Inlet design   79
2.6.2   Inlet parameters and sample considerations   80
2.6.3   Temperature   82
2.6.4   Carrier flow   82
2.7   Auxiliary sampling devices   83
2.7.1   Gas and liquid sampling valves   83
2.7.2   Headspace autosamplers   85
2.8   Conclusions   90
References   90

3   Advances in column technology   91
ALLEN K. VICKERS and DEAN ROOD

3.1   Introduction   91
3.2   Modern capillary columns   92
3.2.1   Capillary column materials   92
3.2.2   Column dimensions   93
3.2.3   Stationary phases   93
3.3   Examples of separations of different types of samples on a variety of columns   95
3.3.1   Petroleum analysis   95
3.3.2   Environmental analysis   97
3.3.3   Life sciences   113
3.3.4   Forensic analysis   121
3.4   Conclusion   121
References   121

4   Detectors for quantitative gas chromatography   122
PAUL LARSON

4.1   Introduction   122
4.2   Pneumatic control   122
4.2.1   Flame ionization detector (FID)   125
4.2.2   Nitrogen phosphorus detector (NPD)   131
4.2.3   Flame photometric detector (FPD)   132
4.2.4   Other sulfur selective detectors   134
4.3   Gas purifiers   135
4.4   Miniaturization   136
4.5   Data path considerations   137
4.6   Summary   138
References   139

5   Detectors for compound identification   140
MARK POWELL

5.1   Introduction   140
5.2   Gas chromatography–mass spectrometry   140
5.2.1   Interface types   141
5.2.2   Ionisation techniques   145

	5.2.3	Analyser types	154
	5.2.4	Selected ion monitoring	167
	5.2.5	GC-MS-MS	168
	5.2.6	Isotopically labelled surrogate standards	170
	5.2.7	Applications	171
5.3	Gas chromatography–atomic emission detection		175
	5.3.1	Theory and instrumentation	175
	5.3.2	Applications	179
5.4	Gas chromatography–Fourier transform infrared spectrometry		181
	5.4.1	Theory and instrumentation	181
	5.4.2	Applications	184
5.5	Applications of combined techniques		186
	5.5.1	GC-MS and GC-AED	186
	5.5.2	GC-MS and GC-FTIR	187
5.6	Future directions and developments		188
	References		189

6   Method validation in gas chromatography   192
R. D. MCDOWALL

6.1	Introduction		192
6.2	Definition of key terms		192
6.3	Equipment qualification and method validation		193
6.4	Equipment qualification (EQ)		193
	6.4.1	The overall EQ process	193
	6.4.2	Equipment qualification to ensure efficient method transfer	194
	6.4.3	Design qualification (DQ)	194
	6.4.4	Installation qualification (IQ)	196
	6.4.5	Operational qualification (OQ)	196
	6.4.6	Modular and holistic qualification	196
	6.4.7	Performance qualification (PQ)	197
6.5	Method validation		197
	6.5.1	Documenting the method	197
	6.5.2	Defining the validation parameters to measure	197
	6.5.3	Analytical reference substances	198
	6.5.4	Scope of validation	198
	6.5.5	Specificity *vs* selectivity	198
	6.5.6	Accuracy	200
	6.5.7	Limit of detection (LOD)	201
	6.5.8	Limit of quantification (LOQ)	201
	6.5.9	Linearity	202
	6.5.10	Range	202
	6.5.11	Robustness	203
	6.5.12	Analyte stability	203
	6.5.13	Revalidation of a method	203
	6.5.14	Internal or external standard method?	204
	6.5.15	Calibration methods	205
	References		205

7   **Faster gas chromatography**   **207**
     CAREL A. CRAMERS, J. G. M. JANSSEN,
     M. M. VAN DEURSEN and J. BEENS

     7.1   Introduction and scope   207
     7.2   Options for increased speed of separation   207
           7.2.1   Resolution normalised conditions   208
           7.2.2   Guidelines for retention time reduction   212
     7.3   Practical consequences of fast GC   227
           7.3.1   Pressure drop   228
           7.3.2   Column loadability   228
           7.3.3   Detection limits   228
           7.3.4   Injection band width   229
           7.3.5   Extracolumn band broadening   229
           7.3.6   Time constant for detection   229
           7.3.7   Elution temperature/maximum column temperature   230
           7.3.8   Coupling to mass spectrometry   230
     7.4   Practical implementations of selected options for fast GC   231
           7.4.1   Narrow-bore columns   231
           7.4.2   Multicapillary columns   237
           7.4.3   Fast temperature programming   240
           7.4.4   Vacuum outlet operation   242
           7.4.5   Comprehensive two-dimensional gas chromatography   249
     References   256

8   **Multidimensional and comprehensive multidimensional**
     gas chromatography   **260**
     PHILIP J. MARRIOTT and RUSSELL M. KINGHORN

     8.1   Introduction   260
     8.2   Gas chromatography as one dimension in multidimensional analytical methods 261
     8.3   Multidimensional separation methods   261
           8.3.1   Scope of conventional capillary gas chromatography: rationale for higher-
                   resolution methods   262
           8.3.2   Approaches to coupled column methods for improved separation
                   in gas chromatography   264
           8.3.3   Conventional multidimensional gas chromatography   268
     8.4   Comprehensive two-dimensional gas chromatography   278
           8.4.1   General benefits of GC × GC   281
           8.4.2   Understanding the comprehensive gas chromatography experiment   281
           8.4.3   Methods for modulation of peak signals between two dimensions   283
           8.4.4   Orthogonality of separation   285
           8.4.5   Presentation of GC × GC data   287
           8.4.6   Applications of GC × GC   288
           8.4.7   Standards and mass spectrometry studies with GC × GC   293
     Acknowledgements   294
     References   295

9   On-line and at-line gas chromatography   298
    TOM LYNCH

    9.1   Introduction   298
    9.2   Why do we need process analysis?   299
          9.2.1   On-line, at-line or central laboratory?   299
    9.3   Trends and recent developments in process gas chromatography   301
    9.4   Recent developments in conventional on-line process gas chromatography
          instrumentation   303
          9.4.1   The application of live switching techniques   304
          9.4.2   The development of plug-and-play gas chromatography hardware 307
          9.4.3   Parallel chromatography   309
    9.5   The development of gas chromatograph-based transmitter instrumentation 316
    9.6   The application of laboratory gas chromatograph instruments as on-line process
          analysis instruments   318
    9.7   The micro gas chromatograph   322
    9.8   Recent developments in at-line gas chromatography   324
          9.8.1   Why do we need at-line gas chromatography?   324
          9.8.2   What are at-line gas chromatographs?   325
          9.8.3   At-line gas chromatography development in BP Amoco   325
          9.8.4   The benefits of at-plant chromatography   326
          9.8.5   The current status of at-line gas chromatography   327
    9.9   Future developments in process gas chromatography   329
          9.9.1   What will drive new developments?   329
          9.9.2   The future for on-line gas chromatography   330
          9.9.3   The future for at-line gas chromatography   332
    Acknowledgements   333
    References   333

Index   334

1 Developments in sample preparation for capillary GC analysis

Pat Sandra and Frank David

1.1 Introduction

Samples are often too complex or too dilute for direct analysis by capillary GC. Preliminary extraction, fractionation/clean-up, concentration and/or derivatisation are needed. A typical flow diagram for a capillary GC analysis is shown in Figure 1.1.

Sampling and sample preparation remain among the more time-consuming, error-prone and contamination-prone aspects of the system. Obtaining a representative sample and proper storage of the sample are important parts of any analysis. Both are often overlooked by analytical chemists, who regard them as self-evident secondary problems, with the chromatographic analysis being of primary interest. However, errors or faults in the sampling protocol cannot be corrected at any point in the analysis. For the analytical data to be meaningful, a plan for acquiring and storing samples should be implemented and, if possible, validated by statistical techniques.

A discussion of sampling is beyond the scope of this chapter and we refer readers to the literature, such as the book by Baiulescu and co-workers [1]. On the other hand, problems of storage occur mostly in the analytical laboratory and

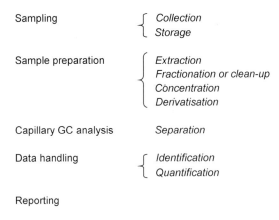

Figure 1.1 Flow diagram for a capillary GC analysis.

are the responsibility of the analyst. Processes that occur in the sample between the time of sample collection and the time of analysis, such as adsorption on the container walls, vaporisation loss, photoreactions, microbial action, etc., can invalidate the data. Samples should be stored in the dark, in brown glass vials or containers, and maintained at 4°C or lower. Especially for aqueous samples, adsorption on the glass wall should be suppressed by adding a polar modifier such as methanol.

In recent years, great effort has been made to develop sample preparation techniques that guarantee high recovery and reproducibility and that moreover are faster, cheaper and easier to automate than older methods.

In this chapter the currently used sample preparation methods for capillary GC analysis are reviewed and new methods are discussed in depth. The subdivision of the chapter is, in the first instance, based on the matrix (gas, liquid, solid) from which the solutes of interest have to be enriched, while clean-up and fractionation, and concentration and derivatisation are treated separately.

The term 'volatiles' used in this chapter cannot be rigorously defined. Volatile compounds may be said to be those boiling up to about 200°C. However, in headspace sampling of liquids it is necessary to consider not only the boiling point of the analyte but its activity coefficient in the liquid matrix. In aqueous solution, organic compounds commonly exhibit very non-ideal behaviour. For example, phenol (b.p. 181.75°C), although readily soluble, has an activity coefficient in water at 25°C of about 60. This means that the vapour pressure of phenol above an aqueous solution is 60 times greater at a given mole fraction in the liquid phase than it would be above an ideal solution and ppm concentrations of phenol in water can be measured by headspace GC. Because of this non-ideality, it is often possible to sample compounds that would not normally be considered to be volatile. The non-ideality of aqueous solutions of organic compounds (and hence the lower limit of detection) can be further enhanced by 'salting out' through the addition of a salt such as potassium carbonate [2–4].

It has to be noted that in recent years the sample preparation strategy has changed drastically. The main reasons are that capillary GC has developed further, for example large volume injection avoiding concentration steps, and that detectors such as mass spectrometers (MS), atomic emission detectors (AED), inductively coupled plasma mass spectrometers, and so on, have partly taken over the selectivity obtained by classical sample preparation methods.

1.2   Enrichment methods

1.2.1   Introduction

Enrichment is normally performed by extraction with a gas, a liquid or a solid in a static or a dynamic mode.

1.2.2 Gaseous (air) samples

1.2.2.1 Introduction

Figure 1.2 presents an overview of the methods currently used for the analysis of gaseous samples. Older methods like universal or selective collection in a liquid or cryogenic concentration are not discussed and the reader is referred to reference 3. The collection of particulates on glass fibre filters is specific for air monitoring. This will be illustrated in section 1.6.1.

1.2.2.2 Adsorptive extraction (AE)

Most of the procedures currently used for the preconcentration of volatile organic compounds (VOCs) in gaseous samples and air are based on adsorption of the analytes of interest on a suitable adsorbent [6] followed by either liquid or thermal desorption. Although CS_2 or CH_2Cl_2 desorption is common practice, thermal desorption is preferred because it guarantees higher detectability. In the thermal desorption method, the desorbed analytes are refocused in a cold trap prior to transfer onto the analytical column. Figure 1.3 compares the 'air' profiles taken in a solvent store room and obtained by enrichment on a charcoal trap followed by CS_2 desorption (Figure 1.3a) and on Tenax followed by thermal desorption (Figure 1.3b). Because of the dilution factor in the first analysis no pollutants are detected. Concentrating CS_2 cannot be applied because the highly volatiles will be partly lost.

Common adsorbents include carbon-based materials such as activated carbon and carbon molecular sieves [7,8] and porous organic polymers such as Tenax and Chromosorb [9]. These are all relatively strong adsorbents giving

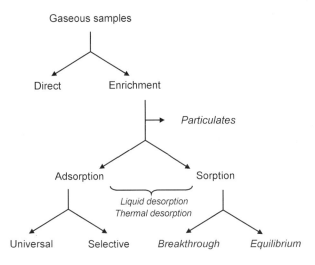

Figure 1.2 Enrichment methods for gaseous samples.

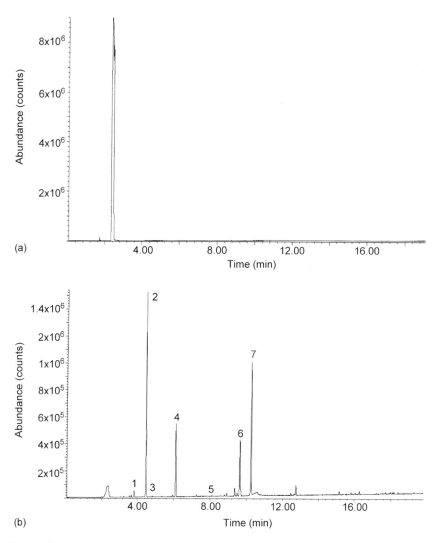

Figure 1.3 Air sampled in a solvent store room, (a) Liquid injection after elution from a charcoal trap with CS$_2$. (b) Sampling on Tenax and thermal desorption. Peaks: 1, carbon tetrachloride; 2, trichloroethene; 3, toluene; 4, 1,1,2-trichloroethane; 5, 3-heptanone; 6, n-nonane; 7, 1,1,2,2-tetrachloroethane.

excellent performance for nonpolar and slightly polar mixtures such as BTX (benzene, toluene, xylenes), polyaromatic hydrocarbons and polychlorinated biphenyls. Unfortunately, their application to the analysis of polar solutes is rather limited. Lack of retention during sampling is generally not the problem, because polar analytes are strongly retained on most adsorbents. This strong retention, however, often precludes rapid and complete desorption, resulting in low recoveries and a severe risk of carry-over. Moreover, the long residence times of the analytes on the hot and active adsorbent surface during desorption might result in reactions with the surface itself or with other adsorbed species. These reactions can result in permanent adsorption and/or in artefact formation, which are clearly undesirable effects [10]. Selective adsorptive extraction or adsorption with reaction can be applied for the enrichment of target compounds. Examples include the determination of aldehydes and ketones in air on a 2,4-dinitrophenylhydrazine-impregnated silica gel adsorbent and of ethylene oxide on a HBr-impregnated adsorbent forming the hydrazones and 2-bromoethanol, respectively. Another complicating factor when working with adsorbents is that the organics that have to be determined can occasionally be formed through degradation reactions from the adsorbent itself. This is the case, for example, with Tenax, where acetophenone and benzaldehyde are easily formed and with Chromosorb, leading to styrene and α-methylstyrene. Alternatives to the classic adsorbents are being investigated for adequate handling of samples containing polar solutes.

1.2.2.3 Static sorptive extraction (SPME and HSSE)

Static sorptive extraction is widely known under the name solid phase microextraction or SPME [11,12]. SPME is applied in headspace analysis. On polymers like poly(dimethylsiloxane) (PDMS) and poly(acrylate) (PA), a partitioning mechanism applies; on divinylbenzenestyrene copolymers or mixtures of Carboxen and PDMS, adsorption controls the enrichment. Passive sampling by means of SPME has been described [13]. A new approach for sorptive enrichment of analytes in air and from the headspace of aqueous or solid samples, referred to as headspace sorptive extraction (HSSE), has recently been developed [14]. The technique involves the sorption of volatile and semivolatile compounds into a large amount of poly(dimethylsiloxane) ($\sim 50\,mg$) placed on a glass rod support (see section 2.4.3). The PDMS-coated glass bar is then thermally desorbed on-line and analysed by capillary GC-MS. With the use of a large amount of sorptive phase, highly volatile as well as semivolatile compounds can be efficiently enriched and compared to solid phase microextraction (SPME) on a $100\,\mu m$ PDMS fibre; a significant increase in sensitivity is achieved. Figure 1.4 shows two chromatograms obtained during passive sampling in a living room equipped with a vinyl carpet. The first rod was placed at ground level and the second at a height of $1.75\,m$. Sampling time was 24 h. The PDMS rods were thermally desorbed on-line with capillary GC-MS on a $30\,m \times 0.25\,mm$ i.d.$\times 0.25\,\mu m$ HP-5MS column. The temperature was programmed from $50°C$ to $250°C$ at $10°C\,min^{-1}$. Phthalates are monitored and the concentrations of the more volatile diisobutyl and di-n-butyl phthalates are higher at $1.75\,m$ compared to ground level, whereas the reverse is noted for the less volatile diethylhexyl phthalate.

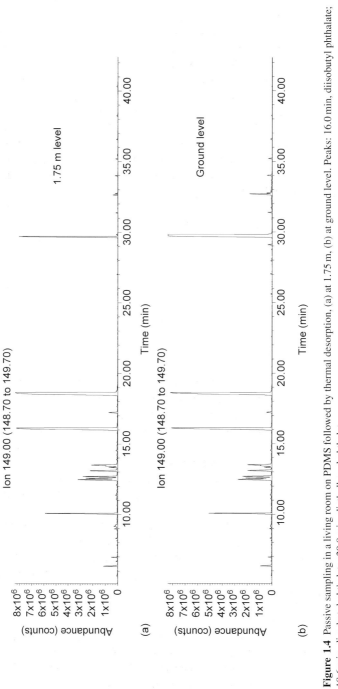

Figure 1.4 Passive sampling in a living room on PDMS followed by thermal desorption, (a) at 1.75 m, (b) at ground level. Peaks: 16.0 min, diisobutyl phthalate; 18.6 min, di-n-butyl phthalate; 29.8 min, diethylhexyl phthalate.

1.2.2.4 Dynamic sorptive extraction (GPE)

Enrichment of organic components from gaseous matrices by dynamic sorptive extraction, also termed gum phase extraction (GPE), opens new perspectives for air monitoring [15,16]. Preconcentration occurs by sorption of the analytes into the bulk of a liquid phase instead of adsorption onto an active adsorbent surface. The most commonly used 'liquid' phase for this purpose is 100% poly(dimethylsiloxane) (PDMS). Preconcentration by sorption has some clear advantages over adsorption onto an active surface. In the sorption mode, polar solutes desorb rapidly at low temperatures owing to the weak interaction of the analytes with the PDMS material. Moreover, PDMS is much more inert than a standard adsorbent, minimising the losses of unstable and/or polar analytes. Another advantage of the PDMS phase is that its degradation products can easily be identified by a mass spectrometric detector as characteristic silicone mass fragments. Peaks originating from the sorbent therefore cannot be mistaken as being from a sampled analyte. For practical purposes, the advantages of PDMS are that, since the analytes are retained in the bulk of this material, retention of the solutes on this phase is more reproducible than in the case of adsorbents. For example, a high water content of the gas sample does not affect retention of the analytes. In addition, poor batch-to-batch reproducibility, such as is sometimes encountered when working with adsorbents, is absent in the case of PDMS.

Traps packed with 100% PDMS particles have been developed to concentrate components from air [15–17]. These packed traps allow sampling flow rates as high as $2.5\,l\,min^{-1}$. Normally gas sampling is stopped before breakthrough occurs, but in equilibrium sorptive extraction it is continued until all compounds of interest are in equilibrium with the sorptive material [18]. Because of the nature of the sorption mechanism, which is basically dissolution, all compounds partition independently into the sorbent (stationary phase) and displacement effects do not occur. This is a great advantage over adsorption materials. Additionally, theory allows the calculation of enrichment factors from literature retention index data. The performance of equilibrium sorptive extraction is illustrated with the analysis of epichlorohydrin in air.

Epichlorohydrin is not easily enriched on classical adsorption materials owing to the easy destruction of the epoxide chain. On a trap containing 0.45 ml PDMS particles, epichlorohydrin has an equilibrium volume of 280 ml. Therefore, 1000 ml air was sampled at $100\,ml\,min^{-1}$ to guarantee equilibrium. Analysis was accomplished on a thermal desoption capillary GC-MS system equipped with a 30 m × 0.32 mm i.d. × 4 μm CP-SIL5CB column (Chrompack, Middelburg, The Netherlands) using the MS in the selected ion monitoring mode (SIM) and positive chemical ionisation (PCI) with methane as reagent gas. The latter was done for added selectivity because the PCI mode allows quantification on the molecular ion (+H), which is much more specific than the lower-mass fragment ions (m/z 63 and 65) generated under electron ionisation (EI)

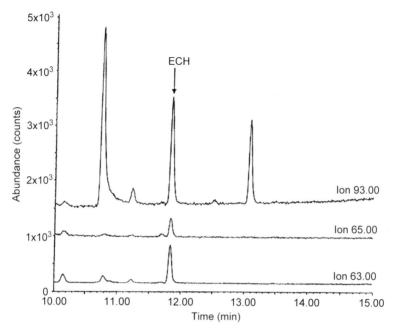

Figure 1.5 Equilibrium sorptive extraction of epichlorohydrin at $0.2\,\mu g\,m^{-3}$ in air.

conditions. Figure 1.5 shows the chromatogram obtained from the enrichment of an air sample spiked with 0.2 ppb epichlorohydrin. Detection limits are around $10\,ng\,m^{-3}$ (ppt).

1.2.3 Liquid (aqueous) samples

1.2.3.1 Introduction
Methods for the analysis of liquid samples are presented in Figure 1.6.

In purity determination of chemicals or pharmaceuticals, simple dilution in a volatile solvent is often applied. Other samples require derivatisation, which will be discussed in section 5. Older techniques include freeze concentration and lyophilisation. In freeze concentration, water is partially frozen and the dissolved substances are concentrated in the unfrozen portion, while in lyophilisation water is completely frozen and removed by sublimation under vacuum. Both methods suffer from loss of volatile organics and concentration of inorganic constituents. The methods applied for aqueous samples—which can be drinking water, waste water, biological fluids, beverages, etc.—can be divided according to the enrichment of the volatiles or of the semivolatiles. Well-known techniques are static and dynamic headspace, which can also be applied for solid

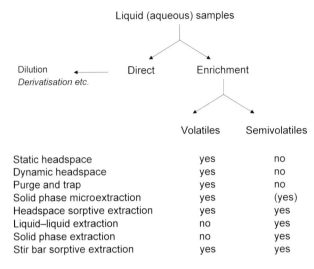

Figure 1.6 Enrichment methods for liquid (aqueous) samples.

samples. The difference between static and dynamic techniques lies mainly in the sensitivity that can be reached. In the static mode, a defined volume of the headspace of a sample at equilibrium is injected; in the dynamic mode, the equilibrium concentration is continuously removed and trapped, which results in higher sensitivity.

1.2.3.2 Gas phase extraction (P and T)

Gas phase extraction from water samples, known as purge and trap (P and T), is in set-up very similar to dynamic headspace sampling. Purge and trap is used for the enrichment of nonpolar volatile organic compounds prior to GC analysis. An inert gas is bubbled through the water sample and the purgeable organics are moved from the aqueous phase to the vapour phase. The volatile compounds are then trapped (solid phase extraction) on an adsorbent such as Tenax or active charcoal. The trap containing the adsorbent is built into a desorption chamber equipped with a heating mechanism that, when activated, permits the desorption (gas phase extraction) of the trapped compounds. This technique has the distinct merit of providing a clean sample, free from its often dirty matrix. A purge and trap device can easily be mounted on a GC equipped with an ECD and PID in series, or with a mass spectrometer. This technique is most appropriate for ppb level analysis of low molecular weight, slightly water-soluble volatile organics with a boiling point below 200°C. A variation of purge and trap is closed-loop stripping analysis [19], which is a combination of gas phase extraction with solid phase extraction in a closed system and allows ppt analysis of pollutants.

1.2.3.3 Micro liquid–liquid extraction (microLL)

A commonly used enrichment method for aqueous samples is liquid–liquid extraction (LLE). LLE may be carried out manually by shaking the water sample with an organic solvent in a separation funnel, or automatically using a continuous liquid–liquid extractor. Continuous LLE is recommended by the US Environmental Protection Agency (EPA) for the enrichment of semivolatiles (base-neutral/acid extractables). Depending on the extraction conditions used, extracts can contain intermediate to low polarity, weakly volatile pollutants (universal extraction for neutral semivolatiles) or acid and base compounds (selective extraction) though adjustment of the pH. Liquid–liquid extraction is very time-consuming and uses toxic solvents. The volume of the extract is usually too large for direct injection and to obtain sufficient sensitivity an additional evaporation–concentration step is necessary (section 4). Particular care needs to be taken in both the solvent extraction and concentration procedures to avoid contamination of the sample. Moreover, solvent impurities will be concentrated as well, often masking the target solutes.

For a number of applications automated micro liquid–liquid extraction in the vial of an injection autosampler can offer a simple and robust alternative. With the use of highly sensitive and selective detectors, eventually in combination with large-volume injection, sub-ppb sensitivities can be reached. Micro liquid–liquid extraction can be used for the analysis of trihalomethanes in drinking water [20].

1.2.3.4 Solid phase extraction (SPE)

This extraction procedure is being used more and more because it is much faster, cheaper, and more versatile than most classical techniques. Moreover, SPE procedures are easily automated using robotics, or can be coupled on-line with capillary GC. The principle of retention (enrichment) is analogous to that of liquid chromatography and is suitable for solutes of low, intermediate and high polarity. SPE media are based on normal-phase, reversed-phase or ion exchange chromatography. The later technique is not often applied for capillary GC analysis because solutes applicable to an ion exchange mechanism are generally very polar or ionic in nature. Some details of SPE sorbents and their applicability are summarised in Table 1.1.

Present designs of SPE devices are the syringe-type cartridge, the Empore discs and the Empore disc cartridges. The syringe cartridge is the most common arrangement and cartridges ranging in mass of sorbent from 50 mg to 10 g are commercially available. Liquid flow-through is typically done using a commercially available vacuum manifold. Permeability is no problem because sorbents are based on irregular silica particles with particle diameter between 30 and 60 μm. In the Empore disc format, particles of 5–10 μm are intertwined with fine threads of Teflon, giving a disc with a thickness of ∼5 mm and a diameter in the range 47–70 mm. Manifolds are also commercially available

Table 1.1 Characteristics of the most important SPE sorbents and their applicability

Mechanism	Sorbent type	Matrix	Eluents	Solutes
Reversed-phase	Octadecyl (ODS) Octyl (OS)	Water Beverages Biological fluids	Methanol Acetonitrile Dichloromethane Hexane	Hydrophobic solutes: pesticides, PAHs, hydrocarbons, PCBs, dioxins and furans, drugs, steroids, sterols, phthalates, preservatives, fat-soluble vitamins, etc.
Normal-phase	Cyanopropyl (CP) Aminopropyl (AP) Diol (DIOL) Silica	Solutions in nonpolar solvents	Hexane/isopropanol Acetone Ethylacetate Methanol	Hydrophylic solutes: amines, phenols, alcohols, acids, etc.
Ion exchange	Benzenesulfonic acid (SCX)	Water Beverages Biological fluids	Alkaline buffers	Ions or ionisable solutes:[a] amines, pyridines, amino acids, etc.
	Quaternary amines (SAX)		Acidic buffers	Phenols, carboxylic acids, amino acids, etc.

[a]Often followed by derivatisation prior to GC analysis.

for multiple sample extraction using Empore discs. The SPE disc allows rapid flow rates of sample and of solvents. One litre of water can be passed through the Empore disc in ~ 10 min, whereas with the syringe cartridge the same volume of water can take more than 1 h. Both approaches need a relatively large volume of desorbing liquid and further concentration is required. In the disc cartridge format, a disc of 10 mm diameter is placed in a cartridge between two polyethylene frits. Only small volumes of solvent are needed for solid phase stripping, eliminating the need for an additional evaporation step, thereby considerably reducing the risk of losses and contamination. The principle of SPE is illustrated with the determination of triazines in water samples applying large volume injection capillary GC-MS.

For the enrichment of triazines from water, a 10 mm/6 ml ODS Empore disc cartridge was chosen. This means that enrichment is based on the reversed-phase mechanism. The operation can be divided into six steps: wetting and stripping the sorbent, conditioning the sorbent, loading the sample, rinsing the sorbent to elute extraneous material, drying of the sorbent and finally elution of the analytes of interest. The disc was first rinsed with 0.5 ml of methanol (wetting and stripping), followed by 1 ml distilled water (conditioning); 25 ml water spiked at the 1 ppb level with 11 triazines was passed through the cartridge at a flow of 50 ml min^{-1} (loading); 5 ml distilled water was then passed through the cartridge to elute polar extraneous solutes (rinsing). After drying under vacuum for 10 min, desorption was performed directly in a 2 ml autosampler vial with 0.5 ml ethyl acetate. Extract (40 µl) was directly injected without further concentration, into the capillary GC system equipped with a programmed temperature vaporising (PTV) injector operated in the solvent vent mode. The optimized injection speed was 100 µl min^{-1}. Analysis was performed on a 30 m $\times$ 0.25 mm i.d. $\times$ 0.25 µm HP-5MS column (Agilent Technologies), with helium as carrier gas. The oven was programmed from 50°C (1 min) to 300°C at a rate of 10°C min^{-1}. The mass spectrometric detector was operated in the scan mode (m/z 20–400). The chromatogram is presented in Figure 1.7.

The detection limits set by the European Community (0.1 ppb) can easily be reached with ion extraction of the mass spectrum, while selected ion monitoring allows determinations down to the ppt or ng l^{-1} level.

1.2.3.5 Sorptive extraction (SPME, GPE and SBSE)

The recently introduced solventless sample preparation techniques of solid phase microextraction (SPME), sorptive or gum phase extraction (GPE) and stir bar sorptive extraction (SBSE) are very powerful methods for the determination of organic solutes in aqueous samples. SPME and SBSE are static in nature while SE is a dynamic procedure. The performance of SPME will be illustrated in section 6.2.

Dynamic sorptive extraction, also called gum phase extraction (GPE), for aqueous samples is a technique that resembles SPE to some extent [21]; the

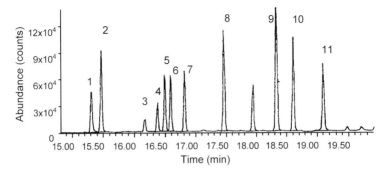

Figure 1.7 Analysis of triazines in water spiked at the 1 ppb level after SPE and large volume injection. Peaks: 1, desisopropylatrazine; 2, desethylatrazine; 3, simazine; 4, atrazine; 5, propazine; 6, terbutylazine; 7, sebutylazine; 8, metribuzine; 9, prometryn; 10, terbutryn; 11, cyanazine.

most important difference is that in GPE retention occurs by dissolution into the bulk of the extractant phase. Pure poly(dimethylsiloxane) (PDMS) is most successfully used as the retaining phase, although the use of an acrylate polymeric sorbent has been described [22]. Owing to the inert nature of PDMS, thermal desorption can be used instead of liquid desorption in SPE, which results in enhanced sensitivity. A drawback of GPE is that, between sorption and desorption, a drying step is mandatory; volatiles are lost in this step. A more universal technique is SBSE.

In stir bar sorptive extraction, a glass-lined magnetic bar is covered with a thick layer of PDMS [23]. Through magnetic stirring of the bar in the sample solution, the components are enriched in the PDMS phase (Figure 1.8). The manipulations in SBSE are illustrated in Figure 1.9.

Figure 1.8 The principle of stir bar sorptive extraction (SBSE).

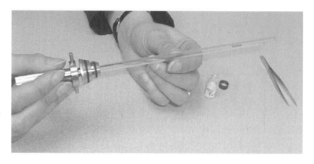

Figure 1.9 Manipulations in stir bar sorptive extraction after SBSE sampling: 1, remove the stir bar with tweezers; 2, rinse briefly in distilled water; 3, dry with lint-free tissue; 4, introduce the bar into the thermal desorber tube; 5, thermally extract; 6, CGC analysis.

After this concentration step, analytes are thermally desorbed (TD) from the stir bar on-line with CGC-MS, CGC-AED or CGC-ICPMS. Major advantages of this technique are ease-of-use, high sensitivity, high accuracy of analysis and reduced risks of contamination compared with other sample preparation techniques. Moreover, the enrichment factors can be predicted from the values of the logarithm of the octanol–water distribution coefficient [24, 25]. Stir bars have recently been commercialised under the name TWISTER and are available from Gerstel (Mülheim a/d Ruhr, Germany). SBSE has been applied for quality control, and trace (ppb) and ultra-trace (ppt) analysis of water samples, beverages, diary products and biological fluids. The performance is illustrated with the analysis of flavour compounds in yoghurt, with the determination of pesticides in wine and with the analysis of polychlorinated biphenyls (PCBs) in human sperm. The instrumentation for thermal desorption, which basically consists of two programmable temperature vaporising injectors in line, has been described in detail [26].

SBSE is very well suited for the analysis of aqueous solutions, but fatty materials like milk, fresh cheese, yoghurt, etc. can also be analysed with SBSE.

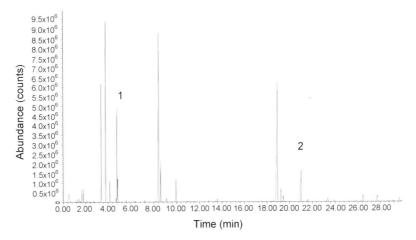

Figure 1.10 SBSE profile of yoghurt with strawberry flavour Peaks:1, ethyl-2-methyl butyrate; 2, γ-decalactone.

A typical example is shown in Figure 1.10, presenting the profile for yoghurt flavoured with strawberries.

For this application, the yoghurt sample was diluted 1:1 with distilled water, and 10 ml was placed in a 40 ml headspace vial. A twister holding 50 mg PDMS was added to the sample and the mixture was stirred for 1 h at 1400 rpm. After extraction, the stir bar was removed by a forceps, washed with water and dried with tissue paper. Subsequently, the stir bar was placed inside an empty conditioned glass tube (187 mm × 4 mm i.d.) for desorption. The stir bar was thermally desorbed by heating from 20 to 240°C at 60°C min^{-1}, with an initial and final time of 1 and 10 min, respectively. Compounds were swept towards the second PTV at 100 ml min^{-1} and trapped at −150°C in an empty baffled liner. When desorption was completed the PTV at a split of 1 to 20 was heated from −150°C at 12°C s^{-1} to 350°C, which was held for 5 min. The TDS-2 was mounted onto a HP 6890 GC/5973 MSD. The MSD was operated in the electron impact mode (70 eV), generating full scan spectra between 50 and 550 amu at 3 scans/s. A 30 m × 0.25 mm i.d. × 0.25 µm Stabilwax capillary column (Restek, Bellefonte, PA, USA) was used. Helium was the carrier gas at 1 ml min^{-1}. The oven was programmed from 40°C (1 min) at 5°C min^{-1} to 240°C. The compounds responsible for the strawberry flavour namely ethyl-2-methyl butyrate and γ-decalactone are clearly present. It is surprising that the lipid matrix did not disturb the SBSE enrichment.

A white wine was analysed under the following conditions. A 25 ml sample was placed in a 40 ml vial and extracted with a stir bar coated with 50 mg of PDMS for 40 min while stirring at 1400 rpm. The stir bar was thermally extracted at 300°C for 10 min. The solutes were cryofocused in the second PTV inlet at $-50°C$. The separation was carried out on a HP-5 MS column. The oven was programmed from $70°C$ (2 min) to $150°C$ at $25°C\,min^{-1}$, to $200°C$ at $3°C\,min^{-1}$ and to $280°C$ at $8°C\,min^{-1}$. Detection was done using an AED by monitoring the chlorine (837 nm), bromine (827 nm) and sulfur (181 nm) emission lines. The thermal desorption GC-AED was operated under retention time locking (RTL) conditions [27,28]. Figure 1.11 shows the Cl-trace and four pesticides are detected in the ppb and sub-ppb level. The main compound (22.1 min) was identified via the RTL pesticide library as procymidone. This was confirmed by analysing the same sample by CGC-MS under RTL conditions. The corresponding spectrum with the library search is shown in Figure 1.12. The concentration was determined by internal standard addition and was 21 ppb. The repeatability of SBSE for this particular pesticide was 6.1 RSD% for $n = 6$.

Human sperm was spiked at the 1 and 10 ppt level with the seven 'Ballschmiter PCBs' [29] and the sample was ultrasonically treated for 3 min at 23 kHz. Sperm (1 ml) was diluted with 9 ml of water–methanol (1:1 by vol) in a 20 ml headspace vial. A stir bar coated with a 50 mg layer of PDMS and conditioned overnight at $325°C$ under a helium stream was placed in the diluted sperm mixture and enrichment was carried out by stirring the twister for 45 min at 1000 rpm. Subsequently, the stir bar was thermally desorbed by heating the glass tube from 20 to $325°C$ at $60°C\,min^{-1}$ with an initial and final time of 1 and 10 min, respectively. Compounds were swept towards the second PTV at $100\,ml\,min^{-1}$ and trapped at $-150°C$ in an empty baffled liner. The PTV was then heated from $-150°C$ at $12°C\,s^{-1}$ to $350°C$, which was held for 5 min. The temperature program started after an additional 30 s at $-150°C$ to allow full flow stabilisation inside the

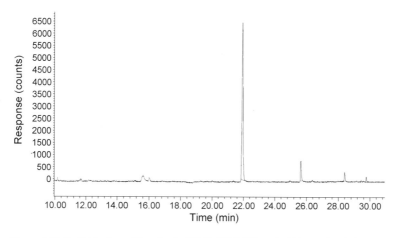

Figure 1.11 Pesticides in a white wine by SBSE-CGC/AED at the 837 nm cl emission line.

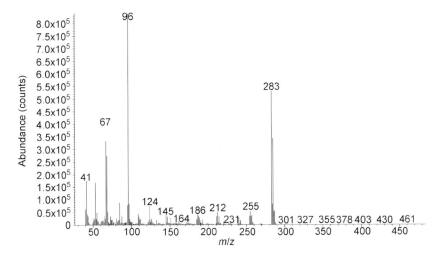

Figure 1.12 Mass spectrum of the pesticide eluting at 22.1 min in Figure 1.11.

PTV-liner after closure of the split valve. The total splitless time was 2.5 min. The TDS-2 was mounted onto a HP 6890 GC/5973 MSD. The MSD was operated in the electron impact mode (70 eV) and in the time-scheduled SIM mode using two ions per congener. An HP-5MS capillary column (30 m × 0.25 mm i.d. × 0.25 μm) was used. Helium was the carrier gas at 1 ml min^{-1}. The oven was programmed from 40°C (2.5 min) at 25°C min^{-1} to 150°C and then to 280°C (3 min) at 10°C min^{-1}. Time-scheduled SIM chromatograms of the sperm sample spiked with 10 ppt (A) and 1 ppt (B) of the PCBs are presented in Figure 1.13. Because of cross-contamination, the limit of quantification was set at 10 ppt. The within-day and between-day reproducibilities at 10 ppt were below 7%.

1.2.4 Solid samples

1.2.4.1 Introduction
Solid samples can be very diverse in nature, for example soil, sediment, pharmaceuticals, polymers, vegetables, fruit, etc. Advancing universal methods is not straightforward but, notwithstanding this, an overview guideline is presented in Figure 1.14.

Direct analysis can be applied, for example, for chemicals and pharmaceuticals that are soluble, perhaps after derivatisation, in organic solvents. Pyrolysis CGC-MS is a powerful technique for the characterisation of macromolecules such as synthetic polymers but can also be applied to fingerprint microorganisms. For the analysis of highly volatile compounds, static and dynamic headspace are established techniques and SPME, HSSE and direct thermal

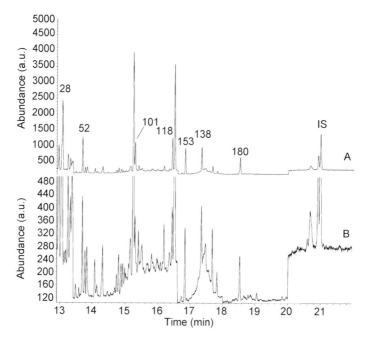

Figure 1.13 PCB analysis of sperm spiked at (a) the 10 ppt level and (b) the 1 ppt level.

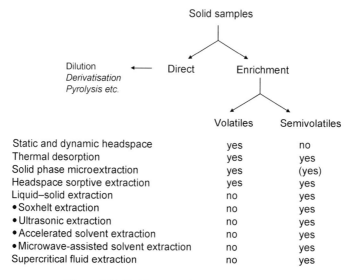

	Volatiles	Semivolatiles
Static and dynamic headspace	yes	no
Thermal desorption	yes	yes
Solid phase microextraction	yes	(yes)
Headspace sorptive extraction	yes	yes
Liquid–solid extraction	no	yes
• Soxhelt extraction	no	yes
• Ultrasonic extraction	no	yes
• Accelerated solvent extraction	no	yes
• Microwave-assisted solvent extraction	no	yes
Supercritical fluid extraction	no	yes

Figure 1.14 Enrichment methods for solid samples.

desorption are gaining in importance. For semivolatiles, older techniques like Soxhlet, shake flask and shake flask with heating are increasingly being replaced by Soxtec (automated Soxhlet), sonication, accelerated solvent extraction (ASE) and microwave-assisted solvent extraction (MASE). Supercritical fluid extraction (SFE), a very powerful technique, seems not able to keep its promises because of the number of parameters that need to be optimised for a particular sample.

1.2.4.2   Gas phase extraction (TD)

Thermal desorption coupled on-line to CGC-MS is a powerful technique not only for the analysis of air volatiles collected on adsorbent and sorbent tubes but also for the analysis of residual monomers in polymers, for the determination of residual solvents in pharmaceutical products and for monitoring volatiles and semivolatiles in soil and sediment. The solid material is directly placed in a thermal desorption tube coupled to a CGC system. The temperature is increased to typically 250–300°C and the released volatiles are transported with the carrier gas stream to a cryotrap. The trap is heated and the solutes are introduced into the column in a narrow injection band. Figure 1.15 shows the analysis of the volatiles in a batch of polystyrene by capillary GC-MS on a conventional HP-5MS column.

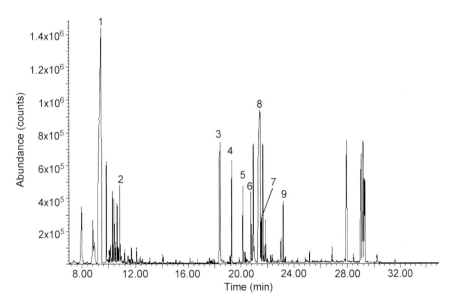

Figure 1.15 Analysis of impurities in polystyrene by direct thermal desorption CGC-MS. Peaks: 1, styrene; 2, C_3-benzenes; 3, butylated hydroxytoluene (BHT); 4, diethylphthalate; 5, 1,3-diphenylpropane; 6, 1,2-diphenylcyclobutane; 7, 1,3-diphenyl-1-butene; 8, 1-phenyltetraline; 9, dibutylphthalate.

1.2.4.3 Sorptive extraction (SPME and HSSE)

The solventless methods of solid phase microextraction (SPME) and headspace sorptive extraction (HSSE) are both ideally suited for the enrichment of volatiles from solid materials. SPME is nowadays widely applied because of its simplicity, while the newcomer HSSE offers higher sensitivity but requires dedicated instrumentation. The analysis of the volatiles emitted by a single rose and analysed on an apolar PDMS fibre and on a polar polyacrylate (PA) fibre are compared in Figure 1.16. The column was a 30 m × 0.25 mm i.d. × 0.25 μm HP-5MS programmed from 40°C to 300°C at 15°C min^{-1}. The identification of the peaks and their abundances on PA and PDMS are given in Table 1.2. The polar compound phenylethylethanol (peak 1) exhibits an increase in abundance by a factor of 10 on the PA fibre; medium polar compounds like peaks 2, 3, and 4 show an increase by a factor of ∼ 1.5, while the apolar sesquiterpenes (peaks 6–13) show an intensity increase by a factor of at least 5 on the PDMS fibre compared to the PA fibre. This discriminative effect of SPME is tiresome in real sampling. Dynamic sampling is much more promising in this respect.

HSSE is similar in principle to SPME but uses a larger amount of sorbent (Figure 1.17). The performance of HSSE is illustrated with the analysis of the aroma carriers in bananas (Figure 1.18). Slices of 1 g of unripe and of ripe banana were put in a 250 ml Erlenmeyer flask and HSSE sampling was

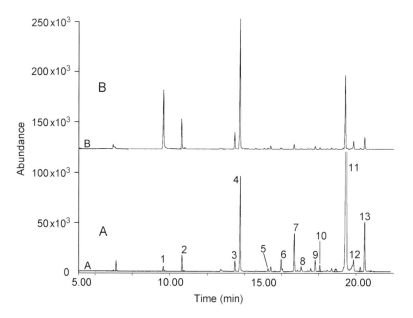

Figure 1.16 Total ion chromatograms of a single rose; (a) SPME with poly(dimethylsiloxane); (b) SPME with polyacrylate. For peak identification see Table 1.2.

Table 1.2 Composition of the rose headspace as observed with SPME on a PA and a PDMS fibre; peak area values relative to 2-phenylethyl acetate as percentage

Peak no.	Compound	Relative area (%)	
		PA	PMSD
1	Phenylethyl alcohol	419	56
2	1-Ethenyl-4-methoxybenzene	155	126
3	2-Phenylethyl acetate	100	100
4	1-(Ethylthio)-2-methylbenzene	869	872
5	3,7-Dimethyl 2,6-octadienoic acid methyl ester		
6	α-Cubelene		81
7	Copaene	29	345
8	Terpene 1		53
9	Terpene 2		113
10	Terpene 3		59
11	Germacrene D	474	4331
12	Terpene 4	61	215
13	Terpene 5	71	465

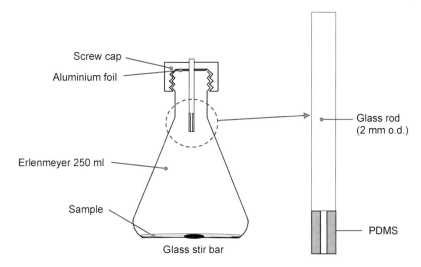

Figure 1.17 Principle of headspace sorptive extraction.

performed for 60 min. Thermal desorption of the HSSE rod was performed in a TDS-2 thermodesorption unit (Gerstel) mounted on a Agilent 6890 GC (Agilent Technologies, Little Falls, DE, USA). For cryofocusing of the analytes prior to injection, a PTV injector with liquid nitrogen cooling was applied. A small plug of Tenax ($\sim$ 20 mg) was placed in the liner of the PTV. Splitless thermal desorption was done by ramping the TDS from 40°C to 250°C at a rate

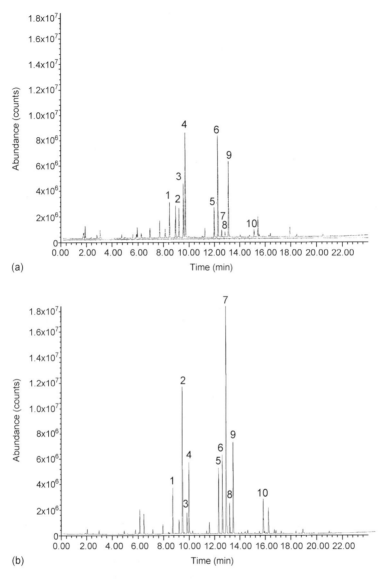

Figure 1.18 HSSE of (a) unripe and (b) ripe banana. Peaks: 1, butyl acetate; 2, 2-pentanol acetate; 3, 1-hexanol; 4, 1-butanol-3-methyl acetate; 5, butyl butyrate; 6, hexyl acetate; 7, 2-methyl-1-methylbutyl propionate; 8, 2-methyl-3-methylbutyl propionate; 9, hexyl butyrate; 10, 2-ethylcyclohexyl butyrate.

of 60°C min^{-1} and holding the upper temperature for 5 min. During thermal desorption, the PTV was cooled at −150°C and then ramped to 250°C at a rate of 600°C min^{-1}. The injector was operated in the split (1/50) mode. Capillary GC analyses were performed on a 30 m × 0.25 mm i.d. × 1 μm HP-1 column

(Agilent Technologies), with helium as carrier gas. The oven was programmed from 30°C (1 min) to 300°C at a rate of 10°C min^{-1}. The mass spectrometric detector was operated in full scan mode (m/z 20–400). The green unripe banana profile (a) changes into the typical banana aroma profile (b) in which 2-pentanol acetate and 2-methyl-1-methylbutyl and 2-methyl-3-methylbutyl propionate play important roles.

1.2.4.4   Liquid phase extraction

Introduction. The extraction of solid samples is most commonly done using traditional liquid–solid extraction methods like shake flask extraction, classical Soxhlet extraction, introduced in 1850(!), and its more modern forms Soxtec and Soxtherm, sonication or ultrasonic extraction. It is not the aim to detail all these methods but rather to concentrate on ultrasonic extraction and the recently introduced methods of accelerated solvent extraction (ASE) and microwave-assisted solvent extraction (MASE).

Ultrasonic extraction (UE). Ultrasonic extraction often provides very good results for solid samples. Advantages over other techniques are simplicity, speed, productivity and low cost. The selection of the solvent is of utmost importance and heating can be applied if needed although self-heating is generated by the sonication process itself. Its application will be illustrated and its performance compared to that of ASE and MASE for the analysis of PCBs in fat samples (section 1.6.3).

Accelerated solvent extraction (ASE). ASE originates from supercritical fluid extraction (SFE) and uses organic solvents at high temperature and high pressure to leach out the organics from solid matrices. Compared to extractions at or near room temperature and at atmospheric pressure, ASE delivers enhanced performance by the increased solubility, improved mass transfer and disruption of surface adsorption. Dionex Corporation in 1995 introduced a fully auto-mated sequential extraction system (Figure 1.19) that was very soon after its introduction recommended by the EPA for the extraction of solid waste [30]. This definitely demonstrates the performance of ASE. Typical conditions for an ASE extraction are temperature 100°C, pressure 2000 psi (13.8 MPa), and extraction time 5 min equilibration and 5 min static extraction with a solvent composed of dichloromethane and acetone in a ratio of 1:1.

The speed of the extraction process is greatly increased compared to that of conventional liquid–solid methods and virtually all organics from the priority pollutant lists can be extracted. Disadvantages of ASE are the lack of selectivity, which means that further clean-up is needed, and the fact that the sample is diluted so that further concentration is required. Recent developments to circumvent these shortcomings are the combination of ASE with SPE and the application of large-volume injection. ASE has been evaluated in our laboratory

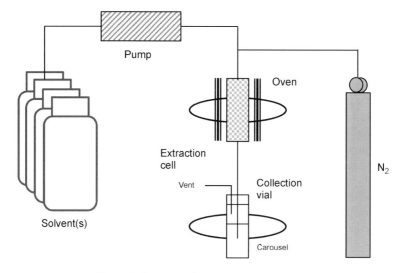

Figure 1.19 Schematic drawing of an ASE system.

for the extraction of PAHs from soil and of PCBs and organometallics from sediment samples. Recoveries were excellent. As an example, dibutyltin chloride and tributyltin chloride were quantitatively extracted with 1:1 hexane-acetone at 1500 psi (10.3 MPa) and 100°C. After derivatisation with methylmagnesium bromide, the analysis was performed with CGC-AED (atomic emission detection) at the Sn 303 nm line.

Microwave-assisted solvent extraction (MASE). MASE utilises electromagnetic radiation to desorb organics from their solid matrices. MASE typically operates at 2.45 GHz. The use of a microwave oven for sample preparation originates from inorganic or elemental analysis. In this case the electromagnetic radiation helps the destruction of inorganic and organic matter using a combination of strong acids and peroxides. The first application of microwaves for the extraction of organics from solid material appeared in 1986 [31]. In recent years, different systems have become commercially available and are based on extraction in a closed vessel with microwave-absorbing solvents in a high-pressure vessel, extraction with a non-microwave-absorbing solvent in an open vessel or extraction with a non-microwave-absorbing solvent in a closed vessel applying a Weflon stir bar that heats the solvent.

The performance of MASE has been compared with that other recently introduced techniques such as ASE and SFE and similar recoveries were obtained for soil and sediment samples [32]. The disadvantages of ASE, namely lack of selectivity and the dilution effect, are the same for MASE. Moreover, care should be taken with solutes that are thermolabile or can rearrange under the influence of electromagnetic radiation. The performance of MASE has also been evaluated for the Belgian dioxin crisis (section 1.6.3).

1.2.4.5 Supercritical fluid extraction (SFE)

The excellent properties offered by a supercritical fluid, namely tuneable solvating power (selectivity), high diffusivity and low viscosity have resulted in analytical usage for many years. Despite early promise, the utility of super-critical fluids for extraction was dormant for many years until the mid-1980s when analytical-scale SFE instrumentation became commercially available. SFE was received with high expectations by the analytical community, but in recent years enthusiasm has declined as the disadvantages have become clearer. The most important disadvantage is definitely the complexity of the extraction procedure, which is matrix dependent and needs careful optimisation. A procedure developed for one particular matrix, for example a sediment from one place, does not automatically work for a sediment sample taken at another place. SFE is therefore not really accepted for routine work but is, in our opinion, definitely a very important research tool [33,34]. The strong point of SFE is the selectivity that can be introduced in the extraction procedure (Figure 1.20).

In the first instance, the extraction selectivity and efficiency can be controlled by the nature of the supercritical medium (Selectivity 1). The reasons for adding a polar or a nonpolar modifier to the CO_2, normally used as the supercritical medium, are threefold: (i) to increase the solubility, (ii) to destroy the matrix effects, and (iii) to enhance diffusion by swelling of the matrix. Modifiers can also be added to retain unwanted solutes but this counteracts (ii) and (iii). The effect of the addition of a polar modifier on the recovery of PCBs has been

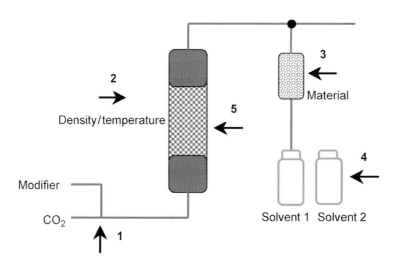

Figure 1.20 Schematic drawing of an SFE instrument with selectivity tools.

described previously [35]. The second opportunity for selectivity concerns the density of the supercritical medium and the temperature (Selectivity 2) [26]. The importance of the latter parameter is often neglected but, as solubility is also controlled by vapour pressure, this can be exploited to introduce selectivity. After leaching of the sample, the extract is collected on a solid trap filled with a nonpolar or polar adsorbent, which can be selected according to the application (Selectivity 3). The trap is then rinsed with a solvent, the polarity of which can be chosen to desorb the solutes of interest in a selective way (Selectivity 4). Last but not least, an adsorbent can be added in the extraction thimble (Selectivity 5). This not only introduces the possibility of extraction liquids by SFE, but also facilitates the retention of unwanted polar solutes (fixation) or the enhancement of recoveries of nonpolar solutes (exsaltation) [26].

1.3   Fractionation and clean-up

1.3.1   Introduction

Solvent distribution and liquid chromatography are mostly applied to further fractionate or clean-up a solvent extract. It is not possible to discuss all the different approaches in detail and only some well-established and newly developed methods are discussed.

1.3.2   Chemical methods

In solvent distribution the principle of 'like dissolves like' is applied and the partitioning is based on total polarity or selective interactions. The separation of apolar/medium polar and polar compounds in, for example, an essential oil can be performed by distribution in a hexane–methanol/water (95:5) mixture, and selective enrichment of polyaromatic hydrocarbons from hydrocarbons in a crude oil can be done with a cyclohexane–nitromethane mixture. Acidic solutes can be extracted with 6% $NaHCO_3$ (C_1 to C_{10} acids and phenols) or 6% KOH (C_{10} to C_{24} acids); alkaline compounds can be recovered with 5% HCl. Alcohols can be fractionated via an isocyanate reaction and aldehydes and ketones with a Schiff base such as phenylhydrazine. A typical scheme for the fractionation of pollutants from particulates collected on a glass fibre filter and Soxhlet extracted using dichloromethane is shown in Figure 1.21.

1.3.3   Chromatographic methods

Liquid chromatography (LC) is frequently used to further fractionate or clean up samples after solvent extraction. The separation mechanism can be based on polarity (normal-phase LC), on hydrophobicity (reversed-phase LC) or on size (size exclusion chromatography). Low pressure LC, high-pressure LC or

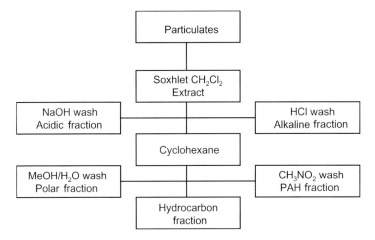

Figure 1.21 Fractionation of pollutants from air particulates.

SPE are all used depending on the aim of the clean-up and the complexity of the sample.

Widely employed adsorbents are silica gel, alumina, Florisil and graphitised carbon. A scheme for the clean-up of dioxins and furans from fats is shown in Figure 1.22. Specific sample preparation stations can now be constructed using, for example, the Gilson autosampler 232 XL.

Fully automated and flexible on-line LC-GC systems for fractionation and clean-up have recently been described [36,37]. The systems can be based on large-volume injection via a cool on-column injector or a programmed temperature vaporization (PTV) injector. The first system allows the enrichment of both volatiles and semivolatiles, while the second is restricted to the analysis of semivolatiles. An on-line LC-PTV-GC system was evaluated for the determination of pesticides in orange oil. The essential oil was fractionated by normal-phase LC, resulting in a separation according to polarity. The fraction containing pesticides was transferred and analysed by capillary GC-NPD (nitrogen phosphorus detector). The LC-GC interface is a modified large-volume sampler (Gerstel, Mulheim a/d Ruhr, Germany) (Figure 1.23).

The heart of the system is a specially designed flow cell that replaces a normal vial. The mobile phase leaving the LC detector is directed via a capillary tube with well-defined dimensions, in a T-shaped flow cell. The cell is equipped with a septumless sampling head through which a syringe needle can be introduced. The sampler is completely computer controlled. To transfer a selected LC fraction, the transfer start and stop times, measured on the LC chromatogram, are introduced in the controller software. The time delay between the LC detector and the flow cell is automatically calculated from the connecting tubing

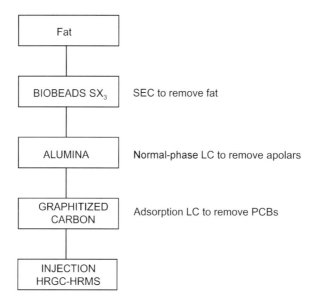

Figure 1.22 Flow diagram for the fractionation and clean-up of dioxins and furans from fat samples. (HR, high-resolution.)

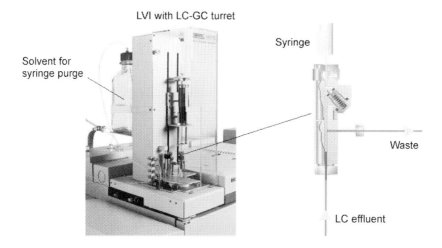

Figure 1.23 Interface based on a PTV injector for LC-GC.

dimensions and the LC flow rate. At the time the LC fraction of interest passes through the flow cell, the syringe needle penetrates the septumless head and samples the LC fraction at a speed equal to the LC flow rate. Volumes up to 2 ml can be collected in the syringe. After collection, the needle is withdrawn from the flow cell, which rotates away from the PTV inlet and a large-volume injection is made using the PTV in the solvent vent mode. Depending on the fraction volume and solvent type, the sample introduction parameters (inlet temperature, vent flow, vent time, injection speed, etc.) are calculated by the PTV calculator program.

For the fractionation of an orange oil sample, the following HPLC parameters were used: column 250 mm $\times$ 4.6 mm i.d. $\times$ 10 µm Lichrospher 100 DIOL, injection volume 20 µl, mobile phase in a gradient from 100% hexane for 10 min, to 40% isopropanol at 20 min (2 min hold) at a flow rate of 1 ml min^{-1} and UV detection at 220 nm. The fraction eluting from 4.4 to 4.9 min (volume = 0.5 ml) was automatically transferred to the GC inlet. The LC-GC interface was programmed to take the sample at a 1000 µl min^{-1} sampling speed (the same as the LC flow rate). The complete fraction of 500 µl was injected in the PTV inlet at 250 µl min^{-1}. This injection speed corresponds to the injection speed calculated by the PTV software program. Capillary GC-NPD analysis was performed on a 30 m $\times$ 0.25 mm i.d. $\times$ 0.25 µm HP-5MS column and the oven was programmed from 70°C (2 min) to 150°C at 25°C min^{-1} and then to 280°C at 8°C min^{-1} (10 min). The detector was set at 320°C with 3 ml min^{-1} hydrogen, 80 ml min^{-1} air and 30 ml min^{-1} helium make-up flow. The LC profile for the orange essential is shown in Figure 1.24. The apolar mono- and sesquiterpenes elute first, followed by the terpenoids and after 16 min the flavanoids are also eluted. These last compounds in particular, give most interference in GC analysis because they have similar molecular weights to the pesticides. Using the same conditions, ethion and chlorpyriphos elute at 4.6–4.7 min. Figure 1.25 shows a comparison of the analysis of a 1 ppm ethion reference sample and the relevant part of the essential oil analysis.

The fraction eluting from 4.4 to 4.9 min was transferred to the CGC-NPD and analysed. Figure 1.26 shows the resulting GC-NPD chromatogram and both ethion and chlorpyriphos (t_R 15.79 min) are detected. The chromatograms are

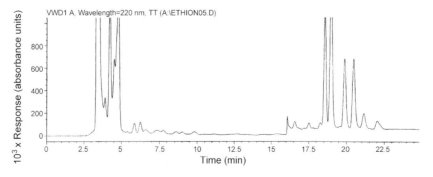

Figure 1.24 LC of an orange essential oil.

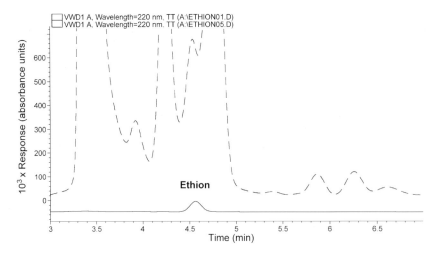

Figure 1.25 Relevant part of the chromatogram of orange essential oil and of ethion.

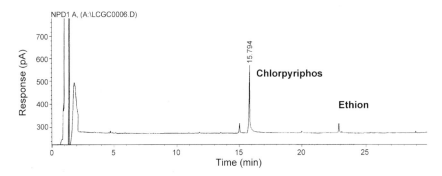

Figure 1.26 CG-NPD analysis of the transferred fraction.

very clean and no interferences are present. This demonstrates the excellent selectivity of the LC-GC combination. The concentrations of ethion and chlorpyriphos were calculated by external standard analysis and are 1.9 ppm for chlorpyriphos and 0.8 ppm for ethion.

1.4   Evaporation

Liquid extraction or desorption techniques frequently need solvent evaporation to concentrate the target solutes. Solvent evaporation can only be applied for concentrating semivolatile compounds and even then care must be taken or solutes of interest may be lost. Moreover, contamination can easily occur and impurities in the extracting solvent can be enriched, hence the need for ultrapure solvents. The most common approaches for solvent evaporation are gas blow-down, rotary evaporation, Kuderna–Danish evaporative concentration, auto-mated evaporative concentration [38], and TurboVap concentration (Zymark Corporation, Hopkinton, MA, USA).

Loss of solutes can be partly overcome by adding a keeper—a small amount of an organic solvent of the same polarity as the extracting solvent but with a higher boiling point, for example adding isooctane to a solution in n-hexane.

To avoid solvent evaporation, micro-methods such as micro liquid–liquid extraction or solid phase extraction on disc cartridges followed by large-volume injection via a cool on-column injection device with precolumn and solvent exit, or via programmed temperature vaporisation injection operated in the solvent venting mode, are becoming more and more frequently applied. An application of large-volume injection without solvent evaporation has been described in section 2.3.4.

1.5   Derivatisation

1.5.1   Introduction

For many applications, it is necessary to derivatise the analytes before they can be properly analysed by capillary GC. A major reason for derivatisation in GC is to 'cap' the polar groups by substituting the active proton so that volatility increases and irreversible or reversible adsorption in the column is avoided. However, derivatisation can also be carried out to increase detectabil-ity, for example introducing a pentafluorobenzoyl group for sensitive ECD detection, or before extraction to improve the extraction yield as well as the chromatographic analysis. The derivatisation procedures applied depend on the functional group to be derivatised and on the purpose of the derivatisation, to improve chromatographic performance, detectability or extraction recovery. Derivatisation reactions should be quantitative, rapid and simple (test tube chemistry) and the excess reagent should be easy to remove or should not interfere with the subsequent chromatographic separation. Some typical reac-tions are listed in Table 1.3. For more information we refer to the book of Blau and Halket [39]. The following applications illustrate the need for derivatisation.

1.5.2   Analysis of phenols in water samples

Because of the toxicity and persistence of phenols, monitoring of these solutes in water samples at the low ppb levels is required. Several methods have been described, but the most straightforward in our opinion is in-situ derivatisation

Table 1.3 Derivatisation reactions for capillary GC

Functional group	Reagent	Derivative
Carboxyl RCOOH	R'OH/HCl R'OH/BF$_3$ R'ONa CH$_2$NH$_2$	Ester ⌈ RCOOR' { RCOOR' ⌊ RCOOR' RCOOCH$_3$
Phenolic ⟨benzene ring⟩—OH	(R'CO)$_2$O	Acyl ⟨benzene ring⟩—OCR' ($\overset{O}{\underset{\|\|}{}}$)
Hydroxyl ROH	CF$_3$—C $\overset{O-Si(CH_3)_3}{\underset{N-Si(CH_3)_3}{}}$	Silyl ether R—O—Si(CH$_3$)$_3$
Amine RNH$_2$	(R'CO)$_2$O	Acyl RNHCR' $\overset{}{\underset{\|\| }{}}$O
Chloro (I, Br, ...) RCl	R'MgBr Na$^+$R'$_4$B$^-$	Alkyl {R—R'
Carbonyl R—C(=O)—R' R—C(=O)—H	CH$_3$ONH$_2$HCl ⟨benzene⟩—NHNH$_2$	Schiff base R R'(H) —C=N—OCH$_3$ R R'(H) —C=N—NH—⟨benzene⟩

prior to extraction. In this way, the extraction yield with SPME is improved and the chromatographic performance is excellent. A typical procedure is as follows. A 10 ml water sample is transferred into a 20 ml headspace vial and 0.5 g potassium carbonate and 0.5 g acetic acid anhydride are added. The phenols are transformed into the phenolates and acylation takes place.

$$\text{2,4,6-trichlorophenol} \quad \xrightarrow{K_2CO_3} \quad \xrightarrow{(CH_3CO)_2O} \qquad (1.1)$$

2,4,6-trichlorophenol

After 15 min stirring, a 100 µm PDMS fibre is exposed to the water for 45 min at a stirring speed of 1000 rpm. The fibre is then desorbed in a split-less injector and the solutes are analysed by capillary GC-MS. Figure 1.27 shows the analysis of a sample contaminated at the ppb level. The solutes are 2-methylphenol (IS1), 3-methylphenol (1), 2,3,5-trimethylphenol (2), 3,4-dichlorophenol (3), 2,4,6-trichlorophenol (IS2), 2,3,5,6-tertachlorophenol (4) and pentachlorophenol (5). The analysis was performed on a 30 m × 0.25 mm i.d. × 0.25 µm SPB-5 column (Supelco, Bellefonte, PA, USA) in a temperature program from 50°C (1 min) to 280°C at 10°C min^{-1} with MS detection. With ion monitoring, the limit of quantification for pentachlorophenol is less than 10 ppt.

1.5.3 Analysis of racemic amino acids in biological fluids

The metabolism of racemic amino acids present in some food supplements is an interesting subject for research. A typical example is the presence of selenomethionine in addition to lipid- and water-soluble vitamins in some food supplements. Selenomethionine is a radical catcher and, in different forms, is consumed in relatively high quantities. When the racemate is taken, it is interesting to know whether both D- and L-selenomethionine are metabolised. The literature is unclear in this respect and some references even claim that the non-natural D-selenomethionine is metabolised to the toxic selenium. For the

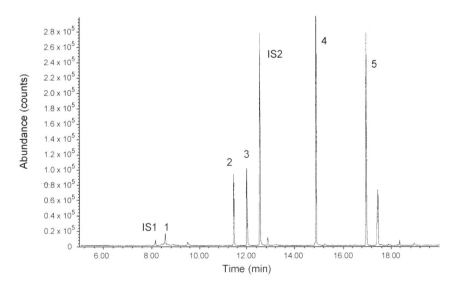

Figure 1.27 CGC-MS analysis of phenols in water after derivatisation and SPME enrichment.

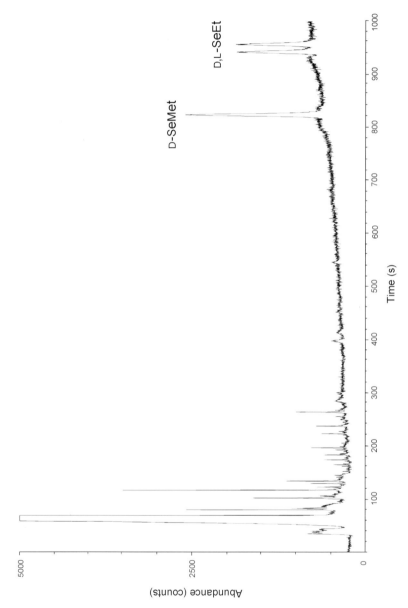

Figure 1.28 CGC-ICPMS analysis of selenoamino acids in urine. (SeMet, selenomethionine; SeEt, selenoethionine.)

sensitive and selective analysis of selenium amino acids capillary GC has been coupled to ICPMS [40].

Isolation of amino acids from aqueous samples like urine is commonly performed by SPE on SCX or SAX (Table 1.1), followed by a two-step derivatisation (esterfication and acylation) before capillary GC analysis. Amino acids can also be derivatised in aqueous medium in a single step by means of ethylchloroformate.

$$
\begin{array}{c}
\text{HOOC}-\text{CH}-\text{NH}_2 \\
\text{H}_3\text{CSeH}_2\text{CH}_2\text{C}
\end{array}
\xrightarrow[\substack{\text{ClCOCH}_2\text{CH}_3}]{\text{CH}_3\text{CH}_2\text{OH H}_2\text{O}}
\begin{array}{c}
\text{CH}_3\text{CH}_2\text{OCCHNHCOCH}_2\text{CH}_3 \\
\text{CH}_2\text{CH}_2\text{SeCH}_3
\end{array}
\quad (1.2)
$$

Figure 1.28 shows the analysis of a urine sample of a person who has taken 100 µg of racemic D- and L-selenomethionine. The urine was collected 5 h after the intake and racemic DL-selenoethionine was added as internal standard. The sample preparation was as follows. To 1 ml urine was added 2 ml of a solvent mixture composed of water–ethanol–pyridine (60:32:8 v/v)and 200 µl chloroformate. After vortexing for 60 s, the derivatised amino acids were extracted with 2 ml chloroform; 1 µl was injected on a 25 m × 0.25 mm i.d. × 0.16 µm Chirasil-Val column (Alltech, Lokeren, Belgium) in a temperature-programmed run from 35°C to 120°C at 4°C min^{-1} and then to 180°C at 8°C min^{-1}. The ^{82}Se line was monitored by ICPMS. The in-situ derivatisation works perfectly and the chromatogram illustrates that L-selenomethionine has been metabolised while D-selenomethionine has not. Note the complete separation of the racemates for D- and L-selenoethionine.

1.6   Some case studies

1.6.1   Fractionation of nitro-PAHs from air particulates

Nitro-substituted polycyclic aromatic hydrocarbons (nitro-PAHs) are an important class of airborne carcinogens. The toxicity and carcinogenic potency of nitrated polynuclear aromatic hydrocarbons is much higher than that of the parent compounds. The concentrations of nitro-PAHs in air are much lower than those of the parent PAHs. Typical concentrations of the carcinogenic benzo[a]pyrene in air samples taken at urban sites are on the order of 1–10 ng m^{-3} whereas concentrations for nitro-PAHs taken at similar sites are generally 10–100 times lower. For reliable assessments of human health risks, analytical methods for the determination of nitro-PAHs have to be extremely sensitive and selective. The target solutes are present in a very complex matrix in which other organic micropollutants, and more especially the phthalates and the PAHs themselves, are present in comparatively very high concentrations.

A method is described for the fractionation of some target nitro-PAHs from co-extracted pollutants that interfere with the capillary GC-MS determinations. The method is based on semipreparative HPLC on aminopropyl silica gel.

Air was sampled on the roof of a university building located close to the center of Ghent, Belgium. Sampling was performed with a Graseby–Andersen high-volume sampler, type GMWL-2000H (Graseby, OH, USA), at a nominal flow rate of 1.13 m^3 min^{-1} for 24 h (1637 m^3 total sampled volume). The collecting filters were 20 cm × 25 cm Whatman Quartz Microfibre QM-A (Vel, Leuven, Belgium). The filters were baked out at 500°C for 24 h before use. Figure 1.29 compares a filter before and after sampling.

The entire filter was introduced to a Soxhlet extractor and extraction was done with 100 ml dichloromethane over 15 h. The extract was concentrated after addition of isooctane as keeper through evaporation by a gentle stream of nitrogen. The solution was filtered through a 0.45 µm pore size Millipore filter, type Millex-HV$_{13}$. Fractionation of the extract was performed on a HP 1100 HPLC system (Agilent Technologies) consisting of a HP 1100 high-pressure pump, a manual injector model G13258H fitted with a 100 µl injection loop, and a variable-wavelength UV-detector, model G1314A. The instrument was equipped with a 25 cm × 10 mm i.d. × 5 µm Adsorbosphere XL amino-propyl column (Alltech, Lokeren, Belgium). The mobile phase flow rate was 5 ml min^{-1}. Gradient elution was applied by programming from 100% n-hexane to 50%/50% v/v n-hexane–dichloromethane in 19 min, then to 100% dichloro-methane in 1 min, followed by an isocratic time of 10 min. The fractiona-tion time, 30 min in total, was followed by 10 min reconditioning with 100% n-hexane before starting the next fractionation. The nitro-PAHs elute in the time window from 8 to 18 min. The collected fraction was then analysed by CGC-MS. A typical LC fractionation pattern, obtained with PAH and nitro-PAH standards,

24h @ 1.13 m^3 min^{-1}

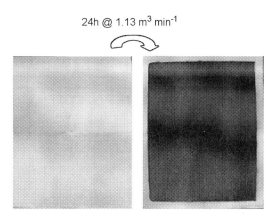

Figure 1.29 Glass fibre filter before and after sampling of air particulates over a period of 24 h.

is shown in Figure 1.30(a). From this chromatogram it could be deduced that the elution window containing the target compounds is between 8 and 18 min. Figure 1.30(b) shows a fractionation of a real air sample extract. The nitro-PAHs are still 'drowned' in many interfering compounds, but, owing to the selectivity of CGC-MS, elucidation and quantification are now feasible.

1.6.2   Versatile robotic system for static (SPME) and dynamic (GPE) sampling

Static and dynamic headspace sampling are complementary techniques. When carefully designed to work together, they can cover a wide range of components from highly volatile to semivolatile compounds or even high boilers, which is of utmost importance in the analysis of complex mixtures such as biogenic emissions. The wider the range of components that an analytical method can cover, the more complete is the emission profile that can be established. Dynamic headspace sampling based on sorption followed by thermal desorption has proved to be a very efficient way to collect and concentrate trace amounts of volatile components emitted by biological materials [41,42]. The number of components that can be enriched depends on the sorbent used. By using a sorbent such as poly(dimethylsiloxane) (PDMS), highly volatile, semivolatile and high-boiling components can be enriched by operating in the breakthrough or equilibrium mode. Solid phase microextraction (SPME) although less sensitive than dynamic headspace sampling has proved to be a powerful static sampling technique for semivolatile compounds. By careful selection of the type and thickness of the film, the range of compounds applicable can be extended from volatiles (PDMS/Carboxen fibre) to high-boiling solutes (7 µm PDMS fibre). Systems for automated SPME are commercially available, such as the FOCUS Robotic Sample Processor (ATAS International BV, Veldhoven, The Netherlands) (Figure 1.31a), the Varian 8200 CX AutoSampler (Varian, Palo Alto, CA, USA) (Figure 1.31b) and the Gerstel Multi Purpose Sampler MPS-3 (Gerstel, Mülheim a/d Ruhr, Germany) (Figure 1.31c). These devices can only employ specifically designed sampling vials, however. The vials also have to be placed in a holder fixed on top of the GC instrument. This limitation is critical in the analysis of biogenic emissions. Sampling units often have different sizes and shapes and are sometimes placed far away from the GC in order to keep the samples under controlled conditions.

An automated system for dynamic and static sampling has been developed (42,43) and the robotic arm for SPME, constructed in our laboratory (Figure 1.31d) from components purchased from FESTO N.V. (Brussels, Belgium) at a cost of only 5000 Euros, is described. The 'Lego'-type construction (Figure 1.32) gives the ability to modify and extend the set-up to readily adapt different requirements in sizes, shapes and locations of the sampling units (Figure 1.33).

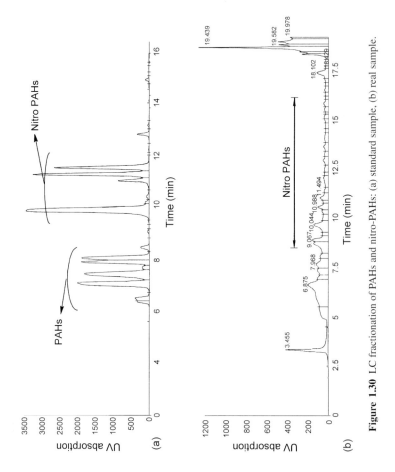

Figure 1.30 LC fractionation of PAHs and nitro-PAHs: (a) standard sample, (b) real sample.

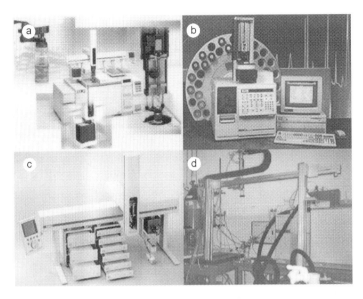

Figure 1.31 Several commercially available automatic samplers with SPME option: (a) ATAS FOCUS Robotic Sample Processor; (b) Varian 8200 CX AutoSampler; (c) Gerstel Multi Purpose Sampler MPS-3; (d) robotic arm constructed in the authors' laboratory.

The total system is based on a GC-MSD (Agilent Technologies) also equipped with a FID. The front inlet is a split/splitless injector with a 0.75 mm i.d. liner for SPME injections. The rear inlet is a thermodesorption system for continuous dynamic sampling (Gerstel). Two identical 30 m × 0.25 mm i.d. × 0.25 μm HP-5 MS (Agilent Technologies) units are installed to connect the inlets to the MSD and FID, respectively. A standard SPME unit i.e. a SPME holder and PDMS fibres was purchased from Supelco. The robotic arm can also operate as an SPME autosampler. For this function a 7000 Ultrorac Fraction Collector (LKB Bromma, Sweden) was used as the sample vial tray (Figure 1.33b). The 7000 Ultrorac was chosen because it moves the tubing racks instead of the drop head, which is common in many other fraction collectors. The 7000 Ultrorac can thus deliver the vials to a fixed sampling point of the robotic SPME in a timely manner. In order to synchronise its operation with the robotic arm, the stepping time of the fraction collector was set to coincide with the total cycle time, that is sampling time plus injection time and moving time, of the robotic SPME device. Headspace SPME analysis of several 'special' Belgian beers was performed using this instrumental set-up. Portions 2 ml of each beer were transferred to 8 ml screwcap vials (Alltech), which were placed into the rack of the fraction collector. A few empty vials were also placed randomly among the beer sample vials to check the blank and completeness of the SPME injections. No carryover was found in the series of robotic injections. Each beer was analysed

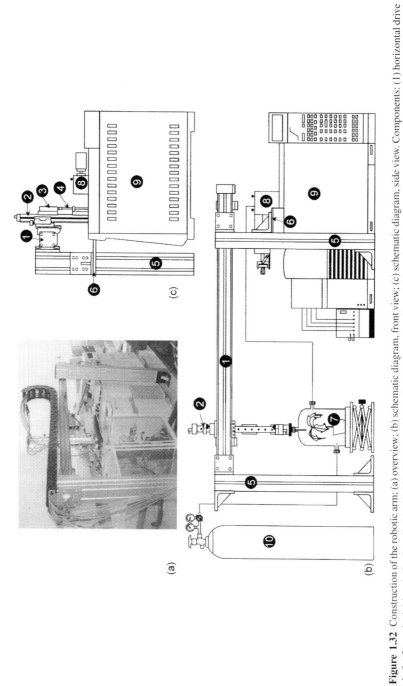

Figure 1.32 Construction of the robotic arm; (a) overview; (b) schematic diagram, front view; (c) schematic diagram, side view. Components: (1) horizontal drive shaft; (2) vertical drive shaft; (3) cylinder; (4) SPME unit attached; (5) robotic arm leg; (6) adapter to the GC; (7) sampling unit; (8) TDSG dynamic sampling; (9) GC; (10) clean air supply.

Figure 1.33 Examples of the sampling units the robotic arm can accommodate: (a) 'in-flow' chamber for simultaneous dynamic and static headspace sampling; (b) 'regular' vial tray using a simple fraction collector; (c) glass bottles for aseptic cultured plants; (d) portable greenhouse.

three times. The reproducibility of the analysis was determined using the peak areas of several characteristic peaks. Relative standard deviations in the range 5–8%, typical for SPME analysis, were found. Figure 1.34 shows the headspace profiles of the beers analysed in this experiment. Identified components are listed in Table 1.4.

The 'regular' beers such as Jupiler and Gulden Draak, do not exhibit any special taste. This corresponds to quite similar headspace profiles with homologous series of ethyl esters of saturated acids. On the other hand, the taste of 'special' beers could be correlated with monoterpenes and monoterpinoids (Timmermans 'Pech') and sesquiterpenes and sesquiterpinoids (St-Bernardus and Mort Subite Framboise) in their headspace profiles. The unique flavours of these specialty beers originate from plant essential oils.

1.6.3   The 1999 Belgian dioxin crisis

In Belgium, 1999 will be remembered as the year of the dioxin crisis. At the beginning of the crisis nobody could imagine the important role separation

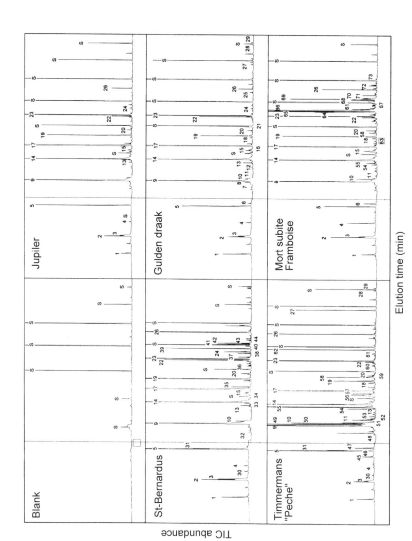

Figure 1.34 Headspace profiles of Belgian 'special' beers analysed by the robotic SPME.

Table 1.4 Volatiles identified in beer SPME profiles

Peak no.	Peak name	Peak no.	Peak name
S	PDMS degradation products	37	Decyl acetate
1	Ethyl acetate	38	Carryophyllene
2	3-Methyl-1-butanol	39	α-Humulene
3	2-Methyl-1-butanol	40	γ-Selinene
4	Ethyl butanoate	41	β-Selinene
5	Isoamyl acetate	42	α-Selinene
6	Styrene	43	α-Amorphene
7	Ethyl methylpentanoate	44	δ-Cadinene
8	β-Myrcene	45	2-Methylethyl butanoate
9	Ethyl hexanoate	46	3-Methylethyl butanoate
10	Hexyl acetate	47	2-Heptanone
11	Limonene	48	Benzaldehyde
12	Ethyl methylhexenoate	49	3-Hexenyl acetate
13	Ethyl heptanoate	50	2-Hexenyl acetate
14	Phenethyl alcohol	51	p-Cymene
15	Octanoic acid	52	Ocimene
16	Ethyl benzoate	53	γ-Terpinene
17	Ethyl octanoate	54	α-Terpinolene
18	Ethyl benzeneacetate	55	Linalool
19	2-Phenylethyl acetate	56	Linolyl acetate
20	Ethyl nonanoate	57	Butanedioic acid, diethyl ester
21	4-Vinyl-2-methoxy-phenol	58	Ethylguaiacol
22	Ethyl 9-decenoate	59	α-Imene
23	Ethyl decanoate	60	5-Pentyldihydro-2(3H)-furanone
24	3-Methylbutyl octanoate	61	Dihydro-β-ionone
25	Perolidol	62	5-Hexyldihydro-2(3H)-furanone
26	Ethyl dodecanoate	63	1,1,4,7-Tetramethylindane
27	Ethyl tetradecanoate	64	Menthadiene
28	Phthalate	65	α-Cedrene
29	Ethyl hexadecanoate	66	α-Ionone
30	Isobutyl acetate	67	Methyl-γ-ionone
31	2-Methylbutyl acetate	68	Di-t-butylphenol
32	Ethyl 2-methyl-2-butenoate	69	2-Methyl-2,6-di-t-butylphenol
33	Heptyl acetate	70	Methyl-α-ionone
34	Ethyl methylheptanoate	71	γ-Isomethylionone
35	Octyl acetate	72	α-Cedrol
36	Isobutyl octanoate	73	Aromadendrene

sciences and especially capillary gas chromatography (CGC) would play to get out of this impasse. Many thousands of food samples had to be analysed before the foods could be released for consumption and this had a tremendous impact on the Belgian economy. This section describes the optimisation procedure for PCB analysis that resulted in a very high throughput and thus high productivity.

The dioxin crisis was in fact a PCB crisis and the analysis of these dioxin precursors was accepted by Belgian and later by European authorities. Beginning in August 1999, the European Community required for export from Belgium certificates for food products with more than 2% fat content! The norm for PCBs in food with more than 2% fat content was set at 200 ppb ($200 \, \text{ng} \, \text{g}^{-1}$ fat) for the sum of seven PCB congeners named according to the nomenclature of Ballschmiter [29] PCBs 28, 52, 101, 118, 138, 153 and 180.

The analytical scheme of the official method for the analysis of PCBs in food consists of steps of sample drying, extraction, clean-up and CGC analysis. Sample drying can be performed by freeze-drying or chemically by sodium sulfate addition. In the second step, the lipophilic contaminants are extracted from the matrix using a nonpolar solvent. The extract contains the lipids, PCBs, PCDDs, PCDFs and other nonpolar solutes such as organochloro pesticides (OCPs), polycyclic aromatic hydrocarbons (PAHs) and mineral oil. Various extraction techniques may be applied: Soxhlet extraction (or the automated versions, Soxtec or Soxtherm), solvent extraction using ultrasonic agitation, accelerated solvent extraction (ASE), microwave-assisted solvent extraction (MASE) and supercritical fluid extraction (SFE). All these techniques perform equally well for the extraction of fat and PCBs, as will be illustrated further. It is obvious that, especially in a crisis situation, selection should be based on sample throughput and cost. Next the PCBs are fractionated from the (co-extracted) fat matrix. For this fractionation, column chromatography on acidic silica gel and aluminium oxide is advised, although other techniques such as gel permeation chromatography (GPC) and solid phase extraction (SPE) may be applied if validated. Both sample extraction and clean-up require a concentration step. Finally, the cleaned extract can be analysed using capillary gas chromatography with electron capture detection (CGC-ECD) or capillary gas chromatography with mass selective detection (CGC-MS). CGC-ECD is extremely sensitive and in most cases sufficiently selective for the detection of PCBs extracted from fat. CGC-MS in the selected ion monitoring mode is somewhat less sensitive but more specific, since the presence of PCBs is confirmed by the detection of several ions per congener in a well-defined relative ratio. For samples found positive by CGC-ECD, MS confirmation was mandatory. The limit of detection for the seven congeners was 5 ppb ($5 \, \text{ng} \, \text{g}^{-1}$ fat per congener). A new and fast sample preparation technique was developed based on ultrasonic extraction followed by clean-up using matrix solid phase dispersion (MSPD). The extract was analysed by CGC-micro ECD. Positive samples were confirmed by CGC-MS. During the crisis, high-speed CGC and the concept of retention time locking (RTL) were implemented and validated. Samples were homogenised using a blender. From fat samples (chicken or pork fat) a 1 g sample was weighed in a 20 ml headspace vial. For eggs, 3 g egg yolk was taken. For animal feed samples or other meat products, a sample size corresponding to 200–500 mg fat was taken. To the sample, 2 g anhydrous sodium sulfate and 10 ml petroleum ether were added. Tetrachloronaphthalene, octachloronaphthalene or the organochloro pesticide Mirex was added as internal standard to the sample at this stage, although external standardisation was also applied. The headspace vial was closed and placed in an ultrasonic bath at 30°C for 30 min. In this step, the fat and PCBs are transferred from the matrix in the petroleum ether phase. The sodium sulfate adsorbs the water present in the sample. After extraction, the sample was allowed to settle. An aliquot (typically 5 ml) was transferred to a test tube and another aliquot (2 ml) was used to determine gravimetrically the fat content of the extract. To the test tube, 2 g of acidic silica gel (44% sulfuric acid) was added and the tube was vortexed for 10 s. This technique is called matrix solid phase dispersion (MSPD). In column chromatography and in SPE, the analytes are eluted through a bed and the fat is retained. In MSPD, the fat matrix is allowed to bind to the adsorbent that is mixed with the sample, while the solutes of interest stay in solution. This method works very efficiently for the fractionation of fat from PCBs. After settlement of the adsorbent ($\sim$ 20 min), an aliquot of the clear solution was transferred to an autosampler vial. The final volume is not important, since the concentrations are calculated to the initial 10 ml solvent and the sample or fat weight. The whole sample preparation takes approximately 1 h and several samples can be prepared in parallel. One technician can handle more than 50 samples per day. The extracts were then analysed by conventional CGC-ECD and CGC-MS on a 30 m $\times$ 0.25 mm i.d. $\times$ 0.25 μm HP-5MS column. Injection of 1 μl is done in the splitless mode.

The new sample preparation method is critical in the analytical scheme. The performance of ultrasonic extraction was in the first instance compared with two other recently introduced sample preparation techniques, namely accelerated solvent extraction (ASE) and microwave-assisted solvent extraction (MASE). The extraction efficiency of the three methods was evaluated with three egg samples contaminated at different levels (low, medium and high). For ASE, a Dionex ASE 200 system (Dionex Corp., Sunnyvale, CA, USA) was used. A 1 g sample was extracted at 100°C and 1500 psi (10.3 MPa) using petroleum ether as solvent. The extraction time was 5 min oven heat-up time, 5 min static extraction and 3 cycles with 60% of the extraction cell volume (22 ml). The extract was then concentrated to 10 ml. For MASE, an ETHOS SEL system (Milestone, Analis, Gent, Belgium) was applied. A 1 g sample was extracted for 20 min at 95°C. The extraction solvent was n-hexane (10 ml in extraction thimble, 10 ml outside thimble) using a Weflon stir bar to absorb the microwave energy in combination with the non-microwave-absorbing solvent. After extraction, the extract was filtered and concentrated to 10 ml to obtain the same final concentration factor as the ultrasonic extraction and ASE. Clean-up and analysis were done in the same way for the three extracts of the three samples. The results, based on duplicate analysis, are summarised in Table 1.5.

For the sample with the lowest concentration (egg 1) the RSD on the PCB sums obtained by the three techniques is 12%; for the two other samples the RSDs are less than 6%. Some small differences are noted for the individual values, but in general these differences are within 10% of the average values. These results clearly demonstrate that there is no statistically significant

Table 1.5 Comparison of ultrasonic extraction (U) (RIC method) with ASE and MASE for PCB enrichment

| PCB | Egg 1 (ppb) | | | Egg 2 (ppb) | | | Egg 3 (ppb) | | |
	U	MASE	ASE	U	MASE	ASE	U	MASE	ASE
118	170	68	147	556	410	550	945	1023	922
153	263	309	210	1142	1051	1184	1811	2032	2042
138	240	320	234	1019	1120	1105	2031	2323	2128
180	111	166	93	696	552	686	1015	1166	1259
Sum	783	863	682	3412	3133	3525	5803	6544	6349

difference between the three techniques and that equally good results are obtained. The ultrasonic method exhibits by far the highest throughput and is extremely cheap compared to ASE and MASE.

The reproducibility and accuracy of the new method was evaluated with the determination of the PCB content in two certified reference materials of the European Community, namely the cod liver oil sample CRM 349 and the mackerel oil sample CRM 350 (IRMM, Geel, Belgium). The analyses of these reference materials were performed by four laboratories having the same CGC-ECD and CGC-MS instrumentation. The results are summarised in Table 1.6 for cod liver oil and in Table 1.7 for mackerel oil. Most values are within 80% and 110% of the certified samples. The values outside these ranges are noted in italic type. For all laboratories and for both techniques, the sum values were always between these limits.

During the interlaboratory study, we noted that the absolute retention times of the different congeners could shift between the laboratories within a window of up to 1 min. The retention time locking (RTL) concept was therefore implemented in the CGC-ECD analysis. The temperature program was changed to 70°C (2 min) to 150°C at 25°C min^{-1}, to 200°C at 3°C min^{-1} and to 300°C (2 min) at 8°C min^{-1} and an initial head pressure of 71 kPa hydrogen was applied. The pressure was then adjusted via the RTL software to obtain a

Table 1.6 Accuracy and reproducibility test for the cod liver oil sample

| PCB | Certified concentration (ppb) | Lab 1 | | Lab 2 | | Lab 3 | | Lab 4 | |
		ECD	MS	ECD	MS	ECD	MS	ECD	MS
28	68	61	65	73	64	*104*	68	*75*	70
52	149	126	*165*	141	164	144	*199*	*159*	148
101	370	333	394	385	404	296	*437*	356	373
118	454	390	467	421	448	479	*508*	397	458
153	938	975	989	886	1010	790	1030	810	1016
180	280	252	295	283	*312*	270	*326*	288	273
Sum	2259	2137	2375	2189	2402	2083	2568	2085	2338

Table 1.7 Accuracy and reproducibility test for the mackerel oil sample

PCB	Certified concentration (ppb)	Lab 1 ECD	Lab 1 MS	Lab 2 ECD	Lab 2 MS	Lab 3 ECD	Lab 3 MS	Lab 4 ECD	Lab 4 MS
28	22.5	25	24	21	21	21	19	19	*16*
52	62	*75*	72	56	56	65	*71*	54	63
101	164	152	*181*	175	175	152	143	150	172
118	142	*163*	152	117	117	138	125	131	134
153	317	337	345	319	319	287	*350*	290	316
180	73	73	79	75	75	76	*83*	70	60
Sum	778.5	825	853	763	739	739	791	714	761

retention time of 26.999 min for p, p'-DDT. The total analysis time under these conditions increased to 36 min, but retention times were now very stable. Figure 1.35 shows the RTL-CGC-ECD profiles of Aroclor 1260 in egg and pork fat recorded with a time interval of nearly one month. As an example, the retention time for PCB 180 varied from 29.230 to 29.234 min (mean 29.232; s 0.002 min; RSD% < 0.01%).

The relatively long analysis times under RTL conditions prompted us to evaluate fast high-resolution capillary GC for PCB analysis [28]. Presently with state-of-the-art capillary GC instrumentation, capillary columns with internal diameters of 0.1 mm and lengths of 10 m can be used, drastically decreasing analysis time while maintaining the resolving power comparison to a conventional capillary column. With the help of the method translation software (MTS), chromatographic conditions of a standard analysis like the RTL-CGC-ECD profiles discussed above could be translated to a high-speed column, keeping the resolution intact. The translated conditions are given in Table 1.8.

The program predicts that the speed gain factor is 4.4 which means an analysis time of 8.2 min. The analyses of an Aroclor 1260 reference sample and an animal feed sample are shown in Figure 1.36.

Compared to Figure 1.35, PCB 180 elutes now at 6.46 min instead of at 29.23 min. The experimental speed gain factor is thus 4.5, corresponding very well with the predicted speed gain factor of 4.4. From the chromatograms, it is clear that the resolution has not been compromised compared with the analysis

Table 1.8 Translation of the chromatographic conditions from a conventional to a fast capillary column

	Conventional	Fast
Column	30 m × 0.25 mm i.d. × 0.25 μm HP-5MS	10 m × 0.10 mm i.d. × 0.10 μm HP-5MS
Pressure	71 kPa hydrogen	233 kPa hydrogen
Program	70°C–2 min–25°C min^{-1}–150°C–3°C min^{-1}–200°C–8°C min^{-1}–300°C (2 min)	70°C–0.45 min–110°C min^{-1}–150°C–132°C min^{-1}–200°C–35.2°C min^{-1}–300°C (0.4 min)
Analysis time	36 min	8.2 min

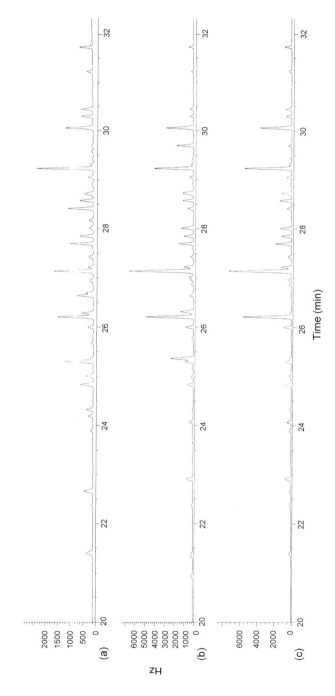

Figure 1.35 Retention time locked CGC-ECD analyses of (a) Aroclor 1260, (b) egg extract and (c) pork fat extract.

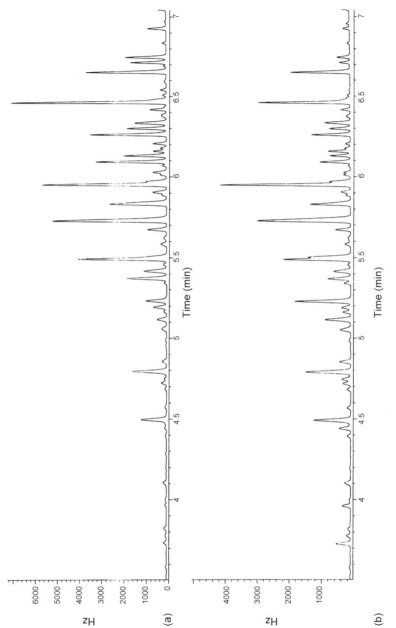

Figure 1.36 Fast high-resolution CGC-μECD analysis of (a) Aroclor 1260 and (b) a contaminated animal feed.

on the conventional columns. It is important to note that splitless injection can be applied in fast high-resolution capillary GC.

In conclusion, with the new sample preparation method, using ultrasonic extraction followed by matrix solid phase dispersion clean-up and fast CGC-ECD or CGC-MS, one technician on one instrument could easily perform 50 analyses per day on a conventional column and, with some help in sample preparation, even a 100 analyses per day with fast high-resolution CGC.

References

1. G. Baiulescu, P. Dumitrescu and P. Zugravescu, *Sampling*, Ellis Horwood, New York, 1991.
2. B. Kolb and L.S. Ettre, *Static Headspace Gas Chromatography: Theory and Practice*, Wiley-VCH, New York, 1997.
3. B. Kolb (ed.), *Applied Headspace Gas Chromatography*, Heyden, London, 1980.
4. H. Hachenberg and A.P. Schmidt, *Gas Chromatographic Headspace Analysis*, Heyden, London, 1977.
5. V. Berezkin and Y. Drugov, *Gas Chromatography in Air Pollution Analyses*, Elsevier, Amsterdam, 1991.
6. (a) K. Ventura, P. Príhoda and J. Churácek, *J. Chromatogr.*, **1995**, *710*, 167. (b) D. Helmig and J.P. Greenberg, *J. Chromatogr.*, **1994**, *677*, 123. (c) R.J.B. Peters and J.A.D.V Renesse van Duivenbode, *Atoms. Environ.*, **1994**, *28* (15), 2413.
7. Y.-Z. Tang, Q. Tran and P. Fellin, *Anal. Chem.*, **1993**, *65*, 1932.
8. T. Knobloch and W. Engewald, *J. High Resolut. Chromatogr.*, **1995**, *18*, 635.
9. F. Jüttner, *J. Chromatogr.*, **1988**, *422*, 157.
10. E. Baltussen, F. David, P. Sandra and C.A. Cramers, *J. Chromatogr. A*, **1999**, *864*, 325.
11. J. Pawliszyn, *Solid Phase Microextraction. Theory and Principles*, Wiley-VCH, New York, 1997.
12. J. Pawliszyn, *Applications of Solid Phase Microextraction*, Royal Society of Chemistry, Hertfordshire, 1999.
13. A. Khaled and J. Pawliszyn, *J. Chromatogr. A*, **2000**, *892*, 455.
14. B. Tienpont, F. David, C. Bicchi and P. Sandra, *J. Microcolumn Sep.*, **2000**, *12*(11) 577.
15. E. Baltussen, H.-G. Janssen, P. Sandra and C.A. Cramers, *J. High Resolut. Chromatogr.*, **1997**, *20*, 385.
16. E. Baltussen, F. David, P. Sandra, H.-G. Janssen and C.A. Cramers, *J. Chromatogr.*, **1998**, *805*, 237.
17. E. Baltussen, A. den Boer, P. Sandra, H.-G. Janssen and C.A. Cramers, *Chromatographia*, **1999**, *49*, 520.
18. E. Baltussen, F. David, P. Sandra, H.-G. Janssen and C.A. Cramers, *Anal. Chem.*, **1999**, *71*, 5193.
19. K. Grob and F. Zurcher, *J. Chromatogr.*, **1976**, *117*, 285.
20. I. Temmerman, F. David, P. Sandra and R. Soniassy, Application Note 228-135, Hewlett Packard, 1991.
21. E. Baltussen, H.Snijders, H.-G. Janssen, P. Sandra and C.A. Cramers, *J. Chromatogr.*, **1998**, *802*, 285.
22. E. Baltussen, F. David, P. Sandra, H.-G. Janssen and C.A. Cramers, *J. High Resolut. Chromatogr.*, **1998**, *21*, 645.
23. E. Baltussen, P. Sandra, F. David, and C.A. Cramers, *J. Microcolumn Sep.*, **1999**, *11/10*, 737.
24. A. Tredoux, H. Lauer, T. Heideman and P. Sandra, *J. High Resolut. Chromatogr.*, **2000**, *23*, 644.
25. W.M. Meylan and P.H. Howard, *J. Pharm. Sci.*, **1995**, *84*, 83.
26. A. Handley, *Extraction Methods in Organic Analysis*, Sheffield Academic Press, Sheffield, UK, 1999.
27. V. Giarrocco, B.D. Quimby and M.S. Klee, Application Note 228-392, Hewlett Packard, 1997.

28. F. David, D.R. Gere, F. Scanlan and P. Sandra, *J. Chromatogr. A*, **1999**, *842*, 309.
29. K. Ballschmiter and M. Zell, *Fresenius Z. Anal. Chem.*, **1980**, *302*, 20.
30. Method 3545 (1995) USEPA SW-846, 3rd edn, Update III, USEPA, Washington DC, 1995.
31. K. Ganzler, A. Salgo and K. Valko, *J. Chromatogr.*, **1986**, *371*, 2999.
32. J. Dean, *Extraction Methods for Environmental Analysis*, Wiley, Chichester, 1998.
33. A. Medvedovici, A. Kot, F. David and P. Sandra, in *Supercritical Fluid Chromatography with Packed Columns* (eds. K. Anton and C. Berger), Marcel Dekker, New York, 1997.
34. P. Sandra, A. Kot, A. Medvedovici and F. David, *J. Chromatogr.*, **1995**, *703*, 467.
35. B. Hawthorne, D.J. Miller, D. Nivens and D.C. White, *Anal. Chem.*, **1992**, *64*, 405.
36. K. Grob, *On-Line Coupled LC-GC*, Hüthig Verlag, Heidelberg, 1991.
37. F. David, P. Sandra, D. Bremer, R. Bremer, F. Roglis and A. Hoffmann, *Lab. Praxis*, **1997**, *21*(5), 33.
38. E.A. Ibrahim, I.H. Suffet and A.D. Sakla, *Anal. Chem.*, **1987**, *59*, 2091.
39. K. Blau and J. Halket (eds.), *Handbook of Derivatives for Chromatography*, Wiley, Chichester, 1994.
40. L. Moens, T. De Smaele, R. Dams, P. Van Den Broeck and P. Sandra, *Anal. Chem.*, **1996**, *15*, 1604.
41. H. Pham-Tuan, J. Vercammen, C. Devos and P. Sandra, *J. Chromatogr. A*, **2000**, *868*, 249.
42. J. Vercammen, H. Pham-Tuan, I. Arickx, D. Van der Straeten and P. Sandra, *Proc. 23rd Int. Symp. Capillary Chromatography* (eds. P. Sandra and A.J. Rackstraw), *J. Chromatogr. A*, **2001**, *912*(1), 127.
43. J. Vercammen, E. Baltussen, T. Sandra, F. David and P. Sandra, *J. High Resolut. Chromatogr.*, **2000**, *23*, 5.

2 Sample injection systems

Tony Taylor

2.1 Introduction

Methods of sample introduction have traditionally been seen as the weakest point in GC analysis. Certainly, thermal decomposition and discrimination of analytes with different volatility can be problematical, but recent advances in sample introduction technology have more than kept pace with advances in capillary column and detector technology. Accuracy and precision of data produced using some of the newest inlets surpass anything that has been seen in GC analysis to date and have expanded the range of samples that can be analysed.

This chapter on GC inlets is intended to present the capabilities, strengths and weaknesses of many common inlets and auxiliary sample introduction devices that are available as part of the armoury of the analyst.

2.1.1 Important concepts

The main function of the GC inlet is to provide accurate, reproducible and predictable transfer of the sample to the column without broadening the band of gas plasma being transferred and without altering the composition of the transferred sample plug relative to that of the original sample.

Inlets are generally divided into two major categories:

- those for use with lower-efficiency packed GC columns
- those for use with high-efficiency capillary GC columns

Packed column inlets are still perhaps the most widely used globally and are arguably the simplest to optimise and operate. These inlets have specific requirements for use while offering a range of particular advantages.

The main difference between packed and capillary inlets is the necessity for the capillary inlet to maintain an efficient sample transfer. Because capillary gas chromatography is a technique that to a large extent is driven by efficiency (capillary columns may generate huge numbers of theoretical plates), the sample inlet must transfer the band of analyte to the column with an efficiency loss that is less than the inherent band broadening exerted by the column. If this is not achieved, the inlet is the limiting factor in the efficiency of the analysis and optimisation of any separation to resolve critical pairs of peaks will be difficult.

Capillary column inlets are further subdivided into the following categories:

- capillary direct (vaporising)
- split/splitless (vaporising)
- programmed temperature vaporiser (vaporising)
- cool-on-column (nonvaporising)
- auxiliary sample introduction devices (headspace, purge and trap, thermal desorption devices and gas and liquid sampling valves)

Vaporising devices rely on a transition of the analyte to the gas phase prior to transfer onto the column, whereas nonvaporising injectors transfer the analyte in the liquid phase directly onto the column.

In general the choice of inlet will very much depend upon the nature of the sample and the analysis goals as well as the column type. After selection of the type of inlet to be used, the following primary variables usually need to be optimised and fixed:

- injection technique
- injection volume
- inlet temperature (initial and/or programmed)
- column type (stationary phase and dimensions)
- column temperature
- liner selection

Where appropriate each of the above variables will be discussed in detail below.

2.2   Split/splitless injectors

The 'split/splitless' injector is the most popular inlet for capillary gas chromatography and is a vaporising injector [1]. The split injector was the first inlet to be developed for capillary gas chromatography and is used to introduce only a small amount of the sample vapour onto the analytical column. Effectively, the split inlet can be used to take an 'on-instrument' aliquot of the sample, whereas in the splitless mode the entire sample is transferred to the column. Split injection guarantees narrow inlet bands owing to the rapid sample vapour transfer, whereas the splitless injector can guarantee high-sensitivity analysis [2].

2.2.1   Split/splitless inlet design (Figure 2.1)

Carrier gas, controlled by pressure regulation devices (forward pressure regulator or flow controller and back pressure regulator), enters the injector at the

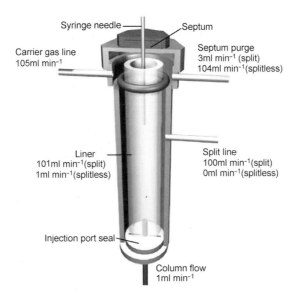

Figure 2.1 Major components of split/splitless injection system.

top and is divided into a number of streams:

1. One stream of carrier is directed past the bottom of the injector septum and is regulated by a needle valve or flow controller. This is known as the septum purge and is designed to carry away the products of septum out-gassing (bleed) and residues from the outside of the injection syringe. This stream helps to reduce baseline noise, will minimise ghost peaks and is fixed by the operator at a nominal flow rate, normally in the range of 1–4 ml min^{-1} depending on instrument type.
2. The other stream of carrier gas is directed into the injection liner and is allowed to mix with the vaporised sample.
3. This mixed stream is then either quantitatively transferred to an analytical column or retention gap (splitless injection), or split between the column inlet and the split vent (split injection), where an on-instrument dilution is effectively achieved for more concentrated samples.

2.2.2   Septa

The septum acts as a barrier between the injection head pressure and atmospheric pressure as well as being a junction through which the injector syringe needle may pass to deliver the sample into the injection port. Two main problems are associated with septa, namely 'bleed' of volatile material into the system at elevated temperatures and leaking, usually as a result of mechanical wear.

Septum bleed occurs when volatile compounds are emitted from the septum; they originate from monomers used during manufacture or from additives used to control characteristics such as thermal stability, mechanical strength or penetrability. When using temperature programmed GC, septum bleed can result in baseline instability or as sets of discrete extraneous peaks that are not associated with the sample or with bleed from the column. The volatile materials produced from the septum collect at the head of the column during the oven cool-down period and initial temperature hold, eluting during the subsequent programmed run. Under isothermal conditions, septum bleed is continuous and will be contributory to a noisy baseline [3] (Figure 2.2).

The use of capillary GC tends to exacerbate the problems of septum bleed owing to the high efficiency of columns, giving rise to discrete peaks rather than a general baseline disturbances. Further, the isothermal hold period associated with splitless injection contributes to sharpening and amplification of the septum bleed products that have not been swept away by the septum purge flow, resulting in significant extra peaks that may interfere with quantification or interpretation of the chromatogram.

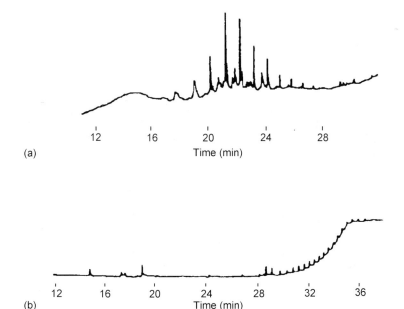

Figure 2.2 Typical septum bleed profile from temperature programmed analysis. (a) Splitless injection. Column: 30 m × 0.53 mm × 1.0 μm (100% PDMS). Oven temperature: 40°C to 300°C at 15°C min⁻¹. Injection/detection temperature: 330°C. Carrier: hydrogen; linear velocity 40 cm s⁻¹. Attenuation: 4 × 10⁻¹¹ AUFS (arbitrary units full scale). (b) Split injection (40 ml min⁻¹ flow rate). Column: 30 m × 0.53 mm × 1.0 μm (100% PDMS). Oven temperature: 40°C to 340°C at 15°C min⁻¹. Injection temperature: 340°C. Carrier: hydrogen; linear velocity 40 cm s⁻¹. Attenuation 8 × 10⁻¹¹ AUFS.

Once the septum loses its sealing capability owing to extensive coring or splitting, the head pressure within the injector (and hence the flow rate of carrier gas) will alter. This may give rise to variable elution times for sample components (affecting peak identification and tracking), and may also result in erratic baselines and/or loss of sensitivity. Eventually the septum will fail and the system will lose pressure, which may be the first time that the problem is identified as being due to the septum!

Figure 2.3 shows a typical chromatographic symptom created by a leak at the septum during injection. The baseline position changes after the elution of a large (usually solvent), peak. This indicates that the septum is not sealing during injection and will therefore leak during injection and for a short time afterwards. This symptom is exacerbated if the diameter of the needle used is too large.

Most manufacturers supply septa that have different temperature ratings (usually septum composition changes for injection port temperatures above 250–300°C) and different properties of robustness to prevent coring and splitting. It is best to identify the type that gives the optimum performance for a particular application. All inlets that employ septa as the sealing mechanism are susceptible to the problems outlined above. Because inlet parameters such as temperature, liner type, oven temperature and injection events differ widely for the split and splitless modes, they will be discussed separately.

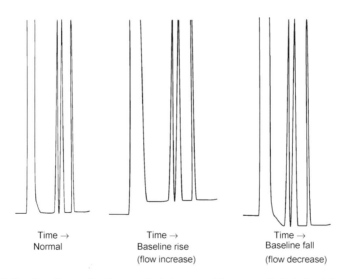

Time →   Time →   Time →
Normal   Baseline rise   Baseline fall
              (flow increase)   (flow decrease)

Figure 2.3 Baseline change after large peak elution caused by septum leak during injection. Split injection ($100\,\mathrm{ml\,min^{-1}}$ flow rate). Column: $30\,\mathrm{m} \times 0.25\,\mathrm{mm} \times 0.25\,\mathrm{\mu m}$ (100% PDMS). Oven temperature: $100°\mathrm{C}$. Injection temperature: $260°\mathrm{C}$. Carrier: hydrogen; linear velocity $50\,\mathrm{cm\,s^{-1}}$.

2.2.3 Split injection mode

Split injection should be chosen for samples that

- are volatile
- are suspected of being thermally unstable
- are permanent gases
- are suspected of containing components without widely differing concentrations, i.e. it should not be used for the analysis of trace components in a major component

These choices are based on the fact that the split injection mode will rapidly transfer the sample to the column (minimising residence time in the high-temperature environment of the injector) and will produce the narrowest initial sample band on the column [4]. Under injection conditions shown in Figure 2.4, for a 1 μl injection of sample in methanol the transfer time of high-volatility sample components onto the column would be

$$(0.525/101) \times 60 = 0.3\,\mathrm{s}$$

(Solvent expansion volume for 1 μl methanol=525 μl gaseous methanol.)

Solutes whose boiling points are significantly higher than that of the solvent may take much longer (several seconds) to fully volatilise and transfer to the analytical column.

2.2.3.1 Split flow rates

As stationary phase capacity is limited in capillary GC, the split flow should be adjusted so that the chromatographic peaks do not show characteristics of column overload as seen in Figure 2.5 below.

A further consideration when choosing split flow rate is the initial sample bandwidth. Reducing the split ratio in Figure 2.4 to 10:1 would result in a transfer time of 3 s, leading to much broader sample bands being introduced into the column and hence broader chromatographic peaks. This effect may be overcome by using focusing techniques that will be discussed later.

Perhaps the most important concern is obtaining a consistent and reproducible split [5]. This is accomplished by calculating, measuring and setting the split ratio in the same manner for a given analysis. The actual split ratio is not critical as long as sufficient compound is introduced into the column without loss of sensitivity (split flow is too high) and the column is not overloaded (split ratio too low), when peak fronting will occur. An excessively low split ratio will often result in peak tailing problems, due to increased sample transfer times. Typical split ratios generally range between 50:1 and 500:1, although specialist applications may require flows outside this range.

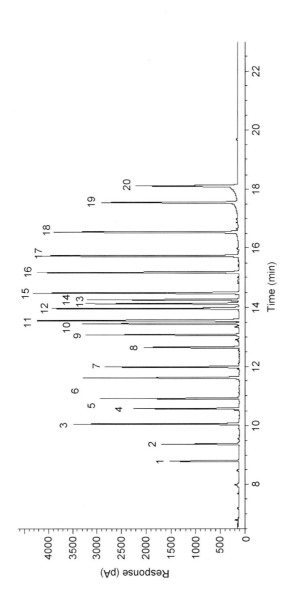

Figure 2.4 PCB congener standard solution analysed by split injection. Oven: 80°C (2 min), 30°C min⁻¹ to 200°C, 10°C min⁻¹ to 320°C (5 min). Inlet: split 300°C (split flow 100:1). Carrier 80°C, helium, 1.3 ml min⁻¹, constant flow. Sample 1 μl. Column HP5-MS 30 m × 0.25 mm × 0.25 μm. Detection: FID 330°C (N₂ make-up, constant column and make-up flow). Peak 1 2,4-dichlorobiphenyl; Peak 20 decachlorobiphenyl. (Courtesy of Agilent Technologies.)

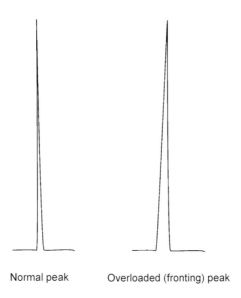

Normal peak   Overloaded (fronting) peak

Figure 2.5 Typical 'shark fin' appearance of an overloaded GC peak.

2.2.3.2   Liners

The liner is used to contain the gaseous plasma of volatilised sample and solvent, and to achieve the required split (dilution), of the sample. Liners with cups, frits, baffles, restrictions or wool packing are available (Figure 2.6), and all are designed with the following key principles in mind.

Ensuring an efficient sample transfer to either the column or the split vent. Split liners tend to have smaller outside diameter (o.d.), as this allows a significant gap between the liner body and the wall of the injection port for the split flow to travel freely. However a reduced inner diameter (i.d.) restricts the amount of sample solution that can be injected due to solvent expansion (see later section on 'Backflash'). Unless specifically designed to do so, the liner should not rest on the bottom plate of the injector, as this will severely restrict the flow of gas and sample components from the liner to the split line outlet. Further, the sample will be more likely to contact the hot metal bottom plate of the injector, increasing the likelihood of thermal decomposition and/or adsorption.

Trapping of particulate matter. Particulate matter from the sample or from septum fragments should be excluded from the column inlet, where it may cause blockage or excessive bleed. Most modern liner designs are very effective at trapping particulates, with no one design being outstandingly more effective than another.

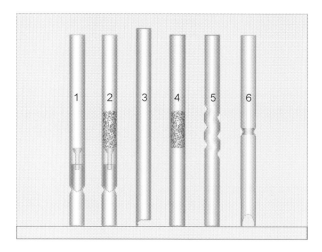

Figure 2.6 Various liner configurations. (1), Jennings cup split liner: improves sample homogeneity and decreases sample discrimination. (2) Packed cup type liner for larger injection volumes and samples containing low volatility components. (3) Small-volume liner. (4) Packed small-volume liner for samples with low-volatility components. (5) Multi-baffled liner: increased surface area gives better volatilisation and sample homogeneity; used extensively for PTV injection. (6) Fritted liner for samples containing particulate matter and low-volatility components.

Reducing the amount of discrimination. Discrimination is the process whereby low-volatility sample components are less effectively transferred to the analytical column than their more volatile counterparts (Figure 2.7). This process may be due to high molecular weight compounds sticking to the syringe needle wall after the solvent and volatile compounds are evaporated or by adsorption onto the liner inner surface and packing when the split flow is poorly optimised. The fast needle injection technique is used by most autosampling devices and ensures that the sample is ejected from the syringe in the liquid form and does not vaporise within the syringe needle leading to selective concentration of the higher-boiling solutes [6] (Figure 2.8).

For manual injection, which can take up to several seconds to complete, the hot needle technique is found to produce the most reproducible level of discrimination, and therefore the best precision for quantitative data, except when dealing with thermally labile analytes [7].

The sample is loaded directly into the syringe barrel (no air gap is left between the plunger and the sample plug), and the sample is withdrawn to leave air in the needle. After insertion into the injection port, the needle is allowed to heat up for 3–5 s, which is sufficient for it to reach the injection port temperature. The sample is then expelled by rapidly depressing the plunger and the needle is quickly withdrawn from the inlet (within 1 s). Injection port discrimination is minimised when the vaporised solvent and analyte components are effectively mixed during residence in the liner and when a surface is provided

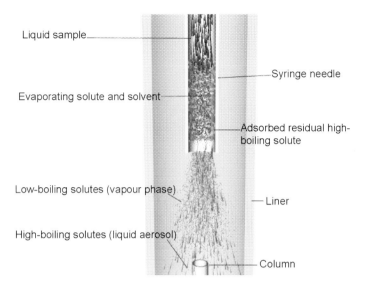

Figure 2.7 High mass discrimination processes.

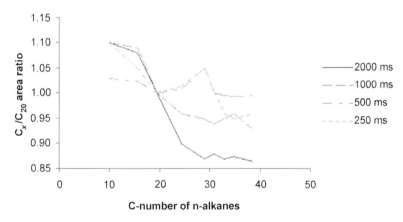

Figure 2.8 Degree of discrimination versus injection port residence time for a 'fast' autoinjector: HP5890 (7673 Autosampler). Column: 10 m × 0.53 mm HP-1. Temperature: 60°C (2 min), 15°C min^{-1} to 310°C (5 min). Injection: splitless. Sample: 1 µl (C_{10} to C_{40} in n-hexane).

for lower-volatility analytes to volatilise completely over an extended period (several seconds). Liners may contain baffles and wool packing to carry out this function as well to trap any particulate matter that may enter the injector.

Ensuring homogeneity of the volatilised sample and optimum sample transfer. Glass wool is one of the most 'active' liner packings and increases the possibility of adsorption and decomposition of the sample through interaction between the wool surface silanol (—SiOH) species and any polar moieties of the analyte molecule. Deactivated quartz or glass wool is recommended for split injection, using an amount sufficient to form a plug within the liner, but care should be taken not to over-pack or form a very dense plug.

Correct placement of the wool packing in the liner is a matter of debate among various manufacturers. The inlet will have a temperature gradient that optimises at the hottest point (closest to the set temperature), somewhere mid-way down the liner, and ideally the wool should be positioned at this point to ensure optimum volatilisation and mixing of the sample. Manufacturers who supply pre-packed liners will generally have located this position and the wool packing should not be moved prior to installation.

If the sample is injected directly into the glass wool (i.e. the needle tip penetrates the packing), mixing tends to be more efficient, high mass discrimination is lowered (the aerosol of lower-boiling components is directed onto a surface where components may volatilise over several seconds) and the tip of the needle is wiped to reduce needle discrimination. The major drawback to this approach is the disruption and breakage of wool strands, which may potentially expose fresh, nondeactivated, silanol species and lead to increased sample adsorption. It may be necessary to move the wool plug away from the needle tip when dealing with particularly polar compounds or compounds that show strong adsorption.

Positioning of the column inlet within the liner is also of vital importance to ensure that the optimum amount of gas plasma is sampled into the column in a smooth and controlled fashion. In this respect most manufacturers will recommend fixed distances for the column inlet from the end of the column nut or ferrule. These distances have been optimised empirically and should be strictly adhered to for optimum inlet performance.

Straight liners are satisfactory for almost all split applications, with baffled or inverted cup liners (Figure 2.6) altering only the relative amounts of sample discrimination. The relative peak size may vary slightly between different liner designs, but the qualitative and quantitative data obtained will be consistent if the same liner design is used for a particular application. It is rare that one particular liner design will give outstandingly better peak shape or area consistency compared to another.

To avoid adsorption of sample components onto the inside surface of the liner, it should be cleaned and deactivated on a regular basis. Liner deactivation regimes are many and varied so one particularly easy and effective procedure is cited for reference [5]. (Care is required not to scratch or etch the liner surface, potentially exposing further silanol species.)

1. Place liner in a screw-cap test tube or other appropriate sealed vessel with PFTE cap.
2. Soak the liner in 0.1 N HCl for 8 h.
3. Rinse with deionised water then methanol.
4. Dry at 100–150°C.

5. Place in a clean test tube and immerse in 10% trimethylchlorosilane in hexane.
6. Replace cap tightly.
7. Soak for 8 h, remove and rinse thoroughly with hexane then methanol.
8. Dry at 100–150°C for 30 min.
9. Handle with tweezers.

2.2.3.3 Temperature

Complete evaporation of the sample via flash vaporisation is necessary to minimise discrimination and to maximise the accuracy and reproducibility of split injections. For an unknown sample, the inlet temperature should be set to 250°C and a scouting oven temperature gradient of 40–300°C (or 10°C below the maximum column temperature) at 10°C min^{-1} implemented. The boiling point of the last major component within chromatogram should be noted and the inlet temperature set at least 20°C above this. Optimisation of the inlet temperature can be achieved by monitoring peak area response at temperatures altered in 20°C steps above and below the initial set temperature. Figure 2.9 shows typical baseline rise before or after the analytical peak that is often indicative of thermal decomposition of the analyte and should obviously be avoided when optimising the inlet temperature.

2.2.3.4 Injection volume

Upon vaporisation, many solvents expand to a volume 10–1000 times their original liquid volume [1]. The volume of a standard split injection liner is usually between 400 and 900 µl. If the vaporised sample volume exceeds the volume of the liner, some of the sample will flow outside the liner zone. A pressure pulse is created on sample vaporisation that often exceeds the carrier gas head pressure and as such the excess vapour volume may travel back up the

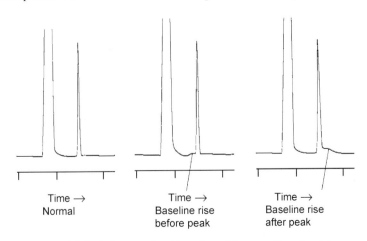

Figure 2.9 Peak shapes attributable to thermal decomposition of analyte.

carrier gas inlet line, leading to two potential problems—peak broadening and injector contamination.

As the solvent 'backflashes' up the carrier gas line, it now occupies a much larger volume and therefore will take longer to transfer to the analytical column. This will cause both solvent and analyte peaks to be broad and tailing, the problem being worse for solvents with large expansion volumes and lower carrier gas flow rates.

Perhaps the more serious problem with backflash is injector contamination, which can be responsible for ghost peaks, carryover and problem baselines. The backflashed compounds may condense in the cool carrier gas line, where they will reside until the next backflash occurs (usually the next injection). The condensed components from the previous injection will be solubilised by the backflash vapours and drawn into the injector body and hence into the analytical column, leading to discrete carryover peaks. Further, the condensed components in the carrier gas line may slowly bleed into the injector during the complete analysis cycle, leading to baseline disturbances and a raised background signal.

Backflash problems may be minimised or eliminated in a number of ways. These include the use of smaller injection volumes, the use of solvents with lower expansion volumes, the use of increased head pressure during the injection phase of the analysis cycle (compressing the vapour) and the use of liners with upper restrictions to limit the amount of vapour that can escape from the top of the liner.

Perhaps the simplest and most effective approach is to ensure that the liner inside diameter is large enough to contain the amount of vapour produced; a list of solvent expansion volumes is given in Table 2.1 for reference to ensure that the user may select an appropriate injection volume for the solvent being used.

2.2.4   Splitless injection mode

Splitless injection is conventionally used for low-concentration samples as the majority of the injected sample is introduced into the column, resulting in much higher sensitivity compared with split injection.

Table 2.1 Expansion volumes for some typical solvents used in GC analysis

Solvent	Boiling point (°C)	Expansion volume (µl)
Acetone	56	290
Acetonitrile	85	405
Carbon disulfide	46	355
Ethyl acetate	77	215
Isooctane	99	130
Methanol	65	525
Methylene chloride	40	330
n-Hexane	69	165
Toluene	111	200
Water	100	1180
For 1 µl injection at 250°C and 15 psig (103.4 kPa)		

Splitless injection is routinely used in areas such as environmental analysis, pesticide monitoring of foods and drug screening. In these application areas, sample preparation requirements are significant and it is not always possible to justify extensive sample clean-up. Therefore, column protection becomes as important as sensitivity.

The hardware used for splitless injection is essentially the same as for split injection. Prior to injection, a solenoid valve is actuated so that the split valve is closed and the split flow is either turned off (with a corresponding reduction in carrier gas), or is diverted through the septum purge line. After injection, the solvent vaporises and is slowly transferred to the column (usually over 30–120 s). Compare the injector residence time of 0.3 s derived in the previous section with a residence time of 31.5 s for the splitless version under identical analytical conditions! After this time the split line valve opens to purge the injector of residual vapours and potential nonvolatile components. This approach helps to protect the analytical column from contaminant species that are more concentrated in splitless injection.

Figure 2.10 illustrates a chromatogram obtained with splitless injection.

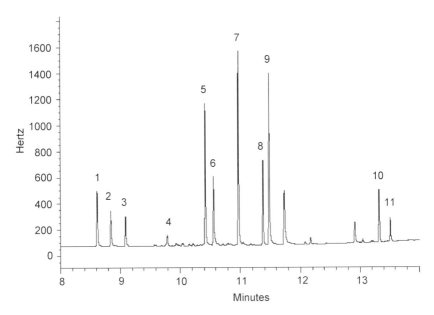

Figure 2.10 Nitroaromatics from forensic wipe sample (solute concentration 40 ppb in ethyl acetate). Oven temperature 50°C (1 min), 20°C min^{-1} to 320°C (0.5 min). Inlet: splitless 250°C (liner: single taper, deactivated). Carrier: hydrogen, 1.4 ml min^{-1}, constant flow. Sample: 1 μl. Column HP5-MS 30 m × 0.25 mm × 0.25 μm. Detection ECD 320°C (argon−5% methane, 20 ml min^{-1}). Peak 1 1-nitronaphthalene; Peak 11 2,7-dinitrofluorene. (Courtesy of Agilent Technologies.)

2.2.4.1   Purge time

If the split flow is not actuated to purge the injector after the sample components have reached the column inlet, significant solvent peak tailing will be observed (Figure 2.11) potentially obscuring early-eluting components and making quantification more difficult [8]. If the purge valve is opened too early, some of the high-boiling components may be ejected before complete transfer to the analytical column, resulting in discrimination. Purge-on times should be optimised for each particular analysis to minimise such injector discrimination.

Appropriate purge delay times are a compromise between the amount of sample transferred to the column and the degree of tailing associated with the solvent front. Optimal purge time is interdependent with all other injector variables and corresponds to quantitative transfer of between 95% and 99% of the vaporised sample onto the analytical column.

Purge delay time should be determined only after all other inlet parameters have been optimised. The following empirical approach may be adopted to optimise the purge-on time.

1. Start by injecting a sample using a long purge time (90–120 s) and measure the area of a solute that elutes with a retention factor > 5. This should correspond to 100% of the sample reaching the column.
2. Reduce the purge time in large increments (30 secs., 20 secs., etc.) and re-inject the sample until a lower area for the peak results.
3. Use smaller time increments to adjust the purge-on time up or down to achieve a peak area reduction of between 1% and 5% from the original (100%) value.

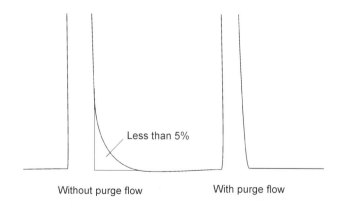

Less than 5%

Without purge flow   With purge flow

Figure 2.11 Purge effects on solvent peak in splitless injection.

For early-eluting analytes it is better to adjust to the shortest purge-on time to ensure that the analytes do not elute on the solvent tail. For late-eluting compounds, the purge-on time should be set as long as possible to maximise analytical sensitivity; however, this approach may potentially lead to increased column contamination.

2.2.4.2 Solvent and temperature choice

As the sample components are transferred very slowly to the column in comparison with split injection, the analytical peaks obtained will be broader. To overcome this problem, 'focusing' techniques are usually employed in splitless injection [9,10] (Figure 2.12).

Stationary phase focusing is only possible with temperature programmed analysis. Retention of analyte species is an exponential function of temperature, therefore as the initial temperature of the column is lowered, the speed at which solutes travel down the column is dramatically slowed. As the vaporised sample moves slowly from the inlet it comes into contact with the stationary phase and is effectively trapped into a narrow zone. The lower the temperature, the more effective the focusing.

Solvent focusing occurs if the polarities of the solvent and the stationary phase are matched. If the initial column temperature is below the solvent boiling point, the solvent will condense on the stationary phase surface, trapping all higher-boiling point sample components. Owing to the high linear gas flow rates and reduced vapour pressure conditions inside the column, the solvent will begin to evaporate from the injector end. As this happens, the solvent forms a continually smaller plug that will contain the more volatile sample components, effectively focusing the dissolved volatile solutes.

Thermal focusing occurs when the initial column temperature is at least 150°C below the boiling point of the sample component. As such, thermal focusing is usually associated with less volatile sample components (i.e. those not associated with the solvent focusing effect). The focusing occurs as a result of the thermal condensation of vapours at the head of the analytical column, narrowing peak volumes during the condensation process. In this sense, thermal focusing does not rely on the chromatographic process but only requires a surface on which vapours may condense. Thermal focusing is usually accompanied by stationary phase focusing.

Injection solvents for splitless injection should be chosen so that the solvent is compatible with the stationary phase in terms of polarity. If the polarity is not matched, the solvent will not form a continuous film upon condensation at the head of the column, but will instead form discrete pools and droplets of solvent, precluding any solvent focusing effect and resulting in broad or split peaks for the early-eluting sample components.

In splitless injection, higher-boiling solvents have several advantages over lower-boiling-point components:

- lower syringe discrimination
- lower pressure pulses associated with sample evaporation

High-boiling components begin to condense via thermal effect

Low-boiling components dispersed in condensed solvent phase

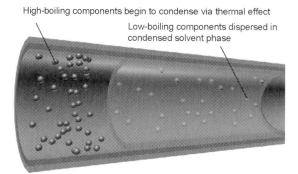

High-boiling components condense at specific column temperature

Low-boiling components focus as solvent evaporates under reduced vapour pressure

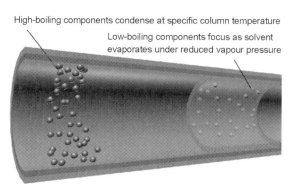

Both high- and low-boiling components now fully focused—temperature increase will begin elution

Figure 2.12 Thermal and solvent focusing effects. Column temperature 20 C below solvent boiling point.

- easier solvent focusing
- higher initial column temperatures (leading to reduced oven cool down times and shorter analytical turnaround)

The initial column temperature should be chosen so that stationary phase and thermal focusing are most effective. This generally occurs when the column temperature is at least 10–20°C below the boiling temperature of the solvent (Figure 2.13).

2.2.4.3   Liners
An important feature of the liner design for splitless injection is the internal volume available for solvent expansion, liners with volumes between 0.25 and 1 ml being usual. Long and narrow inserts are best to ensure minimal sample dilution during the slow vapour transfer, but this may restrict the injection volume owing to solvent expansion (backflash) problems.

Since the contact time between the sample and the liner is very long, liner activity can be a problem in terms of sample adsorption, poor peak shape and decomposition of labile compounds (Figure 2.8). Even deactivated liners and liner packing will become more active over time, so frequent cleaning and replacement of any packing materials is necessary. Wool packing is required for fast autoinjection to maximise reproducibility and to help retain nonvolatile sample components.

2.2.4.4   Temperature and injection volume
The inlet temperature must be high enough to completely vaporise all sample components and minimise residence time in the inlet. However, the lowest

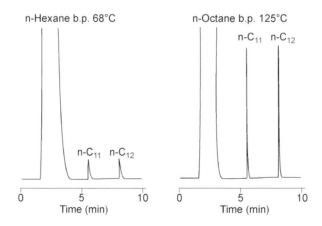

Figure 2.13 Effect of column temperature and solvent boiling point in focusing experiments. Column OV-101 Glass WCOT. Temperature: isothermal 115°C. Injection 2 μl (100 ng C_{11} and C_{12}). Carrier: helium, 0.8 ml min^{-1}.

temperature that achieves these aims should be chosen so as to minimise thermal degradation and reduce the chance of backflash. Solvent volumes must be kept to well below the liner internal volume so that none of the vapour is carried away into the split line, and the full expansion volume is contained within the liner.

If the inlet temperature is too low, this will prevent higher boiling solutes from reaching the column and inlet discrimination will occur. The proportion of the high-boiling solutes reaching the column will decrease as a function of their boiling points.

2.3 Packed-column inlets

Packed-column inlets are traditionally used with glass or stainless steel columns (1/4 inch, 6.35 mm o.d.) packed with silica or a similar inert support material coated with a polymeric liquid stationary phase. As these columns are packed they are inherently less efficient than capillary columns and therefore the efficiency requirements associated with capillary column inlets do not apply. This type of inlet is therefore traditionally used for separations where high efficiency separations are not required [11].

2.3.1   Inlet design

Figure 2.14 shows a typical configuration of a packed inlet including the configuration of inserts that may be used to adapt the inlet for use with columns of varying outer diameter. Carrier gas enters the side of the inlet body and will heat up as it passes up the inlet between the inlet body and the column or insert wall. The hot carrier will then flow down the column to carry the vaporised sample onto the column. In fact, where the inlet is used without inserts the volatilisation of the sample occurs in an unpacked region near the top of the column.

The inlet may also be configured with a septum purge analogous to that used in split/splitless and other inlet types. Gas pressures are controlled via either forward or back-pressure regulated flow controllers to maintain a constant column pressure. More advanced GC systems may also use differential pressure controllers to maintain constant flow (and hence a constant baseline level) during temperature programmed analysis.

Inlet liners can be used in the inlet, the column (1/4 inch, 6.35 mm or 1/8 inch, 3.2 mm o.d.) being attached using nut and ferrule fittings to the base of the inlet. Liners can be packed or straight-through and manufactured from glass or stainless steel lined with glass. The liners are used to protect the column inlet from contamination; thus may be easily cleaned and, if glass, will reduce the amount of sample adsorption and decomposition when using stainless steel packed columns.

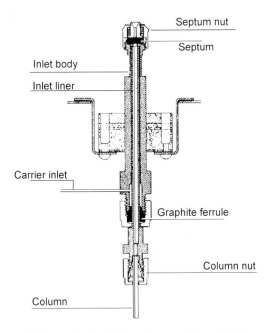

Septum nut

Septum

Inlet body

Inlet liner

Carrier inlet

Graphite ferrule

Column nut

Column

Figure 2.14 Packed column inlet. (Courtesy of Agilent Technologies.)

2.3.2   Inlet parameters and sample considerations

The inlet temperature should be chosen to be well above the boiling point of the sample solvent or that of the major solutes of interest, so that they are efficiently transported through the inlet region of the column. Excessive temperatures may lead to increased discrimination, backflash and degradation of the sample components. The temperature should be set to the minimum value that does not cause excessive band broadening or causes reduced peak areas (due to incomplete volatilisation). When analysing novel solutes, a range of inlet temperatures should be investigated (in steps of $+50°C$ from the solvent boiling point) until constant area is achieved for peaks of later-eluting sample components.

Flow rates for packed columns usually lie in the range 10–50 ml min^{-1} (typically 5–20 cm s^{-1}) but will vary depending upon the carrier gas used (nitrogen is the most commonly used and most efficient for the range of linear velocities suggested). Manufacturers' literature should be consulted for optimum flow rates depending on column and carrier gas choice or may be empirically determined.

Packed column inlets are active towards polar components and as such may cause peak tailing or sample decomposition when the inlet is particularly active. Activity is caused by polar (silanol) species on the glass (quartz) wool packing or the glass walls of the column or inlet liner, giving rise to a secondary (unwanted), mechanism of interaction between the polar silanol species and polar moieties of the solute. The low internal volume of the inlet and the relatively fast flow rates of carrier used, reducing the residence time of the solute species in the inlet, reduce solute tailing to some extent.

The low internal volume of the inlet makes the system more susceptible to backflash into the carrier gas line and onto the septum, leading to problems similar to those outlined in the previous section.

2.4   Capillary direct interfaces (Figure 2.15)

Packed-column inlets may be easily adapted for use with wide-bore capillary columns through the use of appropriate liner and coupling technologies. This approach may be taken because flow rate ranges used for wide-bore (commonly called 'mega'-bore) capillary columns (optimum is around 3–5 ml min^{-1} for 0.53 mm i.d. capillary columns), can be handled by most packed inlet liners although some may require manufacturer-supplied restrictors in the GC pneumatics to stabilise flow at lower flow rates. Capillary direct inlets rely on flash vaporisation and therefore are also susceptible to sample discrimination and decomposition.

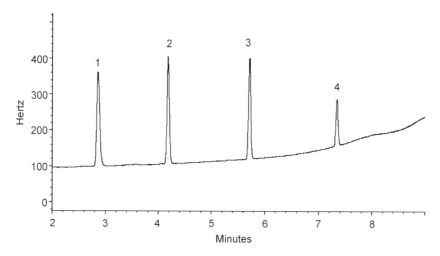

Figure 2.15 Capillary direct injection used in the analysis of trihalomethanes in water (25 µg l^{-1} sample concentration). Oven temperature: 35°C (1 min), 10°C min^{-1} to 125°C. Inlet capillary direct 250°C (liner: double taper, deactivated). Carrier: hydrogen, 5 ml min^{-1}, constant flow. Sample: 1 µl. Column HP5-MS 30 m × 0.53 mm × 0.25 µm. Detection: µECD 340°C (argon−5% methane, 30 ml min^{-1}). Peaks: 1 chloroform; 2 dichlrobromomethane; 3 dibromochloromethane; 4 bromoform.

2.4.1 Inlet design

Capillary direct inlets are usually packed column inlets modified through the use of special liners and inlet base column connectors. Carrier gas flows are configured as for packed column inlets (Figure 2.14).

2.4.1.1 Inlet parameters and sample considerations

Liner volumes should be chosen to be larger than the vapour volume resulting from injection to avoid backflash. Typical liner configurations are shown in Figure 2.16.

The straight liner is very effective for efficient sample transfer at high carrier gas flow rates, although liner volume is limited. The double taper liner has a larger internal volume; the tapers also help to contain sample vapours within the liner and are normally used for extracolumn direct injection. The single taper liner may be used in two orientations. With the expansion volume at the top of the inlet, extracolumn direct injection is possible, and with the taper at the top of the inlet the column can be sealed against the taper to facilitate on-column direct injection. Labile samples are more accurately analysed using intracolumn injection owing to limited exposure to active sites on-column relative to a glass inlet liner. On-column injection volumes should be kept to below 1 μl and flow rates should be kept as high as is practically possible to prevent stationary phase overloading. Slow injection (over a period of a few seconds) helps to

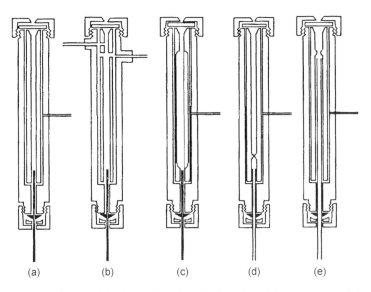

(a)   (b)   (c)   (d)   (e)

Figure 2.16 Liners for direct injection. (a) Open liner. (b) Open liner with septum purge. (c) Open liner with expansion volume. (d) Open liner with constriction. (e) Direct on-column liner.

reduce flashback, decreases peak widths and improves resolution of early-eluting components.

Dirty samples should be analysed in extracolumn mode so that the liner may adsorb high boiling and nonvolatile sample components to protect the analytical column. Alternatively, a retention gap may be used that can be easily replaced following contamination [12].

Band broadening can be a problem with capillary direct inlets because of the lower flow rates compared with packed columns. Stationary phase focusing is normally required to achieve efficient chromatography, and solvent focusing may also be usefully employed, yielding an analysis similar to that of traditional splitless injection (without injection purge). All precautions associated with this mode of injection should be employed, including matched solvent/stationary phase polarity and limited injection volumes. If peak splitting or excessive band broadening are encountered, the use of a retention gap should be considered.

2.5   Cool on-column inlets

When properly optimised, cool on-column injection can provide the most accurate and precise results of all the available capillary column inlets [13]. It has the following advantages over vaporising inlets:

- eliminates sample discrimination—syringe discrimination may be completely eliminated if correct techniques are employed and inlet-related discrimination is overcome as the liquid sample is directly introduced onto the analytical column
- elimination of decomposition—thermally labile samples can be introduced without degradation and rearrangement reactions are eliminated
- solvent focusing of early eluting analytes is possible
- quantitative accuracy and precision are high

However, there are several important restrictions associated with cool on-column inlets:

- sample injection volumes are limited
- peaks eluting prior to the solvent are difficult to analyse
- capillary columns can be easily overloaded leading to poor peak shapes
- inlet variables require careful optimisation
- column contamination is possible with dirty sample matrices

Figure 2.17 shows a cool on-column analysis.

2.5.1   Inlet design

Inlet selection and design depend largely on whether the injection is to be manual or automatic. Manual cool on-column injection can be carried out

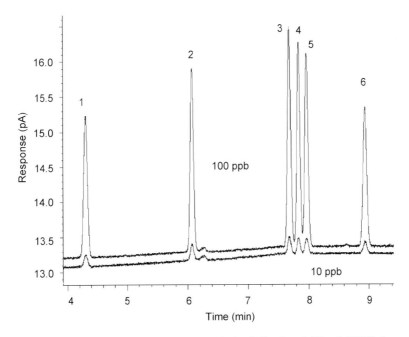

Figure 2.17 Cool on-column ambient headspace analysis of 10 ppb and 100 ppb BTEX (benzene, toluene, ethylbenzene, xylene) in water. Column: 30 m × 0.53 mm × 1.0 μm HP-INNOWax. Oven temperature: 40°C (2.5 min), 10°C min^{-1} to 90°C (7 min). Inlet cool on-column, oven track. Detector: FID, 350°C. Injection volume: 50 μl (headspace vapour). Peaks: 1, benzene; 2, toluene; 3, ethylbenzene; 4, *p*-xylene; 5, *m*-xylene; 6, *o*-xylene.

using capillary columns with internal diameters greater than 0.2 mm, whereas automatic injection is usually carried out with 0.32 mm i.d. columns or more traditionally with 0.53 mm i.d. columns.

Manual injection syringes are usually silica capillaries with outside diameters smaller than the inner diameter of the capillary column (typically 105 mm long × 0.14 mm o.d.), which allows direct introduction of the liquid sample into the column. The inlet is designed to guide the delicate needle into the column while providing a pressure seal during the injection cycle, and to provide accurate thermal control for heating and cooling.

A simple design for manual cool on-column injection is shown in Figure 2.18. The inlet has a low thermal mass to provide rapid and accurate heating and cooling and a duckbill is used to provide pneumatic sealing. A needle guide is usually employed to part the duckbill surfaces and prevent contact between the silica needle and the valve. Once the needle is in position in the column, the needle guide is released and the duckbill seals around the upper part of the needle for injection. The syringe plunger is rapidly depressed to inject the sample, and the needle is immediately withdrawn. For

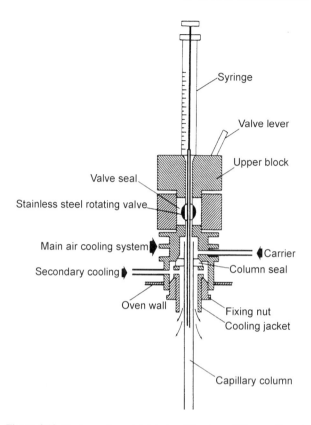

Figure 2.18 Cool on-column inlet design. (Courtesy of Thermo Quest.)

several seconds after injection the sample liquid migrates to form a flooded zone on the inner surface of the capillary. The inlet and oven temperature are then ramped to successively vaporise the sample components and initiate chromatography.

Automatic cool on-column injection is usually carried out using 26-gauge syringes and 0.53 mm i.d. columns or 32-gauge syringes and 0.32 mm i.d. columns. For automatic injection, the duckbill is usually replaced with a septum and nut to ensure a pressure seal, and septum purge gas is usually employed to prevent ghosting. Autoinjection into narrow-bore columns is usually carried out by direct sample introduction into a 0.53 mm i.d. retention gap that is butt-connected to the narrow column using a reducing union.

Cryogenic cooling can be employed for either manual or automated systems to reduce sample run times associated with oven and inlet cooling and to facilitate the use of solvents with particularly low boiling points.

2.5.2 Inlet parameters and sample considerations

Sample preparation is important for on-column injection because of the possibility of column overload, contamination, solvent–stationary phase mismatching and oven temperature considerations. Sample sizes are restricted to the range 0.5–2.0 µl with the ideal volume being governed by the column internal diameter, the compatibility of the sample solvent and the stationary phase, the sample concentration, the stationary phase film thickness and the carrier gas flow rate. A good general rule of thumb is to keep the injection volume as small as possible to satisfy sensitivity requirements.

Proper injection using cool on-column inlets requires the syringe plunger to be depressed as rapidly as possible to prevent the sample from sticking to the outside of the needle via capillary action of the liquid plug between the syringe outer wall and the capillary inner wall. The use of fast autoinjectors is recommended, where the sample plug will be sprayed into the column away from the needle, so reducing discrimination. This concept is outlined in Figure 2.19.

If injection volumes are too large or if the carrier flow is too low, the sample may back up the capillary and be lost via the septum purge lines. Excessive sample volumes or mismatch between the polarity of the stationary phase and the solvent may also lead to peak distortion or splitting that may be overcome using a retention gap.

2.5.3 Retention gap

A retention gap is a piece of column (usually uncoated) that accommodates the liquid sample but will not interact with the solvent or solute once vaporised. The primary function of the retention gap is to reduce the length of the flooded zone

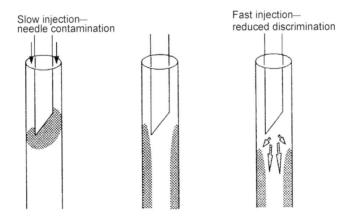

Slow injection—
needle contamination

Fast injection—
reduced discrimination

Figure 2.19 Reduced discrimination using fast injection with cool on-column injection.

created after liquid injection, but it is equally important in protecting the analytical column from nonvolatile sample components. A retention gap is usually employed with cool on-column inlets when peak splitting or broadening occurs.

As the solvent evaporates, all solutes will move freely at the carrier gas velocity to the head of the analytical column where they are focused by the solvent effect (solutes $k < 5$) or by stationary phase interactions (solutes $k > 5$). It is important that the retention gap is properly deactivated to minimise the length of the initial flooded zone and to reduce the possibility of peak tailing or sample degradation.

The polarity of the solvent should be matched with the polarity of the retention gap material to keep the flooded zone as narrow as possible. The more compatible the solvent and stationary phase, the shorter the retention gap may be (i.e. 25 cm per 1 µl sample injected using hexane rather than 2 m per 1 µl methanol using a silica retention gap).

Loss of column efficiency is usually caused by column contamination or degradation of the stationary phase at the column inlet. This will generally lead to broad, tailing or split peaks and can be avoided using well-deactivated retention gap material.

2.5.4 Temperature

The inlet temperature during injection should be below the boiling point of the solvent with the oven temperature matching that of the inlet because the flooded zone within the column may extend out of the inlet. After injection, a stable flooded zone must be created and this is normally achieved by increasing the inlet temperature at a faster rate than the oven program, which will help to focus the higher-boiling components through rapid inlet transfer and stationary phase focusing.

2.5.5 Flow rates

High flow velocities (30–50 cm s^{-1}), are recommended for cool on-column injection to ensure that the sample plug is carried well away from the needle tip during the injection phase. Septum purge flows are usually set in the range 5–10 ml min^{-1}.

2.6 Programmed temperature vaporising (PTV) inlets

PTV inlets combined the flexibility of split/splitless injection systems with the low discrimination associated with cool on-column inlets. The sample is usually injected into a cool liner, after which the temperature can be increased in a controlled fashion to vaporise the sample. Vent times and temperatures can be programmed to achieve cold split, cold splitless or solvent elimination injection.

Advantages of PTV injection include the following:

- reduced needle and inlet discrimination
- ability to handle very large injection volumes
- removal of solvent and low-boiling point components
- removal of nonvolatile materials prior to the analytical column
- excellent accuracy and reproducibility
- cold trapping of gas and sampled fluids

2.6.1 Inlet design

PTV inlets are available from a variety of manufacturers and, in comparison with a split/splitless injection system, the PTV will have [14,15]; lower thermal mass, capability of rapid heating and cooling, low internal volume, and the ability to time heating and split vent events.

The PTV inlet shown in Figure 2.20 contains a 6 cm long 0.15 cm i.d. glass liner packed with silylated glass wool. The carrier flow into and around the inlet is very similar to that of the split/splitless inlet. The inlet is cooled before and during injection using Peltier, forced air or cryogenic cooling. When cryo-cooling is used, it is possible to thermally focus gases introduced from external sampling devices such as headspace or thermal desorption systems. After injection the inlet is heated using electrical heaters that usually are capable of imposing controlled temperature ramps within the inlet.

PTV injection techniques are very useful for the analysis of samples within dirty matrices as most of the nonvolatile material will be trapped within the liner, resulting in a higher number of injections between injector or column maintenance.

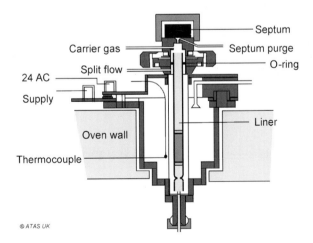

Figure 2.20 Diagram of an OPTIC PTV inlet. (Courtesy of ATAS, Cambridge, UK)

Liners for PTV inlets have a smaller internal volume than those used in traditional split/splitless inlets since there is no flash vaporisation occurring. However, liner internal volume will still govern the maximum injection volume in cold split and cold splitless injection as well as the rate of liquid introduction for solvent elimination injection. Liner activity and internal volume remain key issues when using PTV inlets. Liners require a packing or a specially shaped or modified surface to hold the liquid samples in the liner during the injection and vaporising phases of the injection. Where glass wool packing may cause degradation or adsorption, more inert liner packing such as Tenax may be used or a deactivated baffled glass liner may be used. Generally liners should be cleaned and replaced on a regular basis and whenever a loss of inlet performance is seen.

2.6.2   Inlet parameters and sample considerations

Cold split injection.   This mode is useful for general analysis and for screening of unknown samples. The liquid plug is injected into a cold vaporising chamber, helping to prevent needle discrimination and an introduction of a more reproducible aliquot of the sample. After the needle is withdrawn, the inlet is heated according to a predefined program and evaporation of the solutes occurs in boiling point order. This allows for larger injection volumes compared to classical split injection as there is a lower solute concentration at the column inlet at any one point during the injection cycle; distorted peak shapes are much less likely.

Cold splitless injection.   This injection mode is usually employed for trace analysis (unless only late-eluting compounds are of interest), and suffers much less discrimination than conventional splitless injection. The injection routine is very similar to classical splitless injection but the inlet is initially cool. Stationary phase and solvent focusing are still required to obtain narrow efficient peaks in this mode. In general, cold splitless injection allows larger injection volumes and, because the vaporisation of the solvent and solutes is sequential, flashback is minimised (Figure 2.21).

Solvent elimination split/splitless injection.   This mode is used to selectively remove solvent from the sample to allow larger injections, or to concentrate dilute samples to achieve higher signal-to-noise ratios. This can be achieved via one large injection or several smaller injections, but the maximum allowable injection size remains a function of the liner volume, inlet temperature and flow rate.

The sample is introduced into the inlet according to the following scheme:

1.   The split vent (and solvent vent line if separate) is off.
2.   The inlet temperature is close to, but below, the boiling point of the solvent.

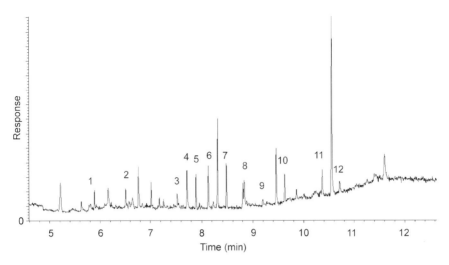

Figure 2.21 PTV-GC analysis of pesticide extract at a solute concentration of 0.1 ppm each. Peaks: 1 methamidophos; 2 acephate; 3 dimethoate; 4 diazinon; 5 chlorothalonil; 6 chlorpyrifos-methyl; 7 chlorpyrifos; 8 thiabendazoe; 9 imazalil; 10 ethion; 11 phosmet; 12 azinphos-methyl. (Courtesy of Agilent Technologies.)

3. The syringe plunger is depressed slowly to prevent flashback (when using bolus injection).

4. After injection, the vent flow (high, up to $1000\,ml\,min^{-1}$), is initiated, which evolves solvent vapours due to the reduced vapour pressures within the inlet.

5. After the major proportion of the solvent has vented through the split line, two options are possible: split line remains open (split solvent-elimination injection) or split line closes (splitless solvent-elimination injection).

Solvent-elimination split injection is rarely used, as the advantage of sample preconcentration would be negated by the on-instrument dilution in split mode.

Solvent-elimination splitless mode improves analytical sensitivity by maximising the amount of higher boiling solutes reaching the column while minimising the initial sample loading on the analytical column. The significant disadvantage of this mode lies in the loss of volatile sample components that are vented with the solvent, and this therefore restricts the technique to analysis of higher-boiling compounds.

Discrimination may occur when analysing highly volatile materials as the solvent venting may cause volatilisation of materials whose boiling points are close to the solvent. Temperatures and vent flow rates need to be carefully optimised to prevent this; however, solutes whose boiling points are within $50°C$ of the solvent are at risk of being partially lost. Analysis of compounds with medium volatility is possible but requires packing of the inlet liner with inert adsorbents such as Tenax, deactivated charcoal or porous polymers. If adsorbents are used, the temperatures required to desorb the sample onto the column can be very high (300– 500°C), when decomposition of the analyte (and the adsorbent) becomes a hazard.

2.6.3   Temperature

PTV inlet temperature is dependent upon the injection mode employed. With cold split or splitless injection, the inlet must be set below the boiling point of the solvent. In cold splitless mode, sample degradation is more likely than with cold split injection because the sample is in contact with the liner and liner packing for extended times. Higher column flow rates, slower temperature ramps and thorough deactivation of the liner and packing should be used when analysing labile samples.

During solvent elimination, the temperature may be raised to just a few degrees below the solvent boiling point to aid selective solvent evaporation.

After the sample is introduced into the liner, the inlet temperature program should be fast enough to ensure that the sample bands are efficiently transferred to the column inlet and should end at a temperature high enough to vaporise all components of interest. The inlet should always be at a higher temperature than the column and the inlet program should not be uncontrolled as this may lead to flashback or column overload if the sample volume is large.

2.6.4   Carrier flow

During cold split and splitless injection, the split flow is off when the sample is loaded into the liner. For cold split injection, the flow rate is adjusted to give the desired split ratio. Cold splitless injection should employ a rapid inlet temperature program so that the sample is not transferred over too long a period, potentially resulting in broad sample bands. Split-on times are optimised as for classical splitless injection.

In solvent-elimination mode, the split vent is on during sample introduction and should be set high enough to clear the inlet of sample vapours so that the liner volume is not exceeded at any point during the sample introduction. Sample introduction rate and inlet temperature are also important in respect to liner overload and therefore these parameters must also be optimised in conjunction with split flow rate.

All PTV injection modes tend to yield wide initial peak widths, so stationary phase and/or solvent focusing is usually required and retention gaps may be usefully employed to prevent peak splitting and distortion in cold splitless injection.

Table 2.2 gives a summary of the main operating modes and instrument configurations for PTV injection.

Table 2.2 Typical PTV inlet settings for three common modes

PTV inlet mode	Cold split	Cold splitless	Solvent elimination splitless
Inlet injection conditions			
Liner temperature	<< Solvent b.p.	<< Solvent b.p.	< Solvent b.p.
Purge/split flow	Off	Off	100–1000 ml min^{-1}
Delay prior to heating	None	None	5–30 s
Inlet heating conditions			
Liner temperature program	Maximum rate	Max. temp. < 80 s	Max. temp. < 80 s
Column temperature	Solute dependent	< Solvent b.p.	< Solvent b.p.
Purge/split flow	Split ratio-dependent	Off until final inlet temp.	Off until final inlet temp.

2.7 Auxiliary sampling devices

When syringe injection is not appropriate, for example when dealing with solid samples, several auxiliary sampling devices are available to introduce the sample into the GC inlet or directly onto the GC column itself. These devices include gas and liquid sampling valves, headspace auto-samplers, thermal desorption devices, and purge and trap autosamplers.

2.7.1 Gas and liquid sampling valves

Valves are usually mechanical devices that are used to introduce a fixed plug of sample into the carrier gas stream. Valves are traditionally used to sample gases or liquids from constant-flow streams in, for example, pipe work carrying liquids or gases, chemical reaction vessels, gas processing plant, waste and effluent streams and distillation equipment.

Gas sampling valves must be appropriately thermostated to prevent condensation of the sample in the valve and this is normally achieved by enclosing the valve system in a box oven heated to an appropriate temperature. Both liquid and gas sampling systems may be either directly connected to a packed column or via a split inlet to a capillary column.

Perhaps the most useful valve system consists of a gas or liquid sampling valve connected via a capillary inlet to the analytical column. Traditionally, a split inlet would have been used to ensure that the sample bands transferred to the column were as narrow as possible; however, PTV type inlet systems are finding very useful applications for sampling large volumes of liquid or gas effluent prior to transfer to high-speed GC systems. Cryogenic cooling systems are usually employed to effectively trap the sample in the inlet prior to transfer into the analytical column, ensure narrow sample bandwidths and allow pre-concentration of dilute samples.

There are several advantages to a valve and inlet system in series, including the following:

- high flow rates from the valve can be maintained to ensure a narrow initial sample band
- potential of splitting to avoid column overload
- dynamic split ratio adjustments to respond to rapidly changing concentrations of analyte in the sampled stream
- possibility of focusing (using cryogenics), to ensure narrow initial sample bands

In most cases the transfer lines from the sampling device to the inlet need to be heated to prevent condensation when sampling gas streams.

If liquids are sampled, all components must be capable of rapid 'flash' vaporisation after the sample plug is introduced into the carrier gas stream to avoid flooding of transfer lines and subsequent ghosting or carry-over. If PTV injection is used, this is not so crucial as the sample may be deposited in the inlet liner in the liquid form. However, all valves and sample transfer lines must be thoroughly washed between injections.

It is possible to use a gas-sampling valve for liquid samples. In this case the valve, loop and transfer line are heated to volatilise the liquid sample prior to its introduction into the capillary inlet. In these cases, Hastelloy connecting tubing and Teflon valve seals are required to ensure that that higher boiling and polar components are not adsorbed onto the transfer system and peak splitting and tailing are avoided.

2.7.1.1   Inlet parameters and sample considerations
Sample loop volume. Injection loop sizes for sampling valves are typically 200–500 µl for gaseous samples and 0.25–0.5 µl for liquids. Initial experiments usually determine the sample volume required for the inlet and column being used and thereafter split ratio or loop pressure may be dynamically adjusted to avoid overloading the column. This procedure avoids the time-consuming process of changing loops to alter the sample volume, which precludes unattended operation.

2.7.1.2   Temperature
Elevated temperatures are important to prevent adsorption of the analyte species to transfer lines and valve components, thereby maximising reproducibility. Gas sampling valves are available in ratings up to 450°C, depending upon the temperature necessary to either volatilise a liquid sample or to keep gaseous samples in that form. The polyimide surfaces used for high temperature rotor seals can be very adsorptive towards polar compounds and care should be taken to ensure repeatability of the sampling method. In general, liquid sampling valves that are not used to volatilise the sample are rated below 75°C.

Setting valve temperatures can be time-consuming and difficult as too high a temperature can lead to decomposition and problems with backflashing samples in the valve system, or leaking and ghosting of sample slugs. The amount of sample contained within the loop is usually inversely proportional to the valve temperature: as the temperature increases the sample amount decreases, as does the loading on the analytical column and the analytical sensitivity. Therefore, careful and reproducible thermostatting of the valve system is crucial to ensure repeatable analysis.

2.7.1.3 Carrier and auxiliary gas flow rates

Peak broadening is a function of the sample volume, the sampling system internal volume and the inlet volume. The longitudinal diffusion of the sample band within the system is reduced in proportion to the gas flow rate through the system and most gas sampling systems will operate most effectively at flows above 20 ml min^{-1}. When operating with split or PTV inlets, this flow may then be distributed through the split and septum purge lines so that an optimum linear velocity may be maintained through the analytical column.

The use of liners of very small internal diameter will minimise peak broadening in the inlet; however, backflash of sample components must be considered and split flow rates adjusted so that the liner volume is not exceeded at any time during the injection cycle.

2.7.1.4 Sample considerations

Changes in baseline position and appearance can be caused by changes in the column flow associated with valve switching operations. As the valve switches the loop contents into and out of the system there is a sudden pressure drop as the rotor seal turns between flow paths. The flow and pressure will then return to operating values once the switch is completed.

Actuator pressure should be high enough to ensure the most rapid and efficient valve movement between the load and inject positions, and a restriction may be included on the sample vent line to maintain the sample loop at system pressure, hence reducing pressure re-stabilisation times.

The amount of sample in the loop is directly proportional to loop temperature and pressure, and any variability in these parameters will lead to variations in peak area obtained. Fitting flow restrictors added to the valve vent lines may help to reduce pressure variability. Temperature stability has already been discussed.

Leaks can of course cause reproducibility problems and the sampling system should be tested regularly for either liquid or gas tightness.

2.7.2 Headspace autosamplers

Headspace analysers are used to introduce a portion of the headspace gas that is evolved by a sample that has been heated in a sealed vessel when vapour has been allowed to equilibrate. Headspace sampling is useful for the analysis of purgeable volatile species in environmental analysis (volatiles in soil, water or waste samples), polymer analysis (additives and residual solvents in formulated polymers), food analysis (aromas and volatiles in food and drinks) and pharmaceutical analysis (residual solvents and volatiles in formulated products including solid dose forms and sterile injectables).

Headspace analysis is used to selectively sample the species of interest while leaving the undesirable components in the vial. Temperature, valve switching times and sample pretreatment are all important in achieving this aim. The main drawback with headspace analysis is the necessity to have a standard contained at the correct concentration in the same matrix as the sample. This overcomes the problem that the sample matrix may attenuate the release of some of the volatile components within the sample. This situation can be overcome by using multiple headspace extraction techniques, in which the sample is repeatedly sampled to gain information on the relative volatility of the sample components. The most volatile species are contained in the first samples and subsequent samplings will contain material that results from a shifted distribution of components within the sample. Overall concentrations decrease with subsequent sampling: more soluble components will decrease the least with each sample and this can provide information to differentiate between polar and nonpolar solutes within a sample matrix.

To ensure narrow sample bands, the flow rates throughout the sampling system must be kept high. This is easily achieved with the column flow rates associated with packed and large internal diameter capillary columns. When using narrow-bore capillaries, the headspace sampling system is normally connected through a split or PTV type injector where the inlet flow rates may be kept high.

Figure 2.22 shows a headspace analysis.

2.7.2.1   Design

Several parameters influence the concentration and distribution of sample components, including vial temperature, equilibration (or heating) time, pressure applied to the vial, loop filling time and flow rate.

The weighed sample is first thermostated to equilibrate the sample; the vial is then pressurised to vent the headspace gases through a gas sampling valve. The sample is injected into the carrier stream and into the column inlet or a capillary or PTV inlet in series with the analytical column. Times, temperatures and pressures at each stage are programmed into the autosampler by the user. Figure 2.23 shows the stages of a typical headspace autosampling protocol.

If vial equilibration time is too short, headspace concentration of volatile species is decreased and sensitivity will be compromised. If vial pressurisation

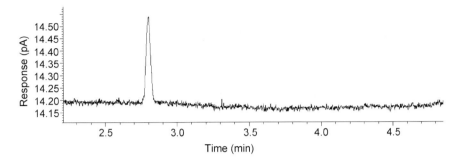

Figure 2.22 Headspace analysis of ethylene oxide (a known carcinogen) from polymer resin beads at 1 ppm. Oven temperature: 30°C. Inlet, see below for headspace autosampler parameters. Carrier: helium at 7.0 psi (48 kPa). Column HP-Wax 30 m × 0.32 mm × 0.50 μm. Detection: FID at 200°C. Headspace autosampler parameters: oven temperature 100°C; valve and loop temperature 105°C; transfer line temperature 105°C; sample equilibration time 60 min; vial pressurisation time 0.5 min; loop fill time 0.15 min; loop equilibration time 0.05 min; injection time 5.30 min; GC cycle time 20 min; loop size 1 ml; vial pressure 10 psi (69 kPa).

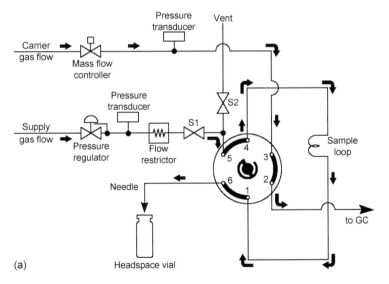

Figure 2.23 Stages in headspace autosampling routine. (a) Standby mode: carrier gas purges the sample loop. (b) Vial pressurisation: probe enters the vial and pressurises. (c) Vent mode: headspace vapour fills sample loop. (d) Injection mode: loop contents swept into GC inlet. (Courtesy of Agilent Technologies.)

time is too short, the headspace gases will not be effectively transferred to the sample loop and again sensitivity will be compromised. Vent time needs to be carefully optimised, as the sample loop needs to contain only sample gas in a nondiffuse form.

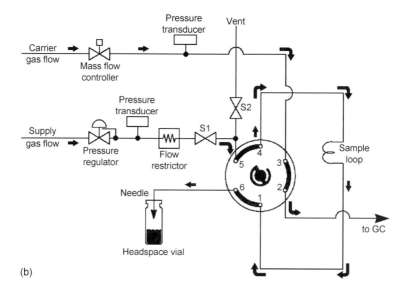

(b)

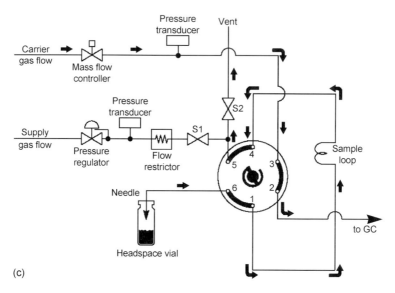

(c)

Figure 2.23 (continued).

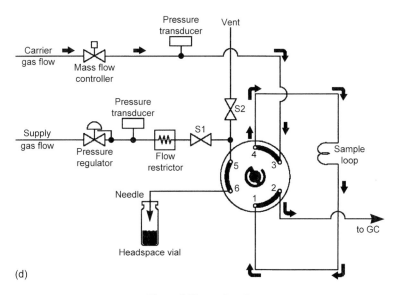

(d)

Figure 2.23 (continued).

2.7.2.2 Inlet parameters and sample considerations

Reproducible headspace data can only be achieved through accurate and repro-ducible sample weighing, well-optimised autosampler conditions and minimised sample matrix effects. Matrix effects may be minimised by homogenising and grinding solid samples; salting out aqueous samples to force polar analytes into the headspace; saturating organic samples with water to decrease adsorptive effects; and adjusting sample pH to promote release of ionic and polar species.

As most headspace analytes are volatile, capillary columns need to have thick stationary phase films in order for focusing to occur effectively. Where stationary phase focusing is not readily achieved, analytes must be trapped in a PTV inlet or on-column using cryogenic methods.

2.7.2.3 Temperature

The sample vial in most modern dynamic headspace sampling devices is heated using either an oven or a thermostated oil or water bath. Temperature control must be stable and reproducible as the headspace concentration of sample com-ponents is directly related to the sample temperature. Increased temperatures drive more volatile species into the vial headspace to increase analytical sen-sitivity but increase the risk of volatilising potentially interfering components. For unknown samples the best approach is to obtain a test chromatogram using a low vial temperature such as $40°C$ and to increase this temperature in stages to increase sensitivity if required. Temperatures used for valve systems and inlet transfer lines should be kept higher than the boiling point of the highest headspace analyte to prevent adsorption or condensation.

2.7.2.4 Flow rate

Gas flows originate from the headspace sampling unit as well as the GC and the magnitude of each flow is governed directly by the type of column being used. For large internal diameter capillary columns and packed columns operating at $20\,ml\,min^{-1}$, $10–15\,ml\,min^{-1}$ will come from the headspace unit and $5–10\,ml\,min^{-1}$ will be supplied from the GC.

Capillary inlets operating at $20\,ml\,min^{-1}$ total flow will take $10\,ml\,min^{-1}$ from the GC and $10\,ml\,min^{-1}$ from the headspace sampling unit. The split ratio is adjusted to achieve the correct sample loading on column in response to analytical sensitivity requirements.

2.8 Conclusions

Sample injection systems and auxiliary sampling devices for GC have been developed to allow the analysis of almost any chemical species that can develop a significant head pressure within the operating temperature range of modern GC columns. As has been discussed, careful selection of the inlet is required, all inlets and sample introduction devices requiring careful optimisation for successful chromatography, with many interdependent parameters affecting response and reproducibility. However, by paying close attention to the points outlined and with a good understanding of the working principles of each injection mode, the chromatographer may successfully analyse by GC a wider range of solutes from a more diverse range of matrices than was previously thought possible.

References

1. P. Sandra, *Sample Introduction in Capillary Gas Chromatography*, Hüthig Verlag, Heidelberg, 1985.
2. K. Grob, *Classical Split and Splitless Injection in Capillary GC*, Hüthig Verlag, Heidelberg, 1986.
3. *A Guide to Minimising Septa Problems*, Restek Corporation, Bellefonte, PA, 1996.
4. I. Chanel, *Analysis of Organochlorine Pesticides and PCB Congeners with the HP 6890 Micro-ECD*, Application note 228–234, Hewlett Packard, Waldbron, Germany, 1997.
5. K. Grob Jr, *J. Chromatogr.*, **1982**, *251*, 235.
6. J.V. Hinshaw, *HRC & CC*, **1993**, *16*, 247.
7. K. Grob Jr and H. Neukom, *HRC & CC*, **1980**, *3*, 627.
8. R.E. Freeman (ed.), *High Resolution Gas Chromatography*, 2nd edn., Hewlett-Packard, Waldbron, Germany, 1981.
9. K. Grob Jr, *J. Chromatogr.*, **1981**, *213*, 13.
10. K. Grob Jr, *J. Chromatogr.*, **1985**, *324*, 252.
11. R. Schill and R.R. Freeman, in *Modern Practice of Gas Chromatography* (ed. R. Grob), Wiley, New York, 1985.
12. M.S. Klee, in *GC Inlets—An Introduction*, Hewlett-Packard, Wilmington, DE, 1994.
13. K. Nauss, J. Fulleman and M.P. Turner, *J. High Res. Chromatogr., Chromatogr. Commun.*, **1981**, *4*, 641.
14. W. Vogt, K. Jacob and H.W. Obwexer, *J. Chromatogr.*, **1979**, *174*, 437.
15. G. Schomburg, *4th Int. Symp. Capillary Chromatography*, Hindeland, Hüthig Verlag, Heidelberg, 1981, p. 921.

3 Advances in column technology

Allen K. Vickers and Dean Rood

3.1 Introduction

Column liquid chromatography has the advantage that both the stationary phase and the mobile phase are capable (at least in theory) of infinite variation. Ironically, this can have the effect of making reproduction of an HPLC separation more difficult. In GC, on the other hand, it is only possible to vary the stationary phase and its operating conditions. It is true that changing the carrier gas can have minor effects on retention and resolution, but in terms of maximum resolving power there is not a great difference between nitrogen, helium or hydrogen if they are used at the optimum flow rates. This is apparent from the well-known family of van Deemter curves for the common carrier gases [1]. There are advantages in the use of hydrogen for fast analysis but this is not really a major concern in the actual resolving power. Thus the manufacturer and user of capillary columns must use such variables as are available, namely

- the composition of the stationary phase
- the film thickness
- the column length
- the column diameter
- the temperature program.

Because the early packed columns were inefficient in terms of numbers of theoretical plates, considerable effort was applied to find a stationary phase that gave the best separation factor, α, for a particular mixture. This was the main reason for the long list of stationary phases that were available for packed columns, although it is clear that many were almost identical in performance. There cannot be much difference in the separations achieved by dinonyl and didecyl phthalate, for example. There were also severe limitations in film thickness and column length. About 1.5% of stationary phase on a solid support represents about the practical lower limit before adsorption effects become severe and 25% was about the maximum amount. Columns were limited to a maximum of about 4 m in length because of the high pressure drop across a long column. In this context it is interesting to note that Scott used a packed column ~15 m long to obtain 30 000 plates for xylene [2]. However, the clear superiority of capillary columns resulted in this development representing an evolutionary dead end.

The use of open tubular (capillary) columns was largely confined to the petroleum industry for separation of complex nonpolar hydrocarbon mixtures on stainless steel columns, although the Grobs, Senior and Junior, did a considerable amount of work on the deactivation and preparation of glass capillary columns. However, it was the development of silica capillary columns by Hewlett Packard in the early 1980s [3] that transformed the situation and caused the rapid displacement of packed columns for the majority of GC separations. The early silica capillary columns were largely coated with poly(methyl siloxane) phases by adsorption and it was only later that chemically bonded phases of other siloxanes were gradually introduced. The limited types of stationary phase available for these early silica capillaries and the large number of theoretical plates available from them led to the philosophy of separating mixtures by 'throwing a large number of plates' into the separation. This was achieved by using capillary columns 100 m or more long. While this approach may give the desired separation, it has the disadvantage of long analysis times of an hour or more. What must represent the near-ultimate in this approach is the separation of a gasoline into 378 peaks on a 1000 ft by 0.02 inch i.d. stainless steel capillary column coated with squalane and programmed up to 140°C at 1°C min^{-1} after an initial hold at 30°C for 20 min [4]. The majority of the peaks were identified by mass spectrometry and, while this separation and identification represented a considerable achievement at the time it was published (1979), the penalty was an analysis that took over four hours to complete with an even greater turnround time before the next sample could be introduced.

3.2   Modern capillary columns

3.2.1   Capillary column materials

Most modern capillary columns are made from synthetic silica of high purity. A protective coating is applied to the outer surface to prevent the development of surface cracks and thereby strengthen the column. In the majority of cases this outer coating is of a polyimide, which is responsible for the brown colour of fused silica columns that often darkens with prolonged exposure to high temperatures. The upper temperature limit for standard polyimide-coated fused silica is about 360°C, but high-temperature polyimide tubing can be used up to 400°C. For even higher temperatures, aluminium-coated capillaries have been marketed with limited success and the favoured approach has now reverted to special stainless steel columns that have been internally deactivated by a chemical treatment or by coating with a thin layer of fused silica. When properly prepared, these stainless steel columns rival those of fused silica for inertness.

The inner surface of fused silica tubing is chemically treated to minimise interaction of the sample with the tubing. The reagents and processes used depend on the type of stationary phase. A silylation process is used for most stationary phases where silanol groups on the inner surface of the silica are reacted with a silane reagent. Typically a methyl or phenylmethyl silyl surface is created for most columns.

3.2.2   Column dimensions

Table 3.1 lists the effect of column internal diameter versus efficiency. The table shows that the smaller the internal diameter, the higher the number of plates per unit length, so a shorter column can be used for a particular separation and the separation can be conducted more quickly. The penalty is that such columns have the smallest sample capacity and therefore components of the sample present in small concentrations may not be detected. At the other end of the scale, the so-called 'wide-bore' columns of 0.53 mm in internal diameter have the lowest efficiency but the highest capacity and are therefore best suited with a thick film for trace analysis or when low-boiling compounds are present. Wide-bore columns operate under conditions not too far removed from those for packed columns and so are ideal for the analyst unused to capillary columns and their particular problems. Because of the low pressure drop and high gas flow rates, wide-bore columns are not suitable for use with mass spectrometric detection.

Column efficiency is related to the square root of the column length so that to double the resolution the column length has to be increased fourfold, which extracts a severe penalty in terms of analysis time, although this may be essential for a complex mixture such as a gasoline.

3.2.3   Stationary phases

As mentioned above, most stationary phases for capillary columns were originally based on silicone polymers and this is still true, although the modern phases are much superior in terms of their range and stability.

Polysiloxanes are the most common stationary phases. They are available in the greatest variety and are the most stable, robust and versatile. Standard siloxanes are characterised by the repeating siloxane backbone, as shown in

Table 3.1 Column efficiency versus internal diameter. Maximum efficiency in each case for a solute with $k = 5$

Internal diameter (mm)	Plates per m
0.01	12 500
0.18	6 600
0.20	5 940
0.25	4 750
0.32	3 710
0.45	2 640
0.53	2 240

Figure 3.1. The type and amount of the groups attached to the silicon atoms distinguish each stationary phase and give it its unique properties. The four most common groups are listed in this figure. The basic material is that which is 100% methyl-substituted. When other groups are present, the amount is indicated as the percentage of the total number of groups so, for example, a 5% phenyl–95% methyl phase contains 5% (di)phenyl groups and 95% (di)methyl groups as substituents. The prefix 'di' indicates that the each silicon atom contains two of that particular group, but it is sometimes omitted even though two groups are present. If the percentage of methyl groups is not stated, it is assumed that it is present in an amount to make up the 100%. Cyanopropylphenyl percentage values can be misleading A 14% cyanopropylphenyl–dimethyl polysiloxane contains 7% cyanopropyl and 7% phenyl with 86% methyl. The cyanopropyl and phenyl groups are on the same silicon atom and thus their amount is summed.

For some stationary phases, a low-bleed version is available with the addition of 'ms' after the main name. These materials incorporate phenyl or phenyl-type groups in the backbone of the siloxane polymer (Figure 3.2) and are called arylenes. The presence of the aromatic ring in the backbone stiffens its structure

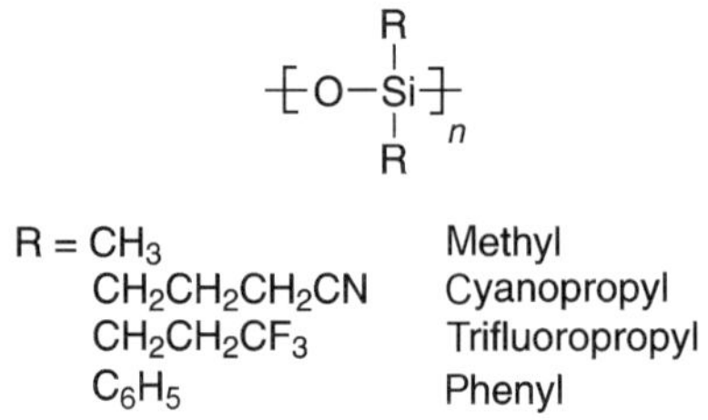

$$\left[O-\underset{\underset{R}{|}}{\overset{\overset{R}{|}}{Si}} \right]_n$$

R = CH$_3$ Methyl
CH$_2$CH$_2$CH$_2$CN   Cyanopropyl
CH$_2$CH$_2$CF$_3$ Trifluoropropyl
C$_6$H$_5$ Phenyl

Figure 3.1 Polysiloxanes.

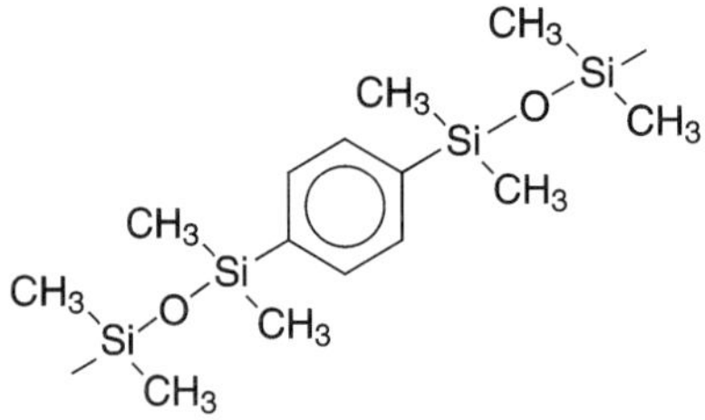

Figure 3.2 Low-bleed phases (arylene).

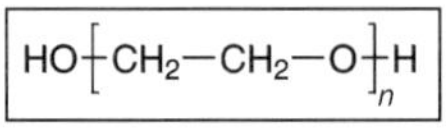

$$HO\left[CH_2-CH_2-O \right]_n H$$

Figure 3.3 Poly(ethylene glycol).

and reduces degradation at high temperatures. In most cases this results in lower bleed rates and higher temperature limits. The arylene phase substitution can be adjusted to give the same separation characteristics as those of the conventional non-arylene version, for example, DB-5 and DB-5 ms. The separations obtained with the two versions are generally very similar. There are some unique low-bleed stationary phases that have no regular equivalent. As the name implies, the 'ms' range of columns is especially suited for GC-MS analysis.

The other main group of stationary phases in addition to the silicones are poly(ethylene glycol)s or PEGs. These have the general structure shown in Figure 3.3. Stationary phases with 'WAX' or FFAP in their title are types of PEGs. Polyglycols are not substituted, so the polymer is 100% of the stated material. They are less thermally and chemically stable than the polysiloxanes, but their unique separation properties make these shortcoming tolerable. There are two types of PEG in common use, one has a higher upper temperature limit (DB-WAXetr) but exhibits a slightly greater tendency to give tailing peaks. The other type (DB-WAX) has lower upper and lower temperature limits and is more reproducible and inert; the two types give slightly different separations. A third type of PEG is the so-called FFAP material (free fatty acid phase) in which the end hydroxyl group on the polyglycol molecule has been esterified with nitroterephthalic acid and, as the name indicates, this phase can be used to analyse mixtures of fatty acids without the usual methylation derivatisation procedure. Finally there is a fourth type (CAM) available for the analysis of basic mixtures.

All these phases are for gas–liquid partition chromatography, but for some samples, especially light gas mixtures, columns using gas–solid chromatography and packed with a variety of active adsorbents are available. An example of a separation with these PLOT (porous-layer open tubular) columns is given below.

3.3 Examples of separations of different types of samples on a variety of columns

3.3.1 Petroleum analysis

Detailed analysis of gasolines. Figure 3.4a shows the modern equivalent of the detailed analysis of a gasoline on a poly(methyl siloxane) 100 m × 0.25 mm column, taking about 90 min to complete; Figure 3.4b shows the same gasoline run on a 40 m × 0.10 mm column where the analysis time has been reduced to about 35 min. If the phase ratio, that is the ratio of column diameter to film thickness, is maintained then the overall chromatography will not change, although it is necessary to use a shorter initial hold time and a faster ramp rate. The main disadvantage is that, because of the smaller sample capacity of the narrow-bore column, some small components may not be detected. However,

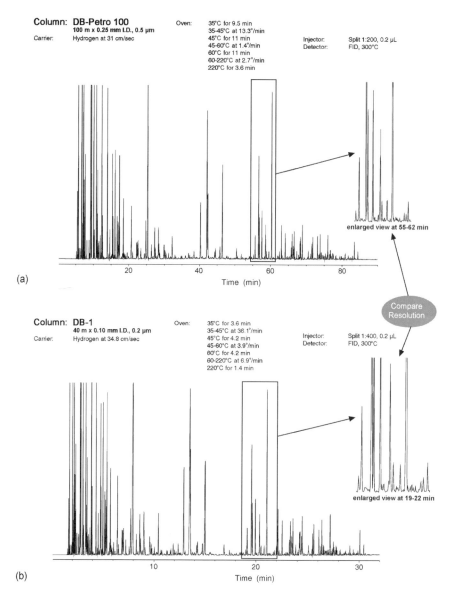

Figure 3.4 Regular unleaded gasoline (California phase 1): (a) 'normal' GC run; (b) 'fast' GC run.

since the majority of components that are of significance in gasoline are in the per cent concentration range, sensitivity in this case is not usually an important issue.

Boiling range of fuels and crude oil. The original ASTM method D-2887 for the determination of the boiling range distribution of petroleum fractions by GC specified a column about 1 m long packed with a methylsilicone phase such as OV-101 and covered the range of compounds with a normal boiling point of 55°C to 538°C, (about n-C_{43}) [5]. However, modern refinery methods now require that the upper boiling range of this method be extended to as high a carbon number as possible. Figure 3.5 shows this extended ASTM D-2887 method with a methylsiloxane-coated capillary programmed up to 400°C and capable of detecting hydrocarbons up to C_{84} and beyond. The column used is made from deactivated stainless steel tubing because the normal polyimide coating degrades rapidly at such high temperatures and the columns become brittle.

Light hydrocarbon gases. Light hydrocarbon gases can be analysed by gas–liquid partition chromatography using a stationary liquid phase but, because of their low boiling points, they have very little retention unless the column is operated at sub-ambient temperatures. This may be acceptable if liquid nitrogen is available as a coolant, but there may be many locations where this is not the case. Classically, gas mixtures have been analysed by gas–solid adsorption chromatography with columns packed with silica gel or alumina. In the past it was difficult to ensure repeatable and consistent performance from such columns because they were very susceptible to the water content of the column packing material. Nowadays deactivated alumina PLOT columns have eliminated this problem and Figure 3.6 shows a chromatogram of a mixture ranging from C_1 (methane) to C_{10} (decane) that does not require sub-ambient operation. Figure 3.7 is a chromatogram obtained on a silica-based PLOT column for the separation of sulfur gases from hydrocarbons and other compounds. Good resolution is obtained, which is important if a flame photometric detector is used since it avoids the possibility of quenching of the detector signal for sulfur compounds by co-eluting hydrocarbons.

3.3.2   Environmental analysis

Table 3.2 shows the EPA list of priority pollutants. With the exception of some of the compounds in the second group, which can be separated by HPLC, all are amenable to GC analysis. As coating techniques have improved and stationary phases with more complex chemistry have appeared, the emphasis is now moving in the direction of tailor-made phases for specific types of mixtures and for particular standard ASTM and EPA methods [6].

Purgeables without cryofocusing. An example of a tailor-made phase is a column coated with DB-624 designed to be used in EPA Method 624 to replace or complement Method 502.2 (Figure 3.8). This column was one of the first designed with a given analysis in mind. The 6% cyanopropylphenyl–methyl polysiloxane provides the dipole–dipole interactions necessary when attempting to resolve analytes of very similar volatility and structure. The excellent selectivity of this phase allows the use of shorter column lengths—75 m rather

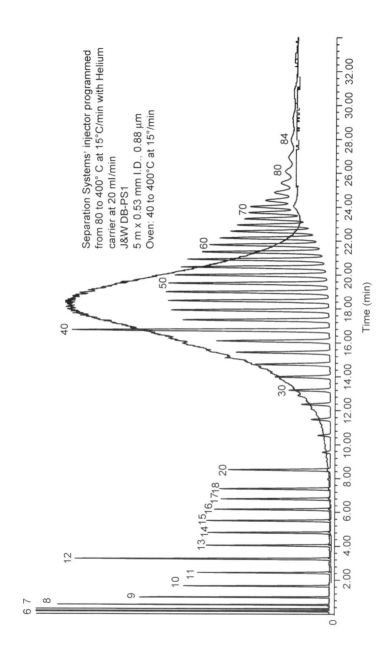

Figure 3.5 Extended method 2887 with MT-60 reference oil. Peak identification denotes n-alkane carbon number. (Chromatogram courtesy of Joaquin Lubkowitz, Separation Systems, Inc., Gulf Breeze, FL, USA.)

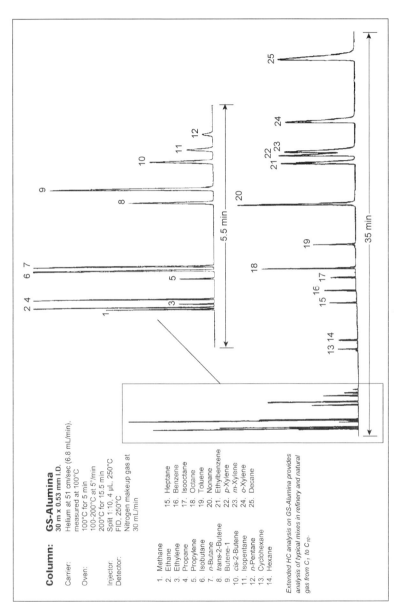

Column: **GS-Alumina**
30 m × 0.53 mm I.D.

Carrier: Helium at 51 cm/sec (6.8 mL/min), measured at 100°C

Oven: 100°C for 5 min
100-200°C at 5°/min
200°C for 15.5 min

Injector: Split 1:10, 4 µL, 250°C
Detector: FID, 250°C
Nitrogen make-up gas at 30 mL/min

1. Methane
2. Ethane
3. Ethylene
4. Propane
5. Propylene
6. Isobutane
7. *n*-Butane
8. *trans*-2-Butene
9. Butene-1
10. *cis*-2-Butene
11. Isopentane
12. *n*-Pentane
13. Cyclohexane
14. Hexane

15. Heptane
16. Benzene
17. Isooctane
18. Octane
19. Toluene
20. Nonane
21. Ethylbenzene
22. *p*-Xylene
23. *m*-Xylene
24. *o*-Xylene
25. Decane

Extended HC analysis on GS-Alumina provides analysis of typical mixes in refinery and natural gas from C₁ to C₁₀.

Figure 3.6 Extended analysis of hydrocarbons.

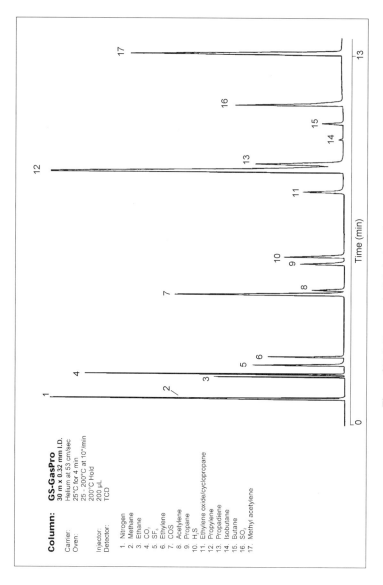

Column: **GS-GasPro**
30 m x 0.32 mm I.D.

Carrier: Helium at 53 cm/sec
Oven: 25°C for 4 min
25 - 200°C at 10°/min
200°C Hold

Injector: 200 µL
Detector: TCD

1. Nitrogen
2. Methane
3. Ethane
4. CO₂
5. SF₆
6. Ethylene
7. COS
8. Acetylene
9. Propane
10. H₂S
11. Ethylene oxide/cyclopropane
12. Propylene
13. Propadiene
14. Isobutane
15. Butane
16. SO₂
17. Methyl acetylene

Time (min)

Figure 3.7 Sulfur gases in light hydrocarbon stream.

Table 3.2 EPA-listed priority pollutants

Volatiles

Acrolein	Chloroethane	trans-1,2-Dichloroethene	1,1,1-Trichloroethane
Acrylonitrile	2-Chloroethyl vinyl ether	1,2-Dichloropropane	1,1,2-Trichloroethane
Benzene	Chloroform	cis-1,3-Dichloropropene	Trichloroethene
Bromomethane	Chloromethane	trans-1,3-Dichloropropene	Trichlorofluoromethane
Bromodichloromethane	Dibromochloromethane	Ethylbenzene	Toluene
Bromoform	1,1-Dichloroethane	Methylene chloride	Vinyl chloride
Carbon tetrachloride	1,2-Dichloroethane	1,1,2,2-Tetrachloroethane	
Chlorobenzene	1,1-Dichloroethene	Tetrachloroethene	

Base–neutral extractables

Acenaphthene	Bis(2-chloroisopropyl) ether	Diethyl phthalate	Indeno[1,2,3-cd]pyrene
Acenaphthylene	4-Bromophenyl phenyl ether	Dimethyl phthalate	Isophorone
Anthracene	Butylbenzyl phthalate	2,4-Dinitrotoluene	Naphthalene
Bezo[a]anthracene	2-Chloronaphthalene	2,6-Dinitrotoluene	Nitrobenzene
Benzo[b]fluoranthene	4-Chlorophenyl phenyl ether	Dioctyl phthalate	N-Nitrosodimethylamine
Benzo[k]fluoranthene	Chrysene	1,2-Diphenylhydrazine	N-Nitrosodi-n-propylamine
Benzo[a]pyrene	Dibenzo[a]anthracene	Fluoranthene	N-Nitrosodiphenylamine
Benzo[ghi]perylene	Di-n-butyl phthalate	Fluorene	Phenanthrene
Benzidine	1,3-Dichlorobenzene	Hexachlorobenzene	Pyrene
Bis(2-chloroethyl) ether	1,4-Dichlorobenzene	Hexachlorobutadiene	2,3,7,8-Tetrachlorodibenzo-p-dioxin
Bis(2-chloroethoxy) methane	1,2-Dichlorobenzene	Hexachloroethane	1,2,4-Trichlorobenzene
Bis(2-ethylhexyl) phthalate	3,3′-Dichlorobenzidine	Hexachlorocyclopentadiene	

Table 3.2 (continued)

Acid extractables

4-Chloro-3-methylphenol	2,4-Dimethylphenol	2-Nitrophenol	Phenol
2-Chlorophenol	2,4-Dinitrophenol	4-Nitrophenol	2,4,6-Trichlorophenol
2,4-Dichlorophenol	2-Metyl-4,6-dinitrophenol	Pentachlorophenol	

Pesticides

Aldrin	4,4′-DDE	Endrin	PCB-1221
α-BHC	4,4′-DDT	Endrin aldehyde	PCB-1232
β-BHC	Dieldrin	Heptachlor	PCB-1242
γ-BHC	Endosulfan I	Heptachlor epoxide	PCB-1248
δ-BHC	Endosulfan II	Toxaphene	PCB-1254
Chlordane	Endosulfan sulfate	PCB-1016	PCB-1260
4,4′-DDD			

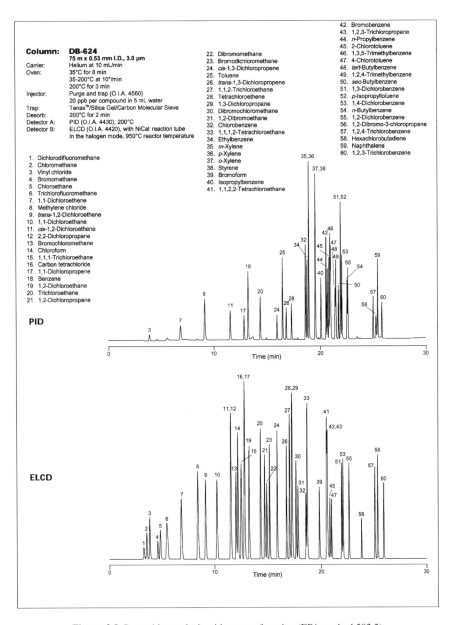

Figure 3.8 Purgeables analysis without cryofocusing (EPA method 502.2).

than the 105 m commonly employed—and as a result the analysis time is reduced from nearly an hour to less than 30 min. The thick stationary phase film of 3 μm is necessary to obtain adequate retention of the most volatile Freons specified in this method. Method 502.2 differs from Method 624 in that it utilises photoionisation (PID) and electrolytic conductivity detectors (ELCD) instead of a mass spectrometer. In Method 502.2, aromatic and halogenated compounds are identified with these selective detectors.

Another phase designed for the analysis of purgeable compounds without the use of cryofocusing is shown in Figure 3.9. This phase, DB-VRX, was designed

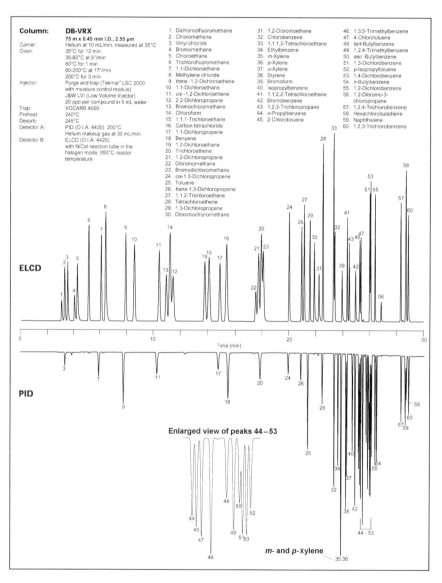

Figure 3.9 Purgeables analysis without cryofocusing (EPA method 502.2).

using computer modelling techniques to give a very high level of selectivity for the volatile organic compounds specified in the EPA 500, 600 and 800 series methods. The strong dipole–dipole interactions induced by this phase provide the selectivity required, and coupling with the PID and ELCD gives an adequate level of identification for routine purposes. The inner diameter of the column has been optimised at 0.45 mm as opposed to the common 0.53 mm to provide improved resolution. The 0.45 mm i.d. of the column also provides another less evident benefit, because the carrier gas is also the desorption gas through the sorbent trap of the sample concentrator. The higher column head pressure associated with the 10 ml min^{-1} flow actually helps to compress the sample band as it is transferred to the column and the result is less band broadening of the early-eluting analytes. This column, like the example above, also uses a thick stationary phase film of 2.55 μm to increase the retention of the lower-boiling compounds.

Polychlorinated biphenyls (PCBs). Another environmental analysis that has achieved considerable importance is that of the separation of PCB congeners. In spite of the fact that manufacture of these compounds ceased worldwide in the late 1970s, their stability is such that they still represent a potential environmental problem. There are 209 chlorine-substituted biphenyl congeners, of which 150 were present at significant levels in commercial mixtures. Fortunately it is not necessary to separate all these compounds and the EPA draft method 1668 specifies measurement of 13 congeners that may be carcinogenic, mutagenic or teratogenic, in particular those with chlorine in one or more of the 2, 3, 7 or 8 positions [7]. The selectivity of DB-XLB (a second-generation arylene polysiloxane phase) has some unique advantages in separating the *ortho*-substituted congeners from those of lesser interest. The numbering in the chromatogram shown in Figure 3.10 corresponds to the Ballschmitter–Zell numbering system [8] and, where different, the IUPAC number is shown in parenthesis. Mass spectrometry is used to detect the PCBs and, because the analytes elute at elevated temperatures where column bleed might play a signif-icant role, the low bleed of DB-XLB minimises spectral interference and helps to increase overall sensitivity by reducing the noise produced by the column. Also, since the pumping capacity of many benchtop mass spectrometers is limited to 1–2 ml min^{-1}, the use of a 0.25 mm i.d. column means that the optimum helium flow rate of about 1 ml min^{-1} for the column can be handled by the mass spectrometer. A 1 m retention gap is included in this set-up to prevent sample residues from reaching the column. As resolution is a function of the square root of column length, a 30 m column was chosen because it allows an acceptable compromise between resolution and run time.

Pesticide and herbicide residues. The utility of cyanopropyl–phenyl siloxane phases for the analysis of pesticides and herbicides is well documented. Many

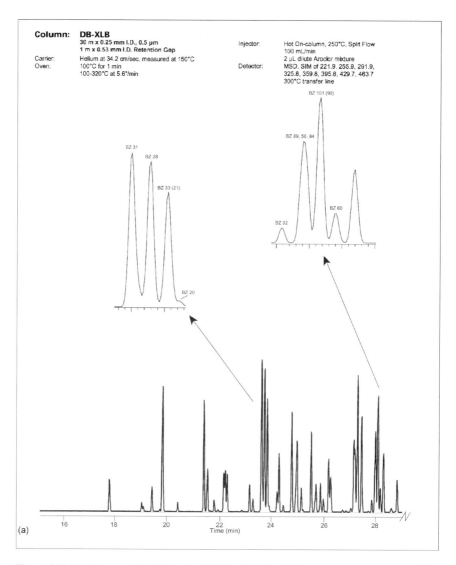

Column: DB-XLB
30 m x 0.25 mm I.D., 0.5 μm
1 m x 0.53 mm I.D. Retention Gap
Carrier: Helium at 34.2 cm/sec, measured at 150°C
Oven: 100°C for 1 min
100-320°C at 5.6°/min

Injector: Hot On-column, 250°C, Split Flow
100 mL/min
2 μL dilute Aroclor mixture
Detector: MSD, SIM of 221.9, 255.9, 291.9,
325.8, 359.8, 395.8, 429.7, 463.7
300°C transfer line

Figure 3.10 (a) Congeners in DIN method PCBs. (b) Extended temperature program resolving congeners 52 and 138.

methods call specifically for the 14% cyanopropyl–phenyl–methyl polysiloxane found in 1701-type phases. The unique selectivity characteristics of this phase have led to its continued use in the environmental field. In the case of phenoxy acid herbicides, it is evident that the sample derivatisation used has a profound effect on the ability of this phase to resolve the analytes of interest. In the example shown in Figure 3.11 the use of a trimethylsilyl (TMS) derivatising

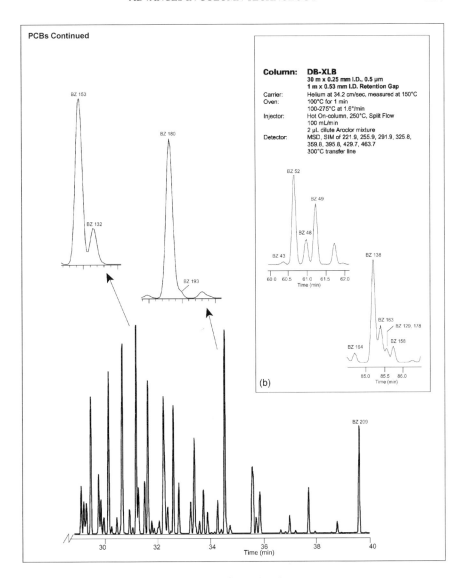

Figure 3.10 (continued).

agent gives the best resolution. With derivatised samples it is recommended that capillary columns with slightly larger internal diameters be used if possible. The 0.32 mm column used for Figure 3.11 has a larger capacity than a smaller-diameter column and has an increased ability to remove matrix-related material. This results in less fouling of the inlet end of the column, giving more consistent performance and requiring less maintenance.

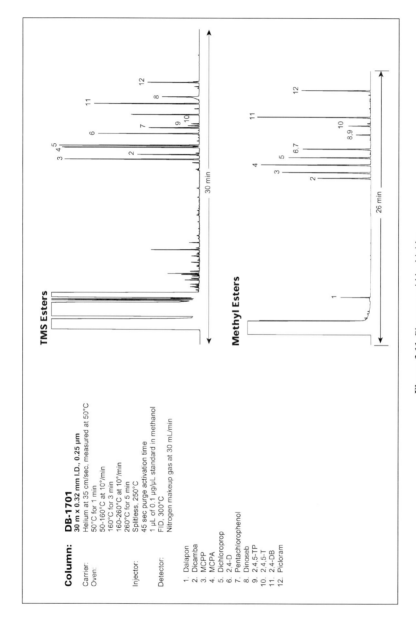

Figure 3.11 Phenoxy acid herbicides.

In many applications the use of a guard column can be advantageous. In the testing of wastewater, sludge or other complex matrices, the guard column is used to trap many nonvolatile residues and prevent them reaching the main column and prolongs column life. However, the connection between the guard column and the main column has always been regarded with some suspicion. Many different types of union have been employed but all have shortcoming of one sort or another, the most important being inconsistent sealing. Indeed, it has been said that the best type of union is one that does not exist! In recent years column manufacturers have taken steps to implement just this and eliminate the union. Using advanced column manufacturing techniques it is now possible to produce a capillary column with a built-in guard column. In essence, a single length of fused silica tubing is now part guard column and part analytical column. Figure 3.12 shows the separation of chlorinated pesticides on such a column coated with a 5% phenyl–methyl polysiloxane; it is evident that the chromatographic quality has been maintained.

In laboratories where GC-MS is not available, the use of a second column with a different stationary phase can offer a higher degree of confidence in peak identification. Figures 3.13a and 3.13b are of a pesticide mixture run on a

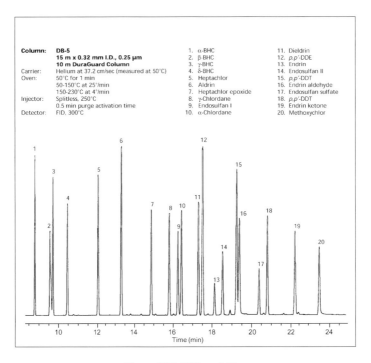

Column:	DB-5		
	15 m x 0.32 mm I.D., 0.25 μm	1. α-BHC	11. Dieldrin
	10 m DuraGuard Column	2. β-BHC	12. p,p'-DDE
Carrier:	Helium at 37.2 cm/sec (measured at 50°C)	3. γ-BHC	13. Endrin
Oven:	50°C for 1 min	4. δ-BHC	14. Endosulfan II
	50-150°C at 25°/min	5. Heptachlor	15. p,p'-DDT
	150-230°C at 4°/min	6. Aldrin	16. Endrin aldehyde
Injector:	Splitless, 250°C	7. Heptachlor epoxide	17. Endosulfan sulfate
	0.5 min purge activation time	8. γ-Chlordane	18. p,p'-DDT
Detector:	FID, 300°C	9. Endosulfan I	19. Endrin ketone
		10. α-Chlordane	20. Methoxychlor

Figure 3.12 CLP pesticides.

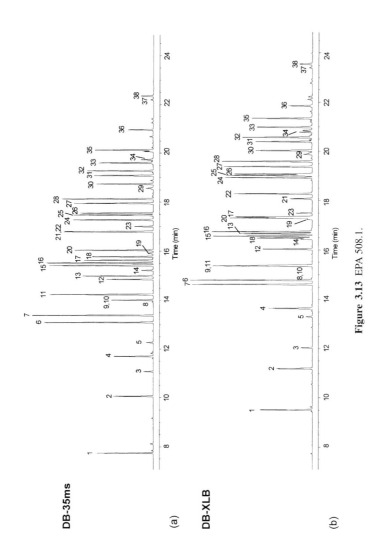

Figure 3.13 EPA 508.1.

DB-35ms

    30m x 0.32 mm I.D., 0.25 µm
    P/N: 123-3832

DB-XLB

    30m x 0.32 mm I.D., 0.50 µm
    P/N: 123-1236

Carrier: Helium at 45 cm/sec (EPC in constant flow mode)
Oven: 75°C for 0.5 min
        75-300°C at 10°C/min
        300°C for 2 min
Injector: Splitless, 250°C
        30 sec purge activation time
        50 pg per component
Detector: µECD, 350°C
        Nitrogen makeup gas
        (column + makeup flow = 30 mL/min constant flow)

1. Hexachloropentadiene
2. Etridiazole
3. Chloroneb
4. Trifluralin
5. Propachlor
6. Hexachlorobenzene
7. α-BHC
8. Atrazine
9. Pentachloronitrobenzene
10. Simazine
11. γ-BHC
12. β-BHC
13. Heptachlor
14. Alachlor
15. δ-BHC
16. Chlorothalonil
17. Aldrin
18. Metribuzin
19. Metolachlor
20. DCPA

21. 4,4´-Dibromobiphenyl
22. Heptachlor epoxide
23. Cyanazine
24. γ-Chlordane
25. α-Chlordane
26. Endosulfan I
27. 4,4´-DDE
28. Dieldrin
29. Chlorobenzilate
30. Endrin
31. 4,4´-DDD
32. Endosulfan II
33. 4,4´-DDT
34. Endrin Aldehyde
35. Endosulfan sulfate
36. Methoxychlor
37. cis-Permethrin
38. trans-Permethrin

Figure 3.13 (continued).

standard DB-35 ms column and on a column coated with another arylene phase, DB-XLB. Not only do the retention times on the two columns differ, but the elution order is also different.

Air analysis. The analysis of volatile organic compounds in air is important in environmental and occupational hygiene studies. EPA Method TO-14 defines the conditions for the analysis of volatiles in air by GC-MS, which requires a thick film column to give adequate separation of the most volatile compounds. While 1,1,2,2-tetrachloroethane (Figure 3.14 peak 36) and *o*-xylene (peak 37) are not completely resolved, these compounds can be distinguished by the *m/z* 83 and 91 peaks from the mass spectrometer detector. The use of a 0.32 mm i.d. column gives greater sample capacity but the MS detector requires a long column in order to obtain a positive head pressure at the required carrier gas linear velocity. A 30 m × 0.20 mm i.d. column would require a head pressure of 5.6 psi (38.6 kPa) to achieve the 25 cm s^{-1} linear velocity of helium required, whereas a 60 m × 0.32 mm i.d. column requires a head pressure of only 3.6 psi (24.8 kPa) to achieve the same linear velocity (Figure 3.14).

Substituted anilines. Anilines are widely used as intermediates in the production of pesticides, dyes and other important chemicals; many are known to be toxic and carcinogenic. Two approaches were adopted for the separation of this mixture. The first was by GC-MS using a DB-5 ms column, a good general-purpose column for trace level analysis (Figure 3.15a). With GC-MS, compounds that are not separated on the column can be resolved by differences in their mass spectra. 4-Chloronitroaniline and 4-nitroaniline co-elute from this column but can be readily separated and quantified by their mass spectra. However, with less-specific detectors, identification by a single elution time

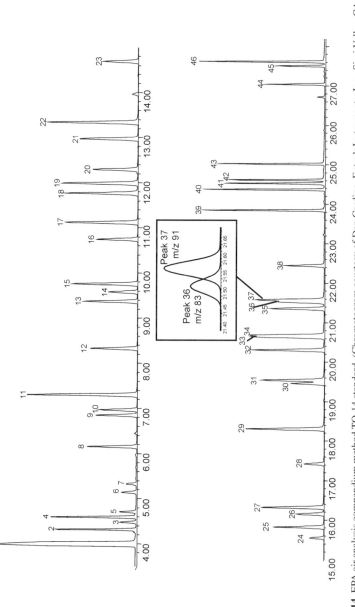

Figure 3.14 EPA air analysis compendium method TO-14 standard. (Chromatogram courtesy of Dan Carding, Entech Instruments, Inc., Simi Valley, CA, USA.)

Column: **DB-1**
    60 m x 0.32 mm I.D., 1.0 µm

Carrier: Helium at 25 cm/sec
           Measured off of CO_2 at 35°C

Oven: 35°C for 5 min
        35°-120°C at 5°/min
        120-220°C at 30°/min
        220°C for 5 min

Injector: Entech 7100 cryogenic
         sample preconcentrator
         400 mL of a 10 ppbV TO-14 standard and
         100 mL of a 20 ppbV IS/SS standard

Detector: HP 5973 MSD,
         full scan of m/z 40-250

1. CO_2
2. Freon 12
   (Dichlorofluoromethane)
3. Chloromethane
4. Freon 114
   (1,2-Dichloro-1,1,2,2-tetrafluoroethane)
5. Vinyl chloride
6. Bromomethane
7. Chloroethane
8. Freon 11
   (Trichlorofluoromethane)
9. 1,1-Dichloroethene
10. Methylene chloride
11. Freon 113
   (1,1,2-Trichloro-1,2,2-trifluoroethane)
12. 1,1-Dichloroethane
13. cis-1,2-Dichoroethene
14. Bromochloromethane (IS)
15. Chloroform
16. 1,2-Dichloroethane
17. 1,1,1-Trichloroethane
18. Benzene
19. Carbon tetrachloride
20. 1,4-Difluorobenzene (IS)
21. 1,2-Dichloropropane
23. cis-1,3-Dichloropropene

22. Trichloromethene
23. cis-1,3-Dichloropropene
24. trans-1,3-Dichloropropene
25. 1,1,2-Trichloroethane
26. Toluene-d8 (SURR)
27. Toluene
28. 1,2-Dibromoethane
29. Tetrachloroethene
30. Chlorobenzene-d5 (SURR)
31. Chlorobenzene
32. Ethylbenzene
33. m-Xylene
34. p-Xylene
35. Styrene
36. 1,1,2,2-Tetrachloroethane
37. o-Xylene
38. 4-Bromofluorobenzene (SURR)
39. 1,3,5-Trimethylbenzene
40. 1,2,4-Trimethylbenzene
41. 1,3-Dichlorobenzene
42. 1,2-Dichlorobenzene
43. 1,4-Dichlorobenzene
44. 1,2,4-Trichlorobenzene
45. 1,2-Dibromobenzene (IS)
46. Hexachloro-1,3-butadiene

Figure 3.14 (continued).

is suspect. Generally, a column with a dissimilar phase must also be used to improve the reliability of the identification. The same set of aniline compounds was run on a DB-17 column (50% phenyl phase) and a DB-1701 column (14% cyanopropylphenyl phase) of the same dimensions. The two columns were part of a dual column assembly in which both columns have a common inlet with a guard column. Ideally the differences in the separation characteristics of the two phases should result in elution order differences and retention time shifts, and this was found to be the case for all but the pair 2-nitroaniline and 3,4-dichloroaniline (Figure 3.15b). The only other option left for positive identification is to use a chlorine-selective detector such as the electrical conductivity detector on one of the channels.

3.3.3 Life sciences

Fatty acids and fatty acid esters. Fatty acids are extremely important in metabolic processes and were among the first group of compounds to be separated using GC by James and Martin in 1952. Fatty acids have traditionally been, and still are (see below), analysed as their methyl esters. This avoids the tailing experienced on most phases. The development of FFAP (free fatty acid phase) by the esterification of a poly(ethylene glycol) with nitroterephthalic acid has resulted in a phase that allows carboxylic acids to be separated without esterification (Figure 3.16). However, for the separation of the key C_{18} saturated and unsaturated acids it is still necessary to analyse the acids as their methyl esters (Figure 3.17). To resolve the close positional isomers, a column with a stationary phase with a large dipole moment is necessary. The stationary phase

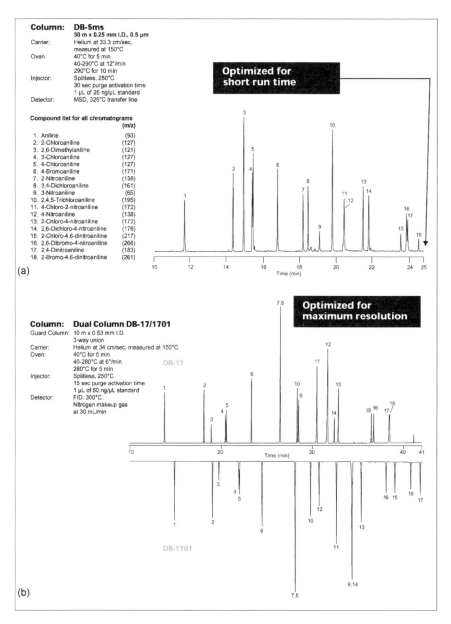

Figure 3.15 Substituted anilines: (a) optimised for short run time; (b) optimised for maximum resolution.

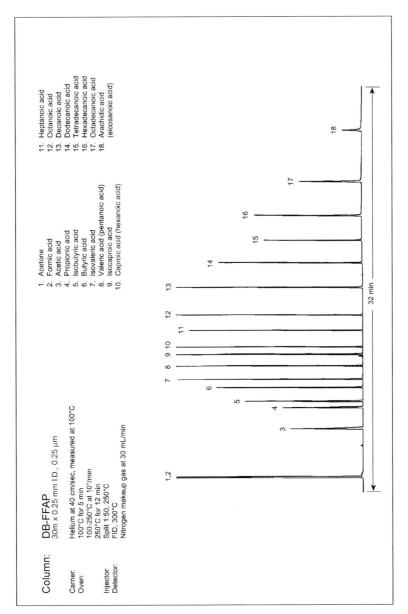

Figure 3.16 Organic acids.

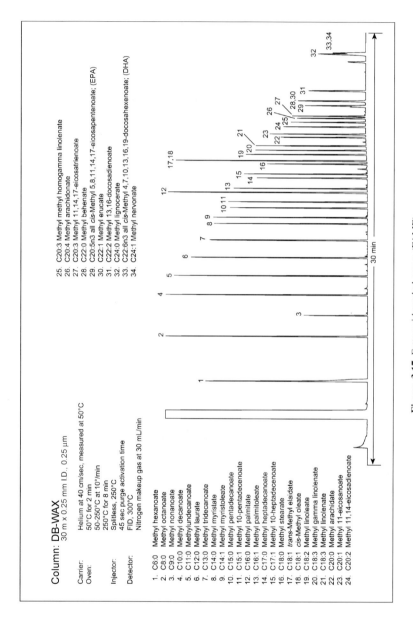

Figure 3.17 Fatty acid methyl esters (FAME).

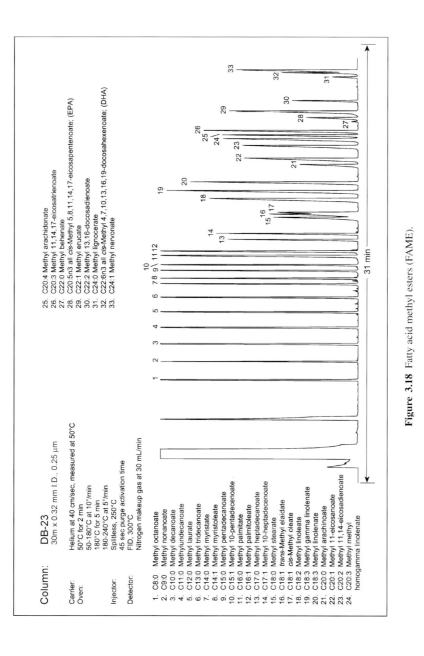

Column: DB-23
30m x 0.32 mm I.D., 0.25 µm

Carrier: Helium at 40 cm/sec, measured at 50°C
Oven: 50°C for 2 min
50-180°C at 10°/min
180°C for 5 min
180-240°C at 5°/min
Injector: Splitless, 250°C
Detector: FID, 300°C
45 sec purge activation time
Nitrogen makeup gas at 30 mL/min

1. C8:0   Methyl octanoate
2. C9:0   Methyl nonanoate
3. C10:0 Methyl decanoate
4. C11:0 Methylundecanoate
5. C12:0 Methyl laurate
6. C13:0 Methyl tridecanoate
7. C14:0 Methyl myristate
8. C14:1 Methyl myristoleate
9. C15:0 Methyl pentadecanoate
10. C15:1 Methyl 10-pentadecenoate
11. C16:0 Methyl palmitate
12. C16:1 Methyl palmitoleate
13. C17:0 Methyl heptadecanoate
14. C17:1 Methyl 10-heptadecenoate
15. C18:0 Methyl stearate
16. C18:1 trans-Methyl elaidate
17. C18:1 cis-Methyl oleate
18. C18:2 Methyl linoleate
19. C18:3 Methyl gamma linolenate
20. C18:3 Methyl linolenate
21. C20:0 Methyl arachinoate
22. C20:1 Methyl 11-eicosanoate
23. C20:2 Methyl 11,14-eicosadienoate
24. C20:3 Methyl methyl
homogamma linolenate

25. C20:4 Methyl arachidonate
26. C20:3 Methyl 11,14,17-eicosatrienoate
27. C22:0 Methyl behenate
28. C20:5n3 all cis-Methyl 5,8,11,14,17-eicosapentenoate; (EPA)
29. C22:1 Methyl erucate
30. C22:2 Methyl 13,16-docosadienoate
31. C24:0 Methyl lignocerate
32. C22:6n3 all cis-Methyl 4,7,10,13,16,19-docosahexenoate; (DHA)
33. C24:1 Methyl nervonate

Figure 3.18 Fatty acid methyl esters (FAME).

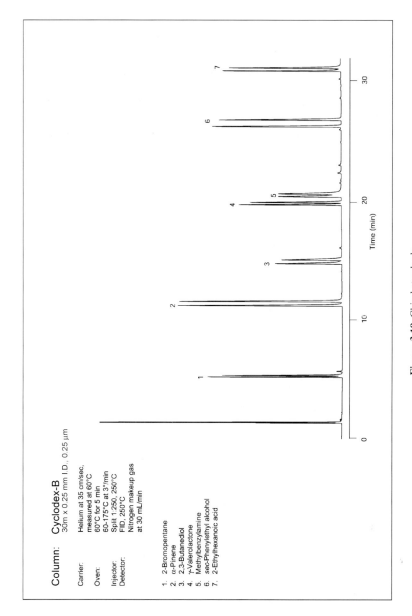

Figure 3.19 Chiral standards.

DB-23 (Figure 3.18) contains 50% cyanopropyl siloxane, which gives good resolution of the positional isomers that are not separated on the DB-WAX column.

Chiral compounds. The separation of chiral mixtures is of increasing importance in the pharmaceutical industry, where only one enantiomer of a drug may biologically active or the other enantiomer may be toxic. The majority of chiral separations are carried out by HPLC since the higher column temperatures involved in GC tend to negate chiral separation and re-form the racemate, which is thermodynamically more stable. However, in favourable instances permethylated β-cyclodextrins embedded in other stationary phases are capable of giving excellent results. Figure 3.19 shows the chromatogram of a synthetic mixture of chiral compounds run on a Cyclodex-B column with excellent results even with a maximum column temperature of 175°C.

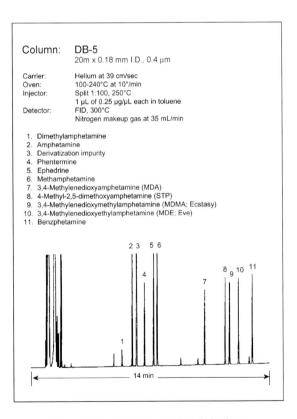

Figure 3.20 Amphetamines: TFA derivatives.

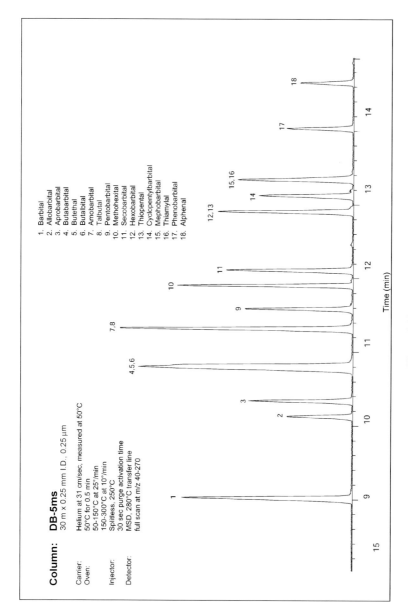

Column: DB-5ms
30 m x 0.25 mm I.D., 0.25 µm

Carrier: Helium at 31 cm/sec, measured at 50°C
Oven: 50°C for 0.5 min
50-150°C at 25°/min
150-300°C at 10°/min
Injector: Splitless, 250°C
30 sec purge activation time
Detector: MSD, 280°C transfer line
full scan at m/z 40-270

1. Barbital
2. Allobarbital
3. Aprobarbital
4. Butabarbital
5. Butethal
6. Butalbital
7. Amobarbital
8. Talbutal
9. Pentobarbital
10. Methohexital
11. Secobarbital
12. Hexobarbital
13. Thiopental
14. Cyclopentylbarbital
15. Mephobarbital
16. Thiamylal
17. Phenobarbital
18. Alphenal

Time (min)

Figure 3.21 Barbiturates.

3.3.4 Forensic analysis

Analysis of amphetamines is now probably the most common task in forensic laboratories (Figure 3.20). Barbiturates were used extensively in the past as sedatives but have now been largely replaced with the benzodiazepines. However, barbiturates still find use as a treatment in epilepsy. Barbiturate overdoses may occur because the effective dose is not too far from the lethal dose; it is therefore imperative that these compounds be detected at very low levels (Figure 3.21).

The use of two nonpolar phases for these analyses may not give the best resolution of the entire family of compounds, but the decreased selectivity is offset by the increased thermal stability, thus allowing for lower limits of detection at the high column temperatures attainable. The low-bleed characteristics of the DB-5 ms phase make it excellent for the detection of low levels of drugs of abuse by mass spectrometry. This is true for even very high-boiling compounds such as steroids.

3.4 Conclusion

It is hoped that the selection of chromatograms presented here, covering a wide range of analytes, shows that it is no longer necessary to use long poly(methyl siloxane) columns in the hope that a large number of theoretical plates will solve a particular separation problem. All of the five operating parameters listed at the beginning of this chapter can be employed to tailor a column for optimum separation in as short a time as possible and to give results compatible with the detectors chosen to complete the information required. It is anticipated that this trend will become even more pronounced in the future.

Acknowledgements

All chromatograms in this chapter are reproduced courtesy of Agilent Technologies, Folsom, CA.

References

1. W. Jennings, E. Mittlefeldt and P. Stremple, *Analytical Gas Chromatography* 2nd edn. Academic Press, San Diego, 1997, pp. 168–247.
2. R.P.W. Scott, in *Gas Chromatography* (ed. D.H. Desty), Butterworths, London, 1953, p. 189.
3. R.D. Dandeneau and E.H. Zerenner, *J. High Resolut. Chromatogr.*, **1979**, *2*, 351.
4. I.M. Whittemore, in *Chromatography in Petroleum Analysis* (eds. K.H. Altgelt and T.H. Gouw Marcel), Dekker, New York, 1979, chapter 2.
5. *Annual Book of ASTM Standards*, vol. 05.02, American Society for Testing and Materials, ASTM, Philadelphia, PA.
6. US Environmental Protection Agency, *Standard Methods of Analysis*, USEPA, Washington DC.
7. G. Frame, *Anal. Chem.*, **1997**, *468A*.
8. K. Ballschmitter and M. Zell, *Fresenius. Z. Anal. Chem.*, **1980**, *302*, 20.

4 Detectors for quantitative gas chromatography

Paul Larson

4.1 Introduction

Quantitative detection in gas chromatography is dependent upon both detector and chromatographic system parameters. These parameters include the detector gas(es) flow rate(s), detector temperature, method of sample introduction, method of sample preparation or collection, choice of column and column type, the ambient temperature and pressure in the laboratory (or the field), the inertness of the system, and signal path and data system considerations.

Another trend in gas chromatography is the movement to performance-based methods from prescriptive methods. ISO (the International Standards Organization) has published ISO International Standard 10723 [1]. This method is for the performance evaluation of on-line analytical systems used for measuring natural gas. OIML (Organisation Internationale de Métrologie Légale) has published International Recommendations R82 and R113 covering both laboratory and portable gas chromatographs [2, 3]. These recommendations give metrological requirements for the gas chromatographs based on such parameters as minimum detection limit and dynamic range for a variety of detectors. R113 also includes performance-based tests for the system.

The biggest changes in gas chromatography over the last decade are in the widespread use of electronic pneumatic control, the development of gas purifiers, and the continuing move towards miniaturization. This chapter will be organized using these topics as the guide for discussion. Results from investigations of the flow space for the flame ionization detector (FID), the nitrogen phosphorus detector (NPD), and the flame photometric detector will be discussed in the section on pneumatic control. The pneumatic control section will include discussion of other sulfur detectors commercialized over the decade. The helium ionization detector (HID) will be discussed under gas purifiers because gas purification is the enabling technology for this detector. The thermal conductivity detector (TCD) and the electron capture detector will be discussed under the topic of miniaturization.

4.2 Pneumatic control

The electronic pneumatic control is a technology enabling reduction in the physical size of the carrier gas(es), and auxiliary controls. The ability to electronically set the pressure or flow rates for the inlets, detectors and auxiliary control has provided a means for easier transfer of methods between different sites with increased precision. This digital control has also allowed ambient temperature and pressure effects to be compensated for [4]. The degree of set point repeatability possible is shown in Figure 4.1, which shows the results of an experiment in which the performance of several different 6890 gas chromatographs with different inlet and detector combinations was characterized by placing the 6890s into a Russells walk-in environmental chamber. This allowed the temperature to be controlled and varied in the sequence 25°C, 22°C, 28°C, 16°C and 34°C. This sequence allowed the temperature difference to increase between each set of runs at a given chamber temperature. The conditions are shown in Table 4.1. The grand relative standard deviation (RSD) was then calculated for the retention times of the different components in the test sample (see Table 4.2 for the composition). The results indicate that the grand RSDs are less than 0.175% for an 18°C temperature range. The effect of barometric pressure changes [4] investigated by testing an instrument at sea level and then at 5000 feet. The effect was shown to be smaller than that of ambient temperature. The results show that for gas chromatographs with correction for ambient pressure and temperature the pneumatic set points for a method can be easily transferred from site to site and instrument to instrument. Electronic pneumatic control will also enable the move to smaller instruments. Use of the manifold technology shown in Figure 4.2 considerably reduces the space for pneumatic control. Pneumatic control has allowed features such as constant flow and compensation

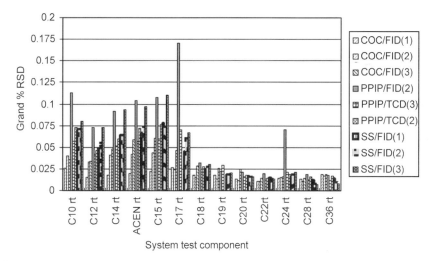

Figure 4.1 Plot of grand RSD for different 6890 inlet detector combinations run in walk-in chamber. RSD calculated as grand average for temperature range 16 C to 34 C. Copyright 2000 Agilent Technologies, Inc. Reproduced with permission.

Table 4.1 HP 6890 operating conditions for system test sample

Parameter/inlet	Split/splitless	COC	PPIP
Column	30 m × 0.320 mm × 0.25 µm HP-5	30 m × 0.320 mm × 0.25 µm HP-5	30 m × 0.320 mm × 0.25 µm HP-5 or 15 m × 0.530 mm × 0.88 µm HP-5
Carrier flow	2 ml min^{-1} (constant flow)	2 ml min^{-1} (constant flow)	5 ml min^{-1} (constant flow)
Inlet temperature	335°C	Oven track	335°C
Detector	350°C	350°C	350°C
Oven	65°C for 1 min 25°C min^{-1} to 140°C 140°C for 15 min 10°C min^{-1} to 320°C 320°C for 14 min	65°C for 1 min 25°C min^{-1} to 140°C 140°C for 15 min 10°C min^{-1} to 320°C 320°C for 14 min	65°C for 1 min 25°C min^{-1} to 125°C 125°C for 12 min 10°C min^{-1} to 320°C 320°C for 14 min

Abbreviations: COC, cool on-column; PPIP, purged packed injection port. Copyright 2000, Agilent Technologies, Inc. Reproduced with permission.

Table 4.2 Composition and concentration of system test sample

Component	Concentration (µg µl^{-1})
n-Pentane (C$_5$)	1000
n-Decane (C$_{10}$)	100
n-Dodecane (C$_{12}$)	300
n-Tetradecane (C$_{14}$)	200
Acenaphthylene (ACEN)	200
n-Pentadecane (C$_{15}$)	200
n-Heptadecane (C$_{17}$)	1.5
n-Octadecane (C$_{18}$)	10
n-Nonadecane (C$_{19}$)	2
n-Eicosane (C$_{20}$)	100
n-Docosane (C$_{22}$)	1000
n-Tetracosane (C$_{24}$)	10000
n-Octacosane (C$_{28}$)	100
n-Hexatriacontane (C$_{36}$)	100
Solvent: isooctane	

Copyright 2000, Agilent Technologies, Inc. Reproduced with permission.

for changes in carrier flow by adjusting the make-up gas flow for the detector (constant = column flow + makeup gas flow). Electronic pneumatic control has also facilitated the study of the response surfaces for detectors such as the flame ionization detector, the nitrogen phosphorus detector, and the flame photometric detector.

Figure 4.2 Detector manifold for a flame ionization detector. Dimensions are ∼85 mm × ∼85 mm × 60 mm. Copyright 2000 Agilent Technologies, Inc. Reproduced with permission.

4.2.1 Flame ionization detector (FID)

The FID is the most widely used detector in gas chromatography. It has many favorable attributes: consistent response for most hydrocarbons, wide linear dynamic range and sensitivity well matched with both capillary and packed-column analysis.

The basis for the FID's consistent response was investigated by Holm using two different approaches and is reported in a series of papers [5–7]. In one approach a flowing stream of hydrogen (to which 0.5% by volume of different compounds were added) passed through a capillary heated to 1400°C. The stream was collected over water in a test tube and then analysed. The second approach used a capillary connected to a mass spectrometer to sample different regions of the flame. The results showed almost complete conversion of the hydrocarbons tested to methane. The relative molar responses for heteroatoms and isotope effects were also studied.

These results are consistent with the experience of chromatographers over the past forty years. Methods have been developed based on normalized response ratios. One such method is ASTM method 5134 [8], which prescribes relative response ratios of 1 for all but benzene (0.90) and toluene (0.95). This allows the response of the method to compounds to be calibrated using a sample with a few components. Thus, a compound such as propane can be used as a model compound for studying the response as a function of different hydrogen and air flow rates.

A 2^3 factorial experiment was used to study how the FID response is affected as a function of propane concentration, hydrogen flow rate and air flow rate. The column used for the study was a 6 foot × $\frac{1}{4}$ inch 10% SP2100 on Chromosorb W HP 80/100. The column flow was set at 30 ml min^{-1}. The oven temperature was set at 50°C with the detector block at 300°C. The jet was the 0.011 inch (0.279 mm) jet. The hydrogen flow rate was set at 24.0, 42.0 or 60.0 ml min^{-1}, while the air flow rate was varied at 200.0, 400.0 or 600.0 ml min^{-1}. The different detector flow conditions were set using electronic pressure control.

Propane and propane mixtures were purchased from Scott Specialty Gases (Plumstead, PA, USA). Samples of the propane and propane mixtures were injected using gas-tight syringes (VICI Precision Sampling, Baton Rouge, LA, USA). Separate syringes were used for each mixture. The data were collected using a Hewlett Packard 3365 ChemStation. The results were analysed using multiple regression on the Statgraphics Plus (Manugistics, Inc., Rockville, MD) statistical software package. The following regression equations were found using stepwise regression. The regression equations are scaled in pA response (height).

For 100% propane.

Response (1 ml, propane) =

$\qquad$ −1547734 − 3162.856 × [hydrogen flow]
$\qquad$ +27609.57 × [air flow] + 158.0558 × [air flow] × [hydrogen flow]
$\qquad$ −702.138 × [hydrogen flow]^2 − 33.0321 × [air flow]^2

$R^2 = 0.9756$

For 1% propane.

Response (0.1 ml, 1% propane) =

$\qquad$ 3733.486 + 305.5687 × [hydrogen flow]
$\qquad$ −3.847408 × [air flow] + 0.122421 × [air flow] × [hydrogen flow]
$\qquad$ −3.95982 × [hydrogen flow]^2 − 0.00105 × [air flow]^2

$R^2 = 0.9936$

For 0.1% propane.

Response (0.1 ml, 0.1% propane) =

$\qquad$ 327.8107 + 32.90859 × [hydrogen flow]
$\qquad$ −0.395417 × [air flow] + 0.01183 × [air flow] × [hydrogen flow]
$\qquad$ −0.41655 × [hydrogen flow]^2 − 0.000065 × [air flow]^2

$R^2 = 0.9773$

These regression equations were input into Statgraphics Plus to generate the response surfaces shown in Figures 4.3, 4.4 and 4.5. The plots and the equations show the response surface for the 1.0 ml injection of propane to be quite different. The amount of propane is near to or exceeds the stoichiometric limit for the air over the range of air flows. The flame is extinguished with the hydrogen set at 24 ml min^{-1} and the air flow at 600 ml min^{-1}. Under these conditions

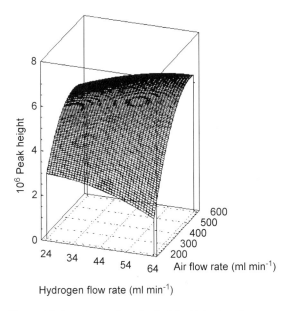

Figure 4.3 Response surface plot for 1.0 ml injection of propane.

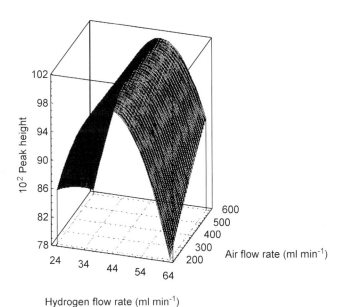

Figure 4.4 Response surface plot for 0.1 ml injection of 1% propane.

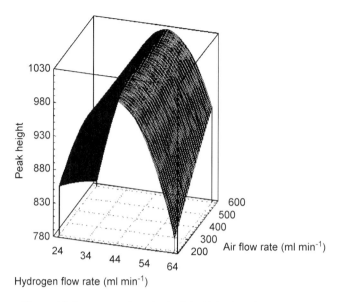

Figure 4.5 Response surface plot for 0.1 ml injection of 0.1% propane.

the propane plug overwhelms the flame chemistry, possibly by cooling the flame. The other two response surfaces have similar characteristics. The regression equations are basically an order of magnitude different (compare the coefficients), with the R^2 values essentially the same. A generalized equation can be generated by including the propane concentration in the equation.

For propane.

Response $(f[\text{propane}]) =$

$C_1 \times [\text{propane}] + C_2 \times [\text{hydrogen flow}] \times [\text{propane}] - C_3 \times [\text{air flow}]$
$\times [\text{propane}] + C_4 \times [\text{air flow}] \times [\text{hydrogen flow}] \times [\text{propane}]$
$- C_5 \times [\text{hydrogen flow}]^2 \times [\text{propane}] - C_6 \times [\text{air flow}]^2 \times [\text{propane}]$

As GC moves to faster analysis using smaller-diameter columns, the data rate setting becomes an important factor in determining peak fidelity and in determining the minimum detection limit (MDL). The minimum detection limit for the FID is defined by the ASTM standard 594 [9] as

$$\text{MDL} = 2\frac{S}{N}$$

where $S = \frac{\text{area}}{\text{mass}}$ and $N = \text{noise}$

The effect of detector flow rate and data rate was investigated using a 10 m × 0.10 mm × 0.20 μm d_f 5% phenyl methyl column and an experimental 0.007 inch (0.178 mm) diameter jet. The combination of the HP Multi-technique ChemStation and 6890 allows two different data channels to be set up with different data rates. This allows a direct comparison of the MDL at 20 Hz and at 200 Hz. The normal flow space was investigated, with the results shown in Table 4.3, which indicate that lower flows of air, hydrogen and make-up gas reduced the noise. The results of reducing the detector fuel gases and the make-up gas flow rates even further are shown in Table 4.4. With the smaller-diameter jet, the lower flows gave a better MDL at 200 Hz. However, the trade-off will be a reduction in the dynamic range of the detector. This will require an increase of the split ratio of the sample, which will reduce the amount of sample reaching the detector.

The consistency of FID response allows other samples to be used for mapping the response of the detector. As part of the 6890 GC development, a system test sample (see Figure 4.6) was used. The sample composition is given in

Table 4.3 MDL values measured at 20 Hz and 200 Hz using a 0.007 inch jet

H_2	Flow (ml min^{-1}) Air	MUG	MDL @ 20 Hz	MDL @ 200 Hz
60	200	5	7.53	42.3
		10	5.5	32.5
		20	4.6	26.1
		30	3.1	17.1
60	400	5	4.96	26.5
		10	4.04	21.9
		20	2.96	15.3
		30	2.65	13.7
60	600	5	4.33	24.3
		10	3.73	19.9
		20	2.73	13.6
		30	2.4	11.6
42.5	200	5	4.22	20.9
		10	3	16.2
		20	2.19	11.7
		30	2.1	9.67
42.5	400	5	3.14	16.5
		10	2.65	12.8
		20	2.22	9.67
		30	1.87	8
42.5	600	5	3.13	14.1
		10	2.51	11.4
		20	2.12	8.45
		30	1.85	7.67

MUG, make-up gas.

Table 4.4 MDL values measured at 20 Hz and 200 Hz using a 0.007 inch jet at low and no make-up flow

Flow (ml min^{-1})				
H$_2$	Air	MUG	MDL @ 20 Hz	MDL @ 200 Hz
6	100	0	2.74	4.81
8	100	0	2.58	4.05
8	200	0	2.57	4.28
10	250	0	2.27	3.97
12.5	100	0	2.45	4.29
12.5	300	0	2.25	3.94
25	600	0	2.23	6.69
12.5	300	10	1.89	3.44
12.5	300	5	1.69	3.26

MUG, make-up gas.

Table 4.2. The range was 3.5 orders of magnitude for the compounds between C$_{17}$ and C$_{24}$ and $\sim$ 5.5 orders of magnitude if the solvent was included. The sample allowed the linearity of the detector to be studied for each injection. This was important because the sample was used to study the effects of the environment upon performance of the instrument. The investigation was carried out using a Russells Technical Corporation, Inc., (Holland, MI, USA) walk-in environmental chamber and by varying the chamber temperature as 25°C, 22°C, 28°C, 16°C and 34°C. The order was chosen to give constantly increasing temperature differences between the chamber test temperatures. The response

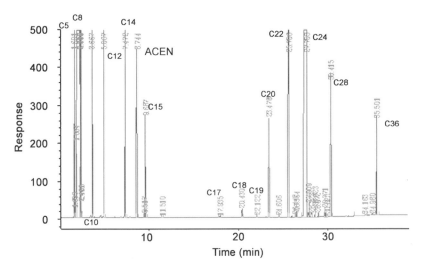

Figure 4.6 Chromatogram of the system test sample. Copyright 2000 Agilent Technologies, Inc. Reproduced with permission.

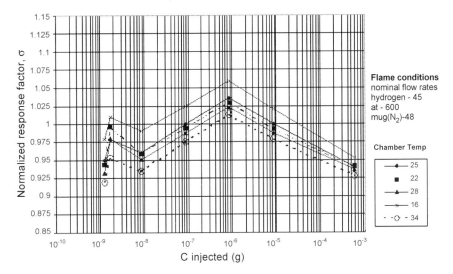

Figure 4.7 Plot of normalized average response factors for system test components as a function of chamber temperature. Flame conditions: nominal flow rates: hydrogen, 45 ml min^{-1}; air, 600 ml min^{-1}; MUG(N$_2$), 48 ml min^{-1}. Abbreviation: MUG, make-up gas.

factor of C$_{20}$ at 25°C was chosen as the reference. A plot of the normalized response factors relative to the C$_{20}$ response factor is shown in Figure 4.7, which shows that the higher chamber temperatures, such as 34°C, tended to give lower normalized response than that at 16°C. This is consistent with the thermal expansion of the sample solution, which would produce a decrease in the volumetric concentration. For most solvents this will be a small effect, a change of less than 0.2% for a 10°C change in temperature.

4.2.2   Nitrogen phosphorus detector (NPD)

The NPD is another detector that has benefited from electronic pneumatic control. Studies by Klein, Larson and Breckenridge used the NPD to illustrate the benefits of maintaining constant flow in the detector by adjusting the carrier and/or make-up gas pressure to compensate for flow changes during temperature-programmed analyses [10]. Another way the NPD has benefited from electronic pneumatic control is in the ability to turn off the hydrogen during the elution of the solvent peak [11]. This facilitates the use of aggressive solvents, such as methylene chloride, when necessary.

It was also possible to study the response surface. A 5890 Series II with a NPD and electronic pressure control connected to the hydrogen, air and make-up gas allowed the flows to be varied easily. The bead current is adjustable manually on the 5890 and was adjusted so that the offset was 20 at the beginning of the experiment. The advantage of the NPD over the FID is the selectivity of the detector for most nitrogen-containing and phosphorus-containing compounds. The selectivity is defined as the response factor of the nitrogen-containing compound or phosphorus-containing compound compared to the response factor of a reference hydrocarbon. As with the FID, an experiment was run that looked at the response surface of the NPD. The nitrogen and phosphorus selectivities for a Hewlett Packard 5890 NPD were studied by varying the hydrogen, air and make-up flows and the results were analysed using stepwise regression. The model generated cannot be considered a general equation for the NPD, but is given to show the complex interactions that are possible with this detector (mug = make-up gas).

$$N_{selectivity} = 279975 - 5107 \times [\text{air flow rate}] + 8895 \times [\text{hydrogen flow rate}]^2$$
$$- 3078 \times [\text{hydrogen flow rate}] \times [\text{mug flow rate}]$$
$$+ 112 \times [\text{mug flow rate}] \times [\text{air flow rate}] + 12 \times [\text{air flow rate}]^2$$
$$R^2 = 0.9478$$

Part of the complexity may be due to cooling effects around the thermionic region of the bead. Because of the possible variation in the NPD bead, a general equation describing the behavior is probably not possible at this time. With the different types of beads available, additional interactions are likely. The surface directly affects the response, while the fuel gases will have an indirect but important effect.

4.2.3   Flame photometric detector (FPD)

The response of sulfur in the FPD is based on the emission of S_2. As cited by Farwell and Barinaga, a wide range of n values have been reported for the response for sulfur [12]. The value of n is the result of fitting the response to the model given in Table 4.5. A series of compounds (2,5-dimethylthiophene, 1,4-butanedithiol, sec-butyl disulfide, dodecanethiol and octyl sulfide) were studied over a range of ~ 50 pg to 10 ng and the peak area was regressed versus the sulfur concentration. The results are given in Table 4.5. The regressions show values of R^2 of 0.99, giving n-values in the range 1.75–2.25. The regression was then forced to a value of $n = 2$. The correlation coefficient is lower with the exception of 1,4-butanedithiol. The regression results show how shallow the optimum is. This indicates that mechanistic conclusions should not be drawn from the $n \neq 2$ regressions, and that $n = 2$ adequately describes the sulfur response.

Another experiment was designed to look at the effects of quenching and flow rates of the detector gases. The quenching was achieved by bleeding propane into the carrier gas in a slug. The carbon concentration was varied by varying the flow rate of the propane (1% propane in nitrogen). An electronic flow controller regulated the amount of propane. This allowed the propane

Table 4.5 R^2 values for best fit and force to $n = 2$ response for different sulfur compounds: response model, area $= C[S]^n$

Compound	log–log regression		$n = 2$ regression	
	n	R^2	n	R^2
Dodecanethiol	1.87	0.9988	2.00	0.9986
sec-Butyl disulfide	1.89	0.9989	2.00	0.9986
Dioctyl disulfide	2.27	0.9952	2.00	0.9929
2,5-Dimethylthiophene	1.90	0.9970	2.00	0.9966
1,4-Butanediol	1.99	0.9990	2.00	0.9990

mixture to be switched on and off at preset runtimes. The elution time of the plug was chosen to coincide with the peak of interest as shown in Figure 4.8. The conditions studied with the experiment are given in Table 4.6. Triplicate injections of the samples were made for each sample and injection volume for each set of flow conditions. The response as a function of the unquenched response is given below, where the amounts of sulfur injected are 0.0031, 0.0156, 0.034 and 0.172 nmol. All the units were converted to a molar basis for regression purposes. The results show that at the lower flow rates of the fuel gases (the higher sensitivity region), the quenching is more prominent than at the higher flow rates. The quenching can be minimized but not eliminated by operating at the elevated flow rates. This can be useful if quenching is suspected. The sample of interest can be run at both sets of flow conditions, then the area for the lower flow response can be divided by the higher flow response for each of the components in the chromatogram. For peaks that are not quenched, the ratio will be about the same, while peaks that are quenched will have a ratio of the areas that is closer to 1.

The other compounds tested responded similarly to the dodecanethiol. The following model for the response was develop for the carbon concentration

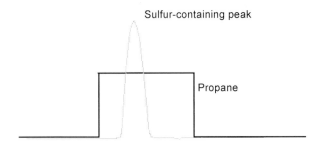

Sulfur-containing peak

Propane

Superimposed slug of propane onto sulfur peak

Figure 4.8 Plot showing co-elution of sulfur-containing peak and propane.

Table 4.6 Detector flow conditions (in ml min^{-1} of component)

Air mode			Oxygen mode			
Hydrogen	Air	1% propane	Hydrogen	Oxygen	Nitrogen	1% propane
50	60	0	40	15	60	0
50	60	2	40	15	60	2
50	60	3.5				
50	60	5				
120	100	0	75	27	80	0
120	100	2	75	27	80	2
120	100	3.5				
120	100	5				

Temperature program 40°C (1 min) to 200°C (5 min) at 10°C min^{-1}.
Column 10 m × 530 mm × 2.65 μm HP-1. Column flow rate 20 ml min^{-1} Nitrogen.

case $[C] = 0$. The results were evaluated and found to predict the response with $R^2 = 0.99$.

$$\text{Area} \propto \frac{[S]^2}{[H_2]^{1/2} - 2[O_2]^{1/2}}$$

where the concentrations are molar concentrations. The term $([H_2]^{1/2}-2[O_2]^{1/2})$ gives an indication of the free hydrogen available in the flame. This model is consistent with the observation that the sensitivity increases as the hydrogen flow decreases. When whole data set was regressed for dodecanethiol, the following equation fitted the whole data set with $R^2 = 0.95$.

$$\text{Area} \propto \frac{[S]^2}{[H_2]^{1/2} - 2[O_2]^{1/2}} \left(1 - \frac{[C]}{[H_2]^{1/2}}\right)$$

There is a self-induced upper limit to the sulfur mode of the FPD. This occurs when the concentration of the solute is great enough to cause self-quenching or self-absorption. This would not normally be seen with the FPD because the photomultiplier tube is normally saturated at this point. However, if the voltage driving the photomultiplier is reduced to near the turn-on voltage for the tube, it is possible to see this phenomenon. Figure 4.9 shows the inverted W-shaped peak. This occurred at concentrations well above the normal working range of the FPD.

4.2.4   Other sulfur selective detectors

Other sulfur detectors have been commercialized over the past decade. The chemiluminescence detectors are based on the interaction of the sulfur with ozone to form SO, which will chemiluminesce. An advantage of this approach is that the second-order response for sulfur is avoided. The atomic emission

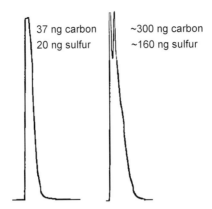

37 ng carbon   ~300 ng carbon
20 ng sulfur   ~160 ng sulfur

Figure 4.9 Self-quenching or self-absorption shown by sec-butyl sulfide.

detector uses the sulfur 183 set of emission lines for the basis of its sulfur detection. Again, the second-order response is avoided. The pulsed flame photometric detector, developed by Amirav and commercialized by several companies under licence, provides a means of separating the carbon emission from the sulfur emission by pulsing the flame on and off and following the emission as a function of time (on the millisecond time scale), but is still based on the S_2 mechanism and will also have the second-order effect [13]. More details of the properties of the sulfur chemiluminescence and the atomic emission detectors can be found in [14]. However, the biggest impact on the quantitative detection of sulfur compounds has been the improvement of the inertness of the sample lines. Tubing passivated with silicon-based compounds to form silica or silicon-based layers on the metal surfaces, such as the SilicoSteel treatment by Restek (Bellefonte, PA, USA), has greatly improved the ability to quantify trace amounts of sulfur compounds by reducing the reaction of the sulfur compounds with the sampling vessel and the transfer lines to the gas chromatograph.

4.3 Gas purifiers

The last decade has been the commercialization of different gas purification products. The availability of these products has enhanced the performance of GC detectors, in particular the helium ionization detector. These gettering products are based on metal alloys, such as the rare gas and nitrogen purifiers available from SAES Pure Gas, Inc. (San Luis Obispo, CA, USA) and gas specific purifiers from Matsen UOP (El Dorado Hills, CA, USA). The gas specific purifiers are available for hydrogen, helium and nitrogen. There are also commercially available air cleaners that significantly reduce the background for the flame ionization and nitrogen phosphorus detectors. For example, during the development of the 6890, the lit flame offset had to be changed from a default value of 2 pA to an adjustable parameter. This was the result of an experiment in which a Zero Air Generator (Agilent Technologies, Wilmington, DE) was used for the air. The system used a 1.2 m × 50 mm column with the flows set to 30 ml min^{-1} hydrogen, 400 ml min^{-1} air, and 25 ml min^{-1} make-up gas. The FID signal level was ~ 0.5 pA with the flame lit. Typically there is reduction of the background by one half. If the normal background is 12 pA, it will be reduced to 6 pA with the air cleaner. The reduction in background signal will also reduce the noise and can lead to better detection limits.

The gettering technology has been an enabling technology for both the helium ionization detector and the atomic emission detector. Wentworth describes the necessity of the getter in his review article on the nonradioactive electron capture detector [15].

VICI Valco Instruments Co. Inc. (Houston, TX, USA) has introduced a series of detectors based on the helium ionization detector. Wentworth and Stearns have been granted several patents. Wentworth *et al.* have published numerous papers, which are summarized in an extensive review covering the development of the detectors [15]. One goal of their work has been to develop a nonradioactive detector to replace the electron capture detector.

Progress with the detector was enabled by the ability to machine sapphire and quartz for use as insulating materials at high temperatures. (The leakage current of insulators is temperature dependent. Amplifying electronics require high impedance and low leakage currents to operated properly.) The helium purity was just as important. It is necessary to begin with 5-nines (99.999%) or 6-nines (99.9999%) helium and to purify it further with a helium purifier (Wentworth and Stearns used a Valco purifier, which is an metal alloy-based purifer). Care must also be taken to avoid other sources of contamination in the detector and carrier gas streams. Two different modes of operation were studied: constant potential and constant current. A linear dynamic range of 10^4 was observed for the constant potential and of 10^6 for the constant current mode with the MDL for carbon tetrachloride near one femtogram (10^{-15} g). Linerization algorithms were also developed.

4.4   Miniaturization

Making the gas chromatograph portable has required reduction in the size of the instrument and, where possible, of the individual components or modules. The actual space required by the pneumatic circuit has been reduced with the introduction of the manifold-based electronic pneumatic modules (see Figure 4.2). The move towards capillary and micro-packed columns also allows possible reduction in analysis time and the size of the column compartment.

Figure 4.10 Thermal conductivity detector for the micro GC. Dimensions of the detector module are ∼ 10 mm × ∼ 10 mm, with cell volume of 240 nl. Copyright 2000 Agilent Technologies, Inc. Reproduced with permission.

The move towards capillary columns has also required changes in the volumes of the different detectors used. One example of the reduction in the physical size and the cell size is the thermal conductivity detector (TCD). The older TCDs had volumes in the range 0.25–1 ml depending on the vintage. The single-filament design reduced the cell volume to 3.5 μl. The silicon-based TCD for the micro GC has a cell volume of 240 nl (see Figure 4.10). This is three orders of magnitude reduction in the cell volume and allows the micro GC TCD to be compatible with 0.20 mm columns with no make-up gas. The TCD is a concentration-dependent detector and its sensitivity is improved when low or no make-up gas can be used. With the Agilent 6890 micro electron capture detector there has been a reduction in cell volume but the size of the entire detector has not been reduced. With the electron capture detector there is a need to control the space charge and this requires that the internal volume and the amount of radioactive material must be carefully considered.

4.5   Data path considerations

The data path can be an important part of the ability of a detector to respond quantitatively to a sample input. The ability to integrate the peaks in the chromatogram can be compromised by inappropriate data path settings. This can occur in two ways. The first is by setting the data acquisition rate too low. With the move towards higher-speed chromatography, there is a need to understand the trade-offs between peak fidelity and, as seen previously, detectability. The limit of the background noise is determined by the shot noise. The peak-to-peak noise is dependent upon the bandwidth (data rate) and the current (for an FID this is the baseline current measured in pA). The shot noise is calculated as a RMS current value from the following equation [16].

$$\text{Noise}_{\text{shot}} = (i_s^2)^{1/2} = (2qI_{\text{dc}}B)^{1/2}$$

where B is the electrical bandwidth in hertz and I_{dc} (the value of the baseline signal for the detector) is the average current in amperes. The actual noise will be the sum of the different noise sources for the detector. However, the equation for the shot noise does point out the trade-off between data rate and MDL. The move to faster chromatography necessitates the move to higher data rates as times of analyses move from the order of 20–30 min to 2–3 min or even faster. As seen earlier in the discussion of gas purification, it is also possible to reduce the level of the baseline value by purifying the gases used in the detector. As the time required for analysis is shortened to seconds, there will be a need for faster data rates to keep peak fidelity. Dyson, in his review article [17] pointed out that good peak fidelity on poor peak shape, such as a tailing peak, may require 350 data points. This indicates that good, quantitative, fast chromatography will also require good chromatography if possible. Referring to the MDL data for the flow space indicates that the ratio for the noise should be $\sim 10^{1/2}$. Examining the noise values at 20 Hz versus 200 Hz will give an indication of other possible noise sources.

Another consideration in the data path is how many significant digits can be handled by the data path. Some of the commercially available gas chromatographs are able to handle 37 bytes in the digital data path to provide 11 orders of magnitude in the signal. This is especially important when detectors such as the flame ionization detector and the helium ionization detectors are used to measure purity in solvents and need to span the dynamic range of the detector and properly measure both the trace components and the solvent purity. A final consideration in the signal path is the bandwidth of the data system being used. Some of the data systems are limited to 100 Hz total for all of the channels being acquired by the data system, that is only one 100 Hz channel could be collected at a time even if 4 or 8 signals were connected. This is an area where the user must exercise care as the speed of analysis increases.

4.6   Summary

As the practice of gas chromatography moves towards faster analyses, the analyst must be aware of the system trade-offs necessary for quantitative results. The fuel gases set points for the FID will depend upon the working range required for the sample. For small-diameter columns, it is possible to reduce the effect of the increased data rate required by shifting to lower flows when working with appropriate split ratios. Full dynamic range work will still require the user to operate closer to the traditional operating region. The use of gas purifiers is required for the helium ionization-based detectors and the atomic emission detector, but has benefits for the other detectors as well by reducing the noise. The miniaturization of the detectors where possible has also provided a means to improve performance and compatibility with capillary columns. For the faster analyses, care must also be taken to match the bandwidth of the data system with the analysis.

References

1. International Standard, ISO 10723, *Natural gas—Performance evaluation for on-line analytical systems*, International Organization for Standardization, Geneva, 1995.
2. International Recommendation R 82, *Gas chromatographs for measuring pollution from pesticides and other toxic substances*, Bureau International de Métrologie Légale (BIML), Paris, 1989.
3. International Recommendation R 113, *Portable gas chromatographs for field measurements of hazardous chemical pollutants*, Bureau International de Métrologie Légale (BIML), Paris, 1994.
4. P. Larson, R. W. Henderson, W.D. Synder and E. E. Wikfors, *The Effect of Environmental Factors on the HP6890 Series Gas Chromatograph's Performance using Manual and Electronic Pneumatics*, Application Note 228–321, Hewlett-Packard, 1995.
5. T. Holm and J.O. Madsen, *Anal. Chem.*, **1996**, *68*, 3607.
6. T. Holm, *J. Chromatogr. A*, **1997**, *782*, 81.
7. T. Holm, *J. Chromatogr. A*, **1999**, *842*, 221.
8. A.W. Drews (ed.), ASTM method 5134-92, Standard Test Method for Detailed Analysis of Petroleum Naphthas through n-Nonane by Capillary Gas Chromatography, *Manual on Hydrocarbon Analysis*, 6th edn, 1998, pp. 786–796.
9. ASTM method E 594-93, *Standard Practice for Testing Flame Ionization Detectors Used in Gas Chromatography*, ASTM, Philadelphia, PA, 1993.
10. K.J. Klein, P.A. Larson and J.A. Breckenridge, *J. High Resolut. Chromatogr.*, **1992**, *15*, 615.
11. K. Meng, Y. Kaplun and R. White, *The New Feature of the HP6890 Nitrogen-Phosphorus Detector to Deal with Hostile Solvents*, Application Note 228-306, Hewlett-Packard, 1995.
12. S.O. Farwell and C.J. Barinaga, *J. Chromatogr. Sci.*, **1986**, *24*, 483.
13. S. Cheskis, E. Atar and A. Amirav, *Anal. Chem.*, **1993**, *65*, 539.
14. R.S. Hutle, *Chromatography in the Petroleum Industry*, Journal of Chromatography Library, **1995**, *56*, pp. 201–229.
15. W.E. Wentworth, J. Huang, K. Sun, Y. Zhang, Lei Rao, Huamin Cai and S. D. Stearns, *J. Chromatogr. A*, **1999**, *842*, 229.
16. G.E. Sullivan, in *Reference Data for Engineers: Radio Electronics, Computer, and Communications*, 7th edn (ed. in chief E.C. Jordan), Howard Sams & Co., 1985, pp. 21–15 to 21–17.
17. N. Dyson, *J. Chromatogr. A*, **1999**, *842*, 321.

5 Detectors for compound identification

Mark Powell

5.1 Introduction

The importance of gas chromatography as an analytical tool for organic compounds results from a combination of the separating ability of modern capillary columns, which remains superior to that of pressure-driven liquid separations, and the range of detectors that can be brought to bear to aid solute identification. Conventional GC detectors, which can be more or less selective towards different compound classes, are covered in Chapter 4. These detectors, however selective they may be, provide only two-dimensional information—a plot of detector signal versus time. This chapter deals with detectors for gas chromatography that provide three dimensions of data. These additional data may be used either on their own, or in combination with other information, such as retention time or the output from an additional detector, to identify unknown compounds. Three spectroscopic detector types will be discussed in detail; all provide information that is useful for solute identification. These are the mass spectrometer, which provides data relating to compound structure; the infrared detector, the output from which is a consequence of chemical functionality, and the atomic emission detector which gives information concerning elemental composition.

5.2 Gas chromatography–mass spectrometry

By far the most popular technique for the identification of unknown compounds in complex mixtures is gas chromatography–mass spectrometry (GC-MS). In recent years, the cost of benchtop GC-MS instruments has fallen steadily and software developments have made instruments easier to use, leading to their ubiquitous presence in routine organic analytical laboratories. J. R. Chapman's book *Practical Organic Mass Spectrometry* [1], is recommended for further reading.

GC-MS systems vary widely in their configuration and capabilities, but they all have certain features in common (Figure 5.1). The heated transfer line maintains solutes in the vapour phase as they pass from the GC to the mass spectrometer. The ion source ionises the solutes and propels them towards the mass analyser, where they are separated according to their mass to charge ratio (m/z). The detector produces a voltage in response to the impact of the resolved

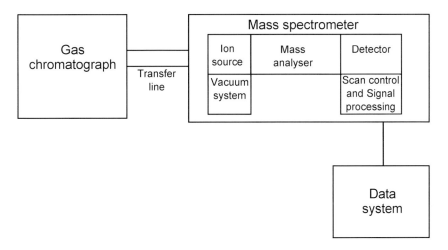

Figure 5.1 Block diagram showing the components of a GC-MS system.

ions, and this signal is then amplified and converted, via on-board electronics and a data system, into usable mass spectral information. The ultimate output from a GC-MS system is known as a mass chromatogram. Each data point in the chromatogram may be displayed as a mass spectrum (Figure 5.2). The vacuum system evacuates the region between the ion source and the detector, ensuring that the chance of collision between ions and neutral species is small. The typical ion source pressure for a mass spectrometer operating in electron impact ionisation mode is 1.3×10^{-4} Pa (10^{-6} torr).

Helium is usually employed as the carrier gas in GC-MS systems, since it forms relatively few ions, does not interfere with common low-mass ions and, owing to its high diffusivity, is rapidly pumped away from the mass spectrometer source region. Some mass spectrometer sources permit the use of hydrogen. Required carrier gas purity is normally 99.95%, with water, oxygen and hydrocarbons present at less than 1 ppm. This is commonly achieved by the use of traps, placed in the carrier gas line between the gas cylinder and the GC, packed with sorbent (for hydrocarbon removal) or reagents that combine with oxygen or water. Higher amounts of air and water can cause mass spectral distortion, reduce filament life and degrade the GC column.

5.2.1   Interface types

The amount of carrier gas that may be introduced into the mass spectrometer through the transfer line is governed by the conductance of the ion source, that is the rate at which carrier gas molecules diffuse out of the source and are removed by the vacuum system. There is an upper limit to the carrier gas flow that can

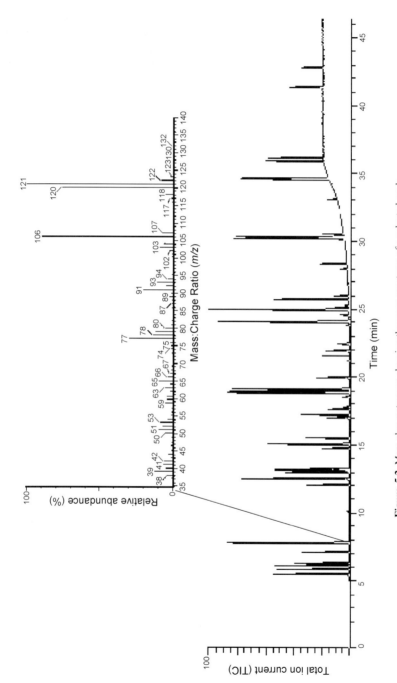

Figure 5.2 Mass chromatogram showing the mass spectrum of a selected peak.

the be introduced into the source region without raising pressure to the point where collisions between ions and neutral species become a problem. GC-MS systems will generally cope with a carrier gas flow up to 1 ml min^{-1}. This means that, in practice, capillary columns of 0.25 mm internal diameter or less may be directly coupled to the instrument. Here, the analytical column is fed through the transfer line, forming a continuous path between the injector and the ion source. Megabore (0.53 mm i.d. capillary) and packed GC columns, however, have typical carrier gas flows of 5 ml min^{-1} or greater. While some instruments are able to cope with the higher carrier gas flows from a direct-coupled megabore column, modification of the interface is usually required to limit the flow of carrier gas into the mass spectrometer. Two devices that are commonly used to reduce the flow of carrier gas into the mass spectrometer are the jet separator and the open split interface.

5.2.1.1 *Jet separator interface*

The jet separator is the more complex and costly of the two options. It consists of an airtight, heated manifold, usually made of glass, connected to a vacuum pump. Two narrow glass jets are mounted coaxially on opposite sides of the manifold with a small gap between them. One of these jets carries the effluent from the GC column, and the other leads, via a transfer line, to the mass spectrometer source (Figure 5.3). As the effluent from the GC column leaves the jet, lighter components (including the bulk of the carrier gas) are pumped away while heavier species have sufficient momentum to travel across the gap and enter the mass spectrometer transfer line.

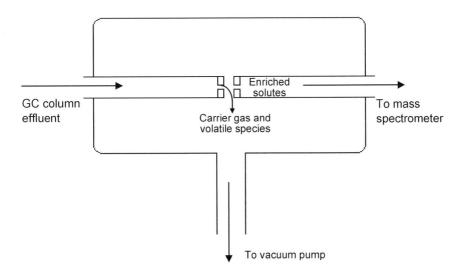

Figure 5.3 Jet separator interface.

The major advantage of this device is that analyte species are enriched with respect to the carrier gas when they enter the mass spectrometer. This is particularly useful in trace analysis, where enrichment confers additional sensitivity. The main disadvantage is that the efficiency of transfer is poor for lighter analytes.

The performance of a jet separator may be characterised by its yield (Y) and degree of enrichment (E) [1]. Yield is calculated as

$$Y = \frac{M_2}{M_1} \times 100\%$$
(5.1)

where M_1 = amount of sample leaving the GC and M_2 = amount of sample entering the MS.

Enrichment is a function both of analyte mass transferred and the ratio of flow rates leaving the GC and entering the MS:

$$E = \frac{M_2 V_1}{M_1 V_2}$$
(5.2)

where V_1 = carrier gas flow leaving the GC and V_2 = carrier gas flow entering the MS.

Chapman [1] quotes typical Y and E values for methyl stearate of 50% and 10, respectively. This means that, although half the methyl stearate is lost, its concentration in the ion source is ten times greater than it would otherwise be.

5.2.1.2   Open split interface

The open split interface consists of an enclosed tubular glass liner into which the analytical column and mass spectrometer transfer line are loosely inserted from opposite ends. The liner is vented to atmosphere via a length of narrow-diameter tubing, and a controlled flow of helium make-up gas may be introduced in order to augment the flow from the GC column (Figure 5.4). In this set-up, the transfer line to the mass spectrometer acts as a flow restrictor. The amount

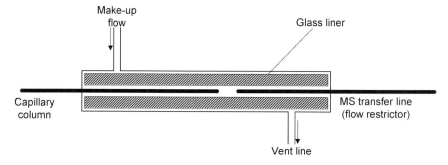

Figure 5.4 Open split interface.

of carrier gas drawn into the mass spectrometer is a function of the length and internal diameter of the transfer line, and the pressure drop between the open split interface and the mass spectrometer source region. It follows that the dimensions of the fused silica transfer line require careful optimisation. A transfer line that is too short or wide will admit too much carrier gas into the mass spectrometer; if the tubing is too long or narrow, a smaller amount of analyte will enter the mass spectrometer, resulting in reduced sensitivity.

The prime function of the make-up gas is to ensure that there is always a high enough gas flow entering the open split interface to replace the gas drawn into the mass spectrometer. This is important in order to avoid the ingress of atmospheric air into the mass spectrometer via the open split vent line. In other words, the direction of flow through the vent line must always be from the open split interface to atmosphere, never the other way around. It is particularly important to guard against this possibility when using capillary columns of 0.25 mm i.d. or less, where the volumetric flow through the column may be less than the flow of carrier gas drawn into the mass spectrometer. When setting the make-up flow on a pressure-regulated temperature-programmed GC system, it should be remembered that carrier gas flow will decrease as column temperature rises, and sufficient make-up flow should be provided to give a small positive flow from the vent line at the GC's maximum temperature.

The make-up flow may also be used, exceptionally, to reduce the amount of solute entering the mass spectrometer by splitting a large proportion of analyte to atmosphere post-column. This is unnecessary when introducing concentrated liquid samples into the instrument, when pre-injection dilution is the obvious solution. When analysing concentrated solid samples for volatile organic compounds (e.g. in investigation of contaminated land) by headspace techniques, however, there is no convenient way to 'dilute' the sample without risking analyte loss, and there is a practical lower limit to the sample mass taken. Under these conditions, providing additional make-up flow to the open split interface can avoid exceeding the detector's linear dynamic range.

Since the column exit is at atmospheric pressure in the open split interface, retention times of solutes ought to be directly comparable with conventional detectors operated at atmospheric pressure, provided that the mobile phase flow rate and stationary phase chemistry are the same. This offers a useful advantage when attempting to identify unknown peaks in a chromatogram produced by, for example, a flame ionisation detector.

5.2.2   Ionisation techniques

The choice of ionisation technique, and the selection of associated ion source variables, such as temperature and pressure, will determine the selectivity of the ionisation process, the abundance of fragment ions and the lowest detectable amount of analyte. Four ionisation techniques will be discussed: electron impact ionisation (EI), chemical ionisation (CI), field ionisation (FI) and hyperthermal surface ionisation (HSI).

5.2.2.1   Electron impact ionisation

In EI, electrons produced by passing an electric current through a tungsten or rhenium filament are accelerated across the ion source by a potential of approximately 70 V. These electrons interact with gaseous analyte molecules entering the ion source from the transfer line, removing an electron to create positive ions. Two magnets, mounted either side of the ion source, collimate the ionising electrons into a narrow beam. At electron energies of around 70 eV, the ion yield for most molecules is maximised. Even so, EI is a relatively inefficient process, with only approximately 1% of molecules that enter the ion source undergoing ionisation [1]. Positive ions are ejected from the ion source by a positive voltage applied to the source, and/or by a negative voltage on ion lenses mounted between the ion source and analyser region. A metal plate, known as the trap, is mounted on the opposite side of the source to the filament. It is held at a positive potential and serves to remove electrons that have traversed the ion source. Figure 5.5 shows the arrangement of components in a typical EI source. The source is commonly heated to around 150–200°C to avoid condensation of gaseous analyte molecules.

Electron impact ionisation produces extremely energetic molecular ions from analyte molecules. A molecular ion is formed by the removal of a single electron

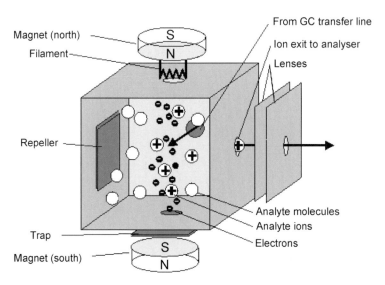

Figure 5.5 Electron impact ion source. (Reproduced by permission of ThermoFinnigan, Hemel Hempstead, UK.)

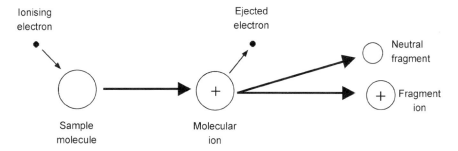

Figure 5.6 Diagram showing the electron impact ionisation process.

from a compound's molecular orbital, resulting in the formation of a radical cation. The energetic molecular ion usually breaks apart, since the breaking of covalent bonds is a convenient way in which the ion can reduce its internal energy. Products of this bond cleavage are fragment ions and neutral species. Fragment ions give rise to an information-rich EI mass spectrum, providing valuable clues as to the structure of the parent molecule. The EI ionisation and fragmentation process is shown graphically in Figure 5.6. EI spectra usually have a large number of different fragment ions. The relative intensity of these ions is subject to thermodynamic control, with the most abundant ions representing the most stable species. The most abundant, and therefore the most stable, ion in a mass spectrum is known as the base peak. Some compound classes, such as polycyclic aromatic hydrocarbons, form exceptionally stable positive ions and the molecular ion is also the base peak. In fact, many compounds of this class are able to support two positive charges. Since $z = 2$ for these ions, this gives rise to a peak in the mass spectrum at exactly half the m/z value of the molecular ion (Figure 5.7). At source temperatures above 200°C, increased fragmentation

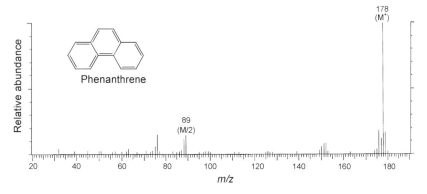

Figure 5.7 Electron impact mass spectrum of phenanthrene, showing the doubly-charged molecular ion at m/z 89.

is likely to occur, owing to the transfer of thermal energy to the gaseous ions and the consequent rise in the ion's internal energy.

Electron impact is the most popular ionisation mode for routine work. Its main advantages are spectral reproducibility, relatively high sensitivity, abundance of fragment ions (which facilitates solute identification), and applicability to a wide range of compound types. Electron impact mass spectra acquired under standard conditions are searchable against mass spectral libraries for the purpose of identifying unknown compounds, usually following the subtraction of background interference caused by column bleed, small air leaks or co-eluting compounds. GC-MS data systems search these libraries using a variety of spectrum-matching algorithms. The simplest methods of comparing sample and library spectra are fit, reverse fit or purity. A 'fit' search measures the degree to which the library entry is contained within the sample. 'Reverse fit' assesses how much of the sample's spectrum is common with the library entry, and 'purity' is a straightforward comparison between library and sample spectra. Another approach is probability-based matching, an algorithm designed to allow both for the presence of mixtures in the sample spectrum and differences in experimental conditions under which the library and data spectra were acquired.

5.2.2.2 *Chemical ionisation*

Chemical ionisation uses a reagent gas to mediate ionisation of analyte species. It is a 'soft' ionisation technique, meaning that the ionisation process is much less energetic than in EI. As a result, the internal energy of the ion is reduced, stabilising the molecular ion and markedly reducing fragmentation. An additional factor, collisional stabilisation, acts to reduce fragmentation in CI still further. Here, ions lose some of their excess internal energy through collisions with reagent gas molecules.

Chemical ionisation may be used to produce either positive ions or negative ions. In both cases, a reagent gas is introduced into the source region of the mass spectrometer, raising the pressure to around 70–130 Pa (0.5–1 torr). Ion sources for chemical ionisation are made with much smaller apertures for electron entry and ion exit than are electron impact sources, so as to maintain a high source pressure without compromising vacuum in the analyser region. It is also common to operate chemical ionisation sources with higher electron energies than used for electron impact work, in order to penetrate more effectively the high partial pressure of reagent gas in the ion source.

Chemical ionisation tends to be less reproducible than electron impact ionisation, in terms both of the appearance of mass spectra and the total ion yield. This is due to difficulty in maintaining stable pressure and temperature conditions in the ion source. Even small increases in temperature can reduce the relative intensity of the molecular ion by increasing its internal energy, while pressure fluctuations serve to increase or decrease the penetration of ionising electrons, resulting in variations in the numbers of reagent gas ions formed. Modern GC-MS instruments equipped for chemical ionisation are usually fitted with electronic reagent gas flow control, establishing a constant pressure regime in the ion source.

Reactions in chemical ionisation can be influenced by the presence of even small amounts of impurities. Reagent gases for chemical ionisation are therefore usually at least 99.9% pure. An additional problem is that, with prolonged use, hydrocarbon-based reagent gases, such as methane or isobutane, form deposits that polymerise on the metal surfaces of the ion source. This results in the build-up of an insulating layer, capable of supporting static charge. Symptoms of source contamination range from altered tuning conditions to loss of sensitivity. The remedy is to dismantle the ion source and remove the contamination with a slurry of fine abrasive powder. This problem is not observed with the use of ammonia, since polymers are not formed from this reagent gas. Care should be taken when connecting ammonia to the reagent gas line because of its corrosive nature. All components in contact with the gas, including tubing and pressure regulator diaphragms, should be made of stainless steel.

In positive CI, cations are produced from the reagent gas by electron bombardment. These ions then react with sample molecules to form a charged species by processes of proton transfer, charge exchange, addition or anion abstraction [1]. The most commonly used reagent gas for positive CI, methane, forms ions (mainly CH_5^+ and $C_2H_5^+$) that react with sample molecules, M, principally by proton transfer:

$$M + CH_5^+, \ C_2H_5^+ \longrightarrow M + H^+ + CH_4, C_2H_4 \qquad (5.3)$$

Proton transfer reactions in positive CI result in the formation of a pseudo-molecular ion $(M + H^+)$, one mass unit heavier than the true molecular ion, often as the base peak. The reactivity of a reagent gas ion is related to its proton affinity. As proton affinity decreases, the protonating ability of the reagent gas increases. It follows that selectivity in positive CI can be tailored through appropriate selection of reagent gas. Proton affinities of three common reagent gases, taken from data compiled by Lias *et al.* [2] are listed in Table 5.1. Methane can protonate practically all organic species, while isobutane and ammonia are much weaker proton donors, unreactive towards n-alkanes and able to protonate only relatively basic compounds. Ammonia, however, is also able to ionise

Table 5.1 Proton affinities of selected reagent gases

Reagent gas	Reagent ion	Proton affinity (kJ mol^{-1})
Methane	CH_5^+	551
Isobutane	$(CH_3)_3C^+$	824
Ammonia	NH_4^+	854

molecules through addition of NH_4^+ to a variety of compound types, forming $M(NH_4^+)$ adduct ions.

Positive CI often yields a molecular ion (or pseudo-molecular ion) with compounds that are too unstable under EI conditions for the molecular ion to be observed. It is therefore commonly used to determine the molecular masses of unknown compounds.

Negative CI is the most selective ionisation method for GC-MS and is applicable mainly to compounds that are capable of stabilising negative charge. These include halogenated species (particularly fluorine-containing molecules), phosphate esters and nitro compounds. The purpose of the reagent gas in negative CI is to slow down electrons entering the ion source in order to facilitate their capture by electronegative compounds. These slow-moving electrons are known as thermal electrons and possess energies in the range 0–2 eV. Excess energy imparted to the ion during electron capture is reduced as a result of collisions with reagent gas molecules. Figure 5.8 illustrates this process. Improved selectivity and sensitivity are the main advantages of negative CI, particularly in 'dirty' matrices that contribute a high background to the mass chromatogram. Figure 5.9 shows two mass chromatograms of the same extract of sewage sludge containing polychlorinated biphenyls, one acquired under EI conditions and one by negative CI. As with positive CI, the molecular ion is commonly observed as the base peak (Figure 5.10), although the electron capture process yields a true molecular ion of the same mass as the parent molecule.

Detector modification is necessary in order to permit recording of negative ions by an electron multiplier. A conversion dynode, a metal plate held at a high

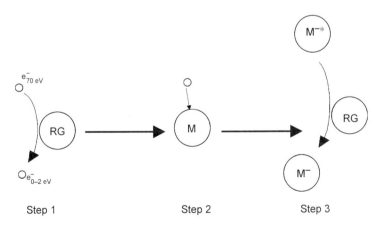

Step 1   Step 2   Step 3

Figure 5.8 Diagram showing the formation of negative ions in chemical ionisation. In step 1, electrons are slowed from 70 eV to thermal energies (0–2 eV). These slow-moving electrons are captured by electronegative species in step 2. Collisions with reagent gas molecules help to stabilise the negative ion. The excited molecular ion is represented by M^{-*} (step 3).

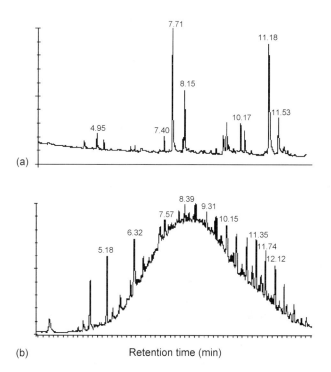

Figure 5.9 Negative chemical ionisation (a) and electron impact (b) mass chromatograms of the same extract of sewage sludge containing polychlorinated biphenyls. (Reproduced by permission of ThermoFinnigan, Hemel Hempstead, UK.)

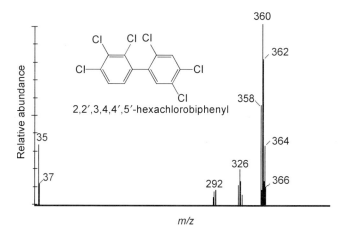

Figure 5.10 Negative chemical ionisation mass spectrum of a polychlorinated biphenyl. (Reproduced by permission of ThermoFinnigan, Hemel Hempstead, UK.)

positive potential (commonly 10–15 kV), is mounted just before the multiplier. A negative ion striking the dynode causes the emission of positive ions, which are then detected by the electron multiplier as if the instrument had been operating in positive ion mode.

5.2.2.3 *Field ionisation*

Field ionisation (FI) is a relatively recent innovation in GC-MS, although the technique has been used as an ionisation method by mass spectroscopists for many years. A high electric field is created between a cathode (commonly held at around $-10 \, kV$) and an anode (the field emitter). The field emitter is usually a wire covered by extremely fine carbon dendrites, at the tips of which field strengths in the order of $10^8 \, V \, cm^{-1}$ are present [1]. Under these conditions, molecules in close proximity to the field emitter are stripped of an electron, forming positive ions. The GC transfer line is arranged so as to terminate close to the field emitter (Figure 5.11). FI is an extremely soft ionisation technique, producing little or no fragmentation. It has significant advantages over positive chemical ionisation, including the absence of potentially interfering reagent gas ions, and the formation of a true molecular ion, free from adducts or additional protons. Figure 5.12 shows a comparison between EI and FI mass spectra of methyl stearate.

5.2.2.4 *Hyperthermal surface ionisation*

Hyperthermal surface ionisation (HSI) occurs when high-velocity analyte molecules approach a hot surface [3]. Ionisation takes place when the difference between the molecule's ionisation potential and the work function of the surface are overcome by the molecule's kinetic energy. In practice, the GC column effluent is mixed with hydrogen or helium make-up gas, and introduced into the differentially pumped ion source through a small nozzle, approximately 0.1 mm in diameter (Figure 5.13). Under these conditions, the velocity of molecules entrained in the gas stream exceeds the speed of sound. The target surface is

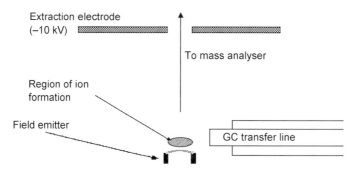

Figure 5.11 Field ionisation source.

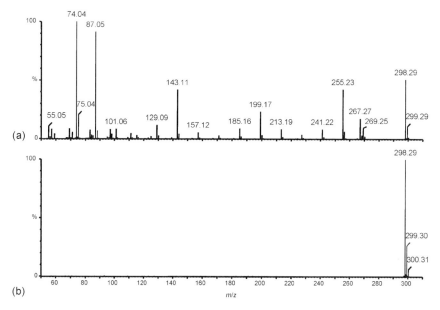

Figure 5.12 Electron impact (a) and field ionisation (b) mass spectra of methyl stearate. (Reproduced by permission of Micromass Limited, Wythenshawe, UK.)

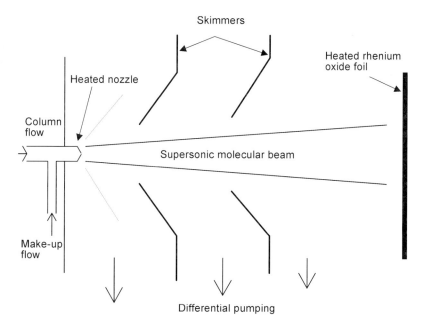

Figure 5.13 Hyperthermal surface ionisation source.

commonly a strip of rhenium oxide foil, which is protected from reduction by a small partial pressure of oxygen (around 1 mPa (10^{-5} Torr)). Ionisation may occur by a variety of mechanisms, including surface-to-molecule electron transfer, molecule-to-surface electron transfer and surface-induced dissociation. HSI is 100 times more efficient as an ionisation technique than EI, making it extremely sensitive. HSI is particularly suited to compounds with low ionisation potential, including polycyclic aromatic hydrocarbons, basic organic compounds and (in negative ion mode) acidic species. Dagan and Amirav claim sub-femtogram limits of detection for deuterated anthracene [4]. The degree of fragmentation observed in HSI spectra increases with molecular beam velocity [5].

5.2.3   Analyser types

Just as sample components are separated in time by the GC column, ions derived from analyte molecules are resolved by the mass analyser. The mass spectrometer separates ions with greater or poorer efficiency depending on the analyser's resolution. Mass resolution, R, is analogous to chromatographic resolution, and is defined as the ability of the analyser to separate two adjacent mass peaks. It is commonly calculated in one of two ways. The most common method is

$$R = \frac{m}{w_{5\%}} \qquad (5.4)$$

where m = mass of ion and $w_{5\%}$ = width of the mass peak at 5% height. Alternatively, resolution may be expressed as 'parts per million':

$$R = 10^6 \times \frac{w_{5\%}}{m} \quad \text{ppm} \qquad (5.5)$$

It should be noted that resolution is mass dependent. If the mass analyser has constant resolution across the mass range, peak width will increase with increasing mass. Similarly, if the analyser operates with constant peak width, resolution will increase with increasing mass. Some mass analysers operate with constant resolution, and some with constant peak width, across their mass range.

With the introduction by manufacturers of capillary columns of 0.1 mm i.d. or less, many analytical laboratories have sought to capitalise on the advantages of fast GC. Many mass analysers are incapable of scanning faster than around 0.1 s per scan without compromising sensitivity or mass resolution. The ability of a mass analyser to scan quickly enough to characterise accurately a narrow GC peak is an important consideration for quantitative analysis. As a rough guide, at least six scans across a chromatographic peak are required in order to establish its profile with reasonable accuracy. The mass analysers most commonly encountered in GC-MS (quadrupole, ion trap, time-of-flight and magnetic sector) are described below.

5.2.3.1   Quadrupole mass analyser

The quadrupole mass analyser comprises four metal rods, of cylindrical or parabolic cross-section, arranged parallel to each other, the same radius (r) apart. Opposite rods are electrically connected to a voltage supply made up of a d.c. component (U) and a r.f. component ($V \cos \omega t$). Here, V is the r.f. voltage and ω its frequency. The voltages applied to each set of rods have opposite signs but equal magnitude (Figure 5.14). Ions entering the mass analyser are made to oscillate between the rods; ions with unstable oscillation paths are neutralised through collision with the rods. The quadrupole mass analyser works as a mass filter, rejecting the majority of ions. At any given moment, only ions of one particular m/z ratio will have stable trajectories through the quadrupole analyser. Figure 5.15 illustrates how ions with unstable paths are removed by the quadrupole mass filter. Quadrupole instruments are normally set up for unit mass resolution, in which ions are baseline-resolved to the nearest whole atomic mass unit. Under these conditions, ion peak width is constant throughout the mass range, and resolution increases with increasing m/z. Mass calibration is achieved through a linear relationship between the r.f. voltage, V, and the m/z of ions with stable trajectories through the analyser. Mass resolution may be adjusted by varying the ratio between U and V. Figure 5.16 shows the effect of varying the $U{:}V$ ratio on the stability of three ions of different m/z. Resolution is limited by the kinetic energy distribution of the ions entering the analyser. It is possible to tune the instrument to achieve resolution below unit mass, but this is accompanied by a corresponding loss in sensitivity.

5.2.3.2   Ion trap mass analyser

The ion trap is arguably the mass analyser that has seen the most development during the last ten years. In its simplest form, it consists of a doughnut-shaped

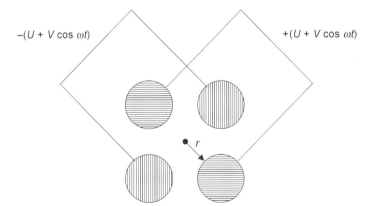

$-(U + V \cos \omega t)$ $+(U + V \cos \omega t)$

r

Figure 5.14 Cross-section through a quadrupole mass filter showing rod spacing (r) and electrical connections.

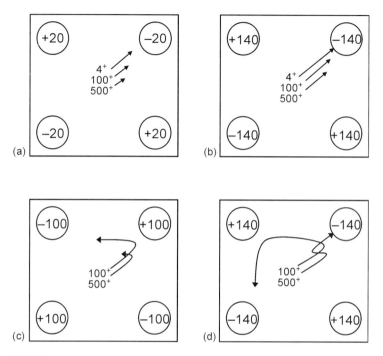

Figure 5.15 Time sequence showing the removal of unstable ions by a quadrupole mass filter. At time 0 (a), four ions (m/z 4, 100 and 500) are introduced into the quadrupole region and start to move towards the negatively charged quadrupole rod. Because smaller ions are more easily deflected in an electromagnetic field, m/z 4 moves fastest and, as the negative charge on the rod increases (b), is neutralised by collision with the rod. The rod polarities now change (c) and the remaining ions are repelled by the nearest rod and start to move towards the nearest centre of negative charge. When the polarities change again, m/z 500 has not moved far enough to avoid being captured by the nearest quadrupole rod and neutralised. m/z 100 continues in a stable trajectory through the mass analyser.

ring electrode, to which r.f. voltage is applied, sandwiched between two cylindrically symmetrical parabolic end-cap electrodes, which are maintained at ground potential (Figure 5.17). In these instruments, ionisation of molecules introduced through the GC transfer line takes place in the analyser cavity, the region bounded by the end-cap and ring electrodes. Electrons produced by an electrically heated filament are permitted or denied access to the analyser cavity by the electron gate. A potential of +150 V is applied to the gate electrode to admit electrons into the trap, via a hole in the top end-cap electrode, and a potential of −150 V is used to 'close' the electron gate. During ionisation the electron gate is 'opened', allowing electrons into the analyser cavity, after which the gate is 'closed' and all the resulting ions are confined in a stable trajectory inside the trap. A small partial pressure of helium carrier gas helps to reduce the kinetic energy spread of the ions by collision, improving mass resolution.

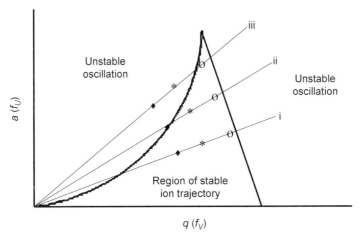

Figure 5.16 Diagram showing the effect of altering the $U{:}V$ ratio on the mass resolution of a quadrupole mass filter. Ions (♦), (∗) and (○) have adjacent m/z values. Under conditions of low $U{:}V$ ratio (i), all three ions have stable trajectories. As the $U{:}V$ ratio is increased (ii), the lowest-mass ion (♦) becomes unstable. As $U{:}V$ is increased still further (iii), only one ion has a stable trajectory through the mass filter.

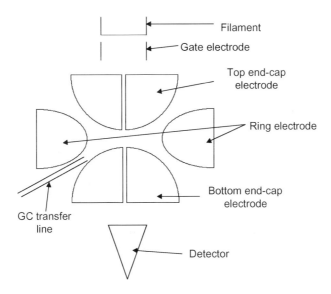

Figure 5.17 Schematic diagram of an ion trap detector.

As the r.f. voltage is ramped, ions are destabilised in order of increasing m/z. Destabilised ions are drawn towards the two end-cap electrodes; some of those travelling downwards pass through holes in the bottom end-cap electrode, and reach the detector. Ions destabilised upwards, towards the filament, are lost.

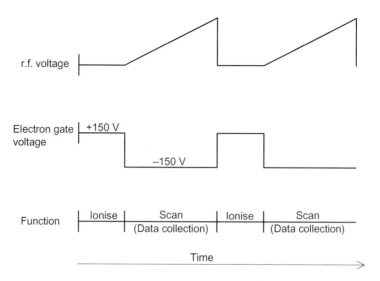

Figure 5.18 Basic ion trap scan function. Ionisation is controlled by the gate electrode, and the r.f. voltage determines the stability of ions in the trap. As the r.f. voltage is scanned, a mass spectrum is acquired.

Figure 5.18 shows the basic ion trap scan function (the relationship between r.f. and electron gate voltages, and data acquisition). As with the quadrupole mass analyser, the ion trap detector is operated with constant peak width across its mass range. For a comprehensive discussion of the theory and operation of ion trap mass spectrometers, refer to *Practical Aspects of Ion Trap Mass Spectrometry* by March and Todd [6].

There is a practical limit to the number of ions that can be stored in the ion trap at any moment. This limit may be exceeded when analysing concentrated or highly contaminated samples, resulting in a phenomenon known as space charging. Space charging is the generation of an electromagnetic field inside the trap cavity by a large number of moving ions. The induced field may be large enough to affect ion trajectories, resulting in loss of mass resolution and spectral distortion. Space charging effects may be greatly reduced by using a variable ionisation time, adjusted as a function of the amount of ionisable material present in the trap. This is achieved in practice by a technique called automatic gain control (AGC). The instrument ionises the contents of the trap for a fixed period of time, before rapidly scanning out all the ions (the AGC pre-scan). The detector records and integrates the signal from this pulse of ions, and calculates the optimum ionisation time to be used for the subsequent analytical scan, based on the number of ions detected during the pre-scan. Analytical ionisation times typically range from 78 μs to 25 ms, shorter times being used as the magnitude of the pre-scan peak increases. The AGC scan function is shown in Figure 5.19.

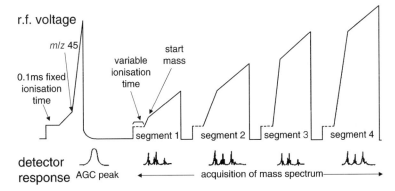

Figure 5.19 The AGC scan function. Following measurement of the AGC peak, a variable ionisation time is set and the analytical scan is performed. Here, full scan data are collected in four sub-scans (segments). (Reproduced by permission of Varian Limited, Warrington, UK.)

Another problem with in-trap ionisation with contaminated or concentrated samples arises as a result of interactions between ions and neutral species. When there is a relatively high partial pressure of neutral molecules in the ion trap, the chance of collisions between ions and molecules is high. This can lead to CI-type proton transfer reactions between ions and molecules, or a more intense molecular ion as a result of collisional stabilisation. This type of spectral distortion is shown in Figure 5.20, which compares quadrupole and ion trap EI spectra of a pesticide. Proton transfer is particularly likely with basic analytes since these have, by definition, a high proton affinity. This phenomenon is not as easily dealt with as that of space charging, since there is no convenient way, with older instruments, of excluding neutral species from the analyser cavity. One solution is to ionise the sample components before they enter the analyser region (see below).

In the early 1990s, manufacturers modified the way in which ion trap instruments worked by applying a r.f. voltage to the end-cap electrodes, of lower frequency and voltage than that supplied to the ring electrode. This technique, known as axial modulation, results in ions of the same m/z being ejected from the trap in a much tighter band, and at a lower ring electrode r.f. voltage than would otherwise be the case, significantly improving mass resolution and increasing mass range. Axial modulation also increases the number of ions that can be stored in the ion trap before spectral distortion through space charging occurs, thus improving sensitivity.

More recently, commercial instruments have been developed that overcome the problems of ion–molecule interactions by ionising solutes outside the analyser cavity. The external ion source produces ions continuously; these are injected into the analyser in packets by a gating lens, in much the same way as electrons were admitted into the trap cavity in the older design. A schematic

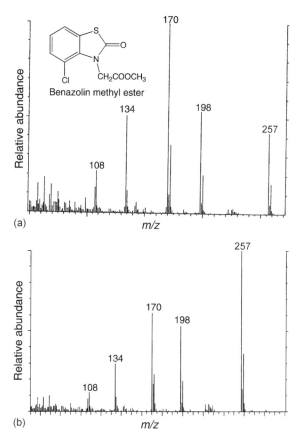

Figure 5.20 Quadrupole (a) and ion trap (b) electron impact spectra of benazolin methyl ester. Note the augmented molecular ion at m/z 257 in the ion trap spectrum.

diagram of an external ion source and analyser arrangement is shown in Figure 5.21. The ion injection time is varied depending on the magnitude of an AGC peak acquired using a 1 ms fixed injection period. As the r.f. voltage is ramped to eject ions from the trap, an axial modulation voltage (also called a resonance ejection waveform) is applied to the end-cap electrodes. The scan function for an external ion source instrument closely resembles the AGC scan function shown in Figure 5.19, with variable ion injection time replacing variable ionisation time. The major advantages of external ionisation are a much-reduced partial pressure of neutral species in the source and easier implementation of CI. Positive CI is possible on older instruments, but requires a modified scan function whereby the reagent gas is first ionised and then allowed time to react with analyte molecules. With external ionisation, a standard CI source design is employed and the analyser sees no difference in its mode of operation.

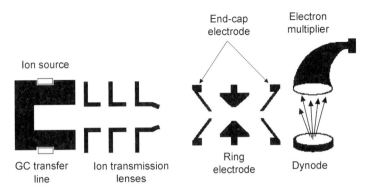

Figure 5.21 Schematic diagram of an ion trap with an external ion source. (Reproduced by permission of ThermoFinnigan, Hemel Hempstead, UK.)

Another significant advance has been the development of instruments that permit GC-MS-MS experiments to be performed with the ion trap (see section 5.2.5 for a fuller discussion of GC-MS-MS). For GC-MS-MS, an r.f. voltage (the isolation waveform) is applied to the end-cap electrodes after ion injection, accompanied by a complementary increase in the ring r.f. voltage. The resulting field is capable of isolating ions of a specified m/z, ejecting all other ions from the trap. Following isolation, a further variable-amplitude voltage (the excitation waveform) is applied to the end-cap electrodes, causing the isolated ions to gain kinetic energy. These fast-moving ions undergo collisions with helium atoms present in the trap, causing fragmentation (collision-induced dissociation). Daughter ions are stored, ejected and detected by scanning the r.f. voltage, as previously. The scan function used in this mode of operation is shown in Figure 5.22. The process of ion isolation followed by collision-induced dissociation may be performed, in theory, a limitless number of times on the same group of ions, increasing the selectivity of the process each time. This technique (known as MS^n) is limited only by the system's ability to detect the much fewer ions left in the trap after multiple isolation and collision events.

Ion traps operating in MS-MS mode vary in the manner in which they impart kinetic energy to ions during the collision-induced dissociation process. The two techniques most commonly used are resonant excitation and nonresonant excitation. Briefly, in resonant excitation, only the ion of interest gains kinetic energy; daughter ions are not excited and undergo no further fragmentation. Nonresonant excitation, on the other hand, imparts kinetic energy to all ions in the trap, resulting in their fragmentation for as long as the excitation field is applied. Resonant excitation may therefore be applied to only one parent ion at once. Non-resonant excitation is used in a technique known as multiple reaction monitoring, which typically allows up to five parent ions to be isolated

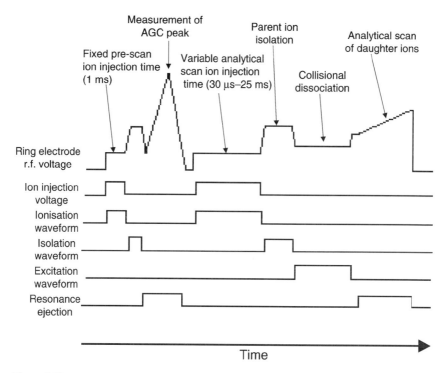

Figure 5.22 Ion trap MS-MS scan function. (Reproduced by permission of ThermoFinnigan, Hemel Hempstead, UK.)

and dissociated, with detection of all the daughter ions produced. This approach is particularly useful for the selective detection and quantification of co-eluting compounds yielding different characteristic fragment ions.

Some ion trap mass spectrometers permit a mode of operation known as selected ion storage (SIS). Here, ions are accumulated in the trap, and unwanted ions are ejected through the application of waveforms to the end-cap electrodes that cause them to resonate and develop unstable trajectories (resonant ejection). There is no excitation step, however, as in MS-MS, and stored ions have the same relative abundances as in an EI mass spectrum. Characteristic ions from a target compound may thus be selectively enriched relative to background or interfering ions. Commercial instruments operating in this mode typically permit the storage of five separate ion ranges, and the rejection of up to five discrete masses within these ranges. SIS can provide additional sensitivity for compounds at low levels in difficult matrices. Figure 5.23 illustrates the application of SIS to the analysis of oestrogenic compounds in liver extract. Andalò *et al.* [7] reported that the performance of SIS was poor at high sample concentrations, owing to protonation of the parent ion under conditions of

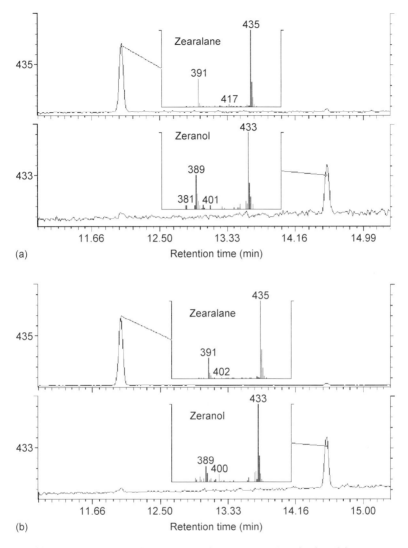

Figure 5.23 Narrow mass range ion trap mass chromatogram (a) and selected ion storage mass chromatogram (b) of oestrogenic compounds in sheep liver extract. Numbers on the vertical axes are the mass/charge values of the extracted ion chromatograms. (Reproduced by permission of Varian Limited, Warrington, UK.)

raised partial pressure. When this occurs, the protonated analyte molecule is ejected from the trap along with other 'undesirable' ions, and is then unavailable for subsequent detection. This is a consequence of in-trap ionisation with this particular instrument.

5.2.3.3 *Time-of-flight mass analyser*

The time-of-flight (TOF) mass analyser has traditionally been used for the analysis of large molecules (including polymers and proteins), but has undergone something of a renaissance recently through its application to small molecules in GC-MS instruments. TOF instruments separate ions according to the amount of time that they take to travel between a pulsed acceleration region and the detector. After ionisation, ions are injected into the acceleration region, where a repelling high voltage pulse (push-out voltage) applied to a metal plate fires them towards the detector down a field-free flight tube. Smaller ions travel faster and arrive at the detector sooner than larger ions. Mass calibration is achieved by establishing the relationship between m/z and the time from the application of the accelerating voltage to arrival at the detector (the 'time of flight'). Theoretically, there is no upper limit to the mass range of a TOF analyser, since even the heaviest ion should arrive at the detector, given sufficient time.

As with quadrupole and ion trap instruments, mass resolution is limited by the spread of kinetic energies and direction of travel as ions enter the analyser. Ions moving towards the repelling electrode are forced to change direction when the electrode is switched on, and will take longer to arrive at the detector than other ions of the same m/z. Conversely, ions moving towards the detector as they enter the analyser will have their direction of travel reinforced by the repelling electrode and will arrive at the detector sooner than other ions of the same m/z. The problem persists despite the use of collimating ion optics between the ion source and the repeller electrode. One way of overcoming this limitation is to use an instrument equipped with a reflectron. A reflectron (also known as an ion mirror) consists of a number of repelling plates or grids placed at the opposite end of the flight tube to the ion source. Ions penetrate the reflectron field and are repelled and reflected towards the detector (Figure 5.24). Faster-moving ions penetrate further and spend longer in the reflectron than slower moving ions of the same m/z, with the result that they arrive at the detector at approximately the same time.

TOF analysers are operated with constant resolution across their mass range. Those equipped with a reflectron are capable of considerably higher mass resolution ($R \approx 5000$) than quadrupole instruments at m/z values up to approximately 500. It is therefore feasible to resolve ions with low to mid-range m/z values to better than one decimal place with a bench-top TOF instrument. The main benefit of this improved resolution is that analyte ions may be separated from background ions of the same unit mass but different elemental composition, provided that their masses differ by more than approximately 0.05 amu.

Another significant advantage of the TOF mass analyser is its very high scanning speed (also known as duty cycle). Scanning speeds in mass analysers that scan r.f. or d.c. voltages are limited by the speed with which these voltages

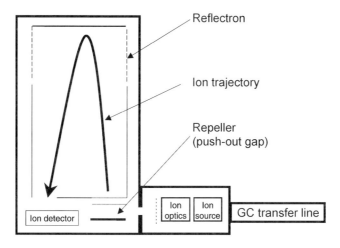

Figure 5.24 Schematic diagram of a time-of-flight mass analyser.

can be changed, and the amount of time that it takes for the voltages to settle between scans. These limitations do not affect TOF instruments, since the push-out voltage is pulsed, not scanned, and the drift tube requires no electromagnetic field for mass separation. TOF instruments typically acquire data at the rate of 50 spectra per second, making them ideal detectors for fast GC [8]. Figure 5.25 shows the GC separation of volatile organic compounds in under 7 s using a TOF mass spectrometer. In addition, fast scanning speeds enable closely eluting peaks to be identified and resolved (Figure 5.26).

5.2.3.4   Magnetic sector and double focusing mass analysers
Magnetic sector mass analysers were used in the first mass spectrometers. In its simplest form, the magnetic sector mass analyser comprises a source in which ions are accelerated at high voltage (commonly 2–8 kV) towards a curved flight

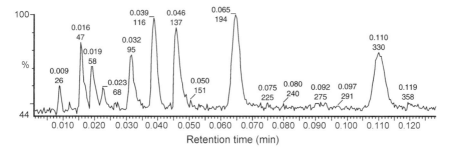

Figure 5.25 GC-TOF-MS chromatogram showing the separation of organic solvents. Most peaks have widths at base of less than 0.3 s. (Reproduced from [8] by permission of John Wiley and Sons Limited.)

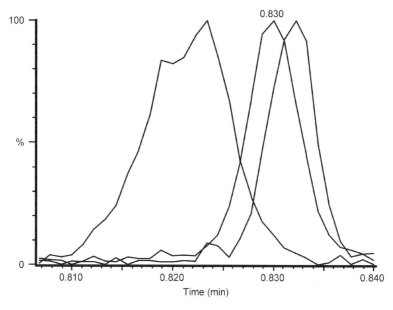

Figure 5.26 Resolution of three closely eluting compounds with different characteristic ions using GC-TOF-MS. (Reproduced from [8] by permission of John Wiley and Sons Limited.)

tube mounted in a magnetic field. Ions with stable trajectories are detected after passing through the tube; ions with unstable trajectories are deflected by the magnet into the walls of the tube (Figure 5.27). Variable-width slits are used to collimate ions entering the analyser, and to admit only a narrow m/z range to the detector. Magnetic sector instruments resolve ions of different momentum according to the equation (5.6), which establishes the conditions necessary for an ion to be detected:

$$mv = Bzer \qquad (5.6)$$

where m = mass of the ion, v = velocity of the ion, B = strength of magnetic field, z = number of charges on the ion, e = charge on an electron, r = radius of curvature of tube.

Ions of different m/z may be brought into focus at the detector sequentially, by scanning either the accelerating voltage or the magnetic field strength.

As with other mass analyser types, mass resolution with the magnet sector analyser is limited by the kinetic energy distribution of ions entering the analyser (ions of the same m/z but different velocities will have different trajectories through the magnetic field). A resolution value of 5000 is common for single-analyser magnetic sector instruments. In combination with an electrostatic mass

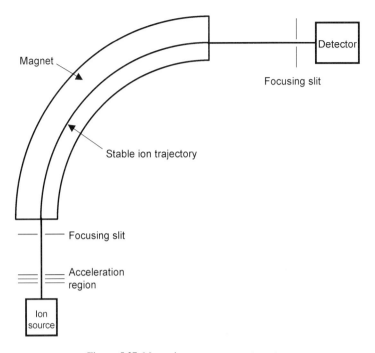

Figure 5.27 Magnetic sector mass spectrometer.

analyser, however, it is possible to compensate for the spread of kinetic energies such that all ions of the same m/z are brought into focus together. These double focusing mass analysers are capable of operating at mass resolutions in excess of 60 000, and have mass ranges up to approximately 10 000 amu. They are the most expensive and sophisticated type of mass spectrometer, requiring skilled staff to maintain and operate them. The highly accurate mass assignments provided by these high-resolution instruments enable the elemental composition of ions to be deduced, since all elements (with the exception of ^{12}C) are either slightly heavier or slightly lighter than their nominal unit masses. For example, carbon monoxide and nitrogen gases have the same nominal mass (28 amu) but different accurate masses (27.9949 and 28.0062 amu, respectively) [1].

5.2.4   Selected ion monitoring

Selected ion monitoring (SIM) (sometimes called 'selected ion recording' or SIR) is used in target compound analysis to improve the limits of detection afforded by some analyser types. Target compound analysis is used to confirm the presence or absence of a particular compound in a sample, usually quantitatively. Ions that are characteristic of the compound(s) of interest are chosen and the instrument is set up to scan only those ions. Improvement in sensitivity arises from increased transmission of analyte ions to the detector, owing to an increase in their dwell time (the amount of time that the analyser spends transmitting these ions to the detector). It follows that the more m/z values that are monitored simultaneously in SIM mode, the poorer will be the resulting sensitivity. A number of different scan windows can be programmed, such that m/z values of the monitored ions change with time as different target analytes elute from the column. An illustration of the enhanced sensitivity possible with SIM operation is shown in Figure 5.28.

The instrument types that benefit most from SIM operation are those with quadrupole or magnetic sector analysers, where ions with unstable trajectories are not detected. In TOF or ion trap instruments, all ions have at least a nominal chance of being detected, and SIM is an unnecessary tool. It is usual to monitor three or more different ions per target compound to reduce the possibility of reporting a false positive result. Compound identity can be confirmed by checking that the ratios of the monitored ions are consistent with those of the target compound.

Although SIM cannot be used to screen samples for unknown compounds, some MS data systems permit the acquisition of alternate scans in SIM and full-scan mode during the same run. The result is two separate data files: a full-scan file for the identification of unknowns, and a SIM file for target compound quantification.

5.2.5   GC-MS-MS

MS-MS is a technique that involves the selective isolation and fragmentation of ions, followed by detection of the resulting charged fragments. It is a powerful technique for improving selectivity and sensitivity in GC detection, and is particularly good at separating analyte signal from a noisy chemical background in target compound analysis. The way in which modern ion traps achieve MS-MS has already been described (section 5.2.3.2). Triple-quadrupole instruments can operate in the same way as ion traps functioning in MS-MS mode, but are also capable of other types of MS experiment.

A triple-quadrupole (or tandem) mass analyser consists of two sets of quadrupole rods, mounted in line, separated by a collision cell. The collision cell is equipped with a set of either quadrupole or hexapole electrodes, to which r.f. voltages of varying amplitude may be applied, and a controlled supply of collision gas (usually argon) (Figure 5.29). The collision cell is known as the Q2 region, and the pre- and post-collision quadrupoles are termed Q1 and Q3, respectively.

MS-MS works on the assumption that parent ions with the same m/z but different structures or chemical compositions are likely to produce characteristic daughter ions of different m/z. Because of the absence of virtually all background

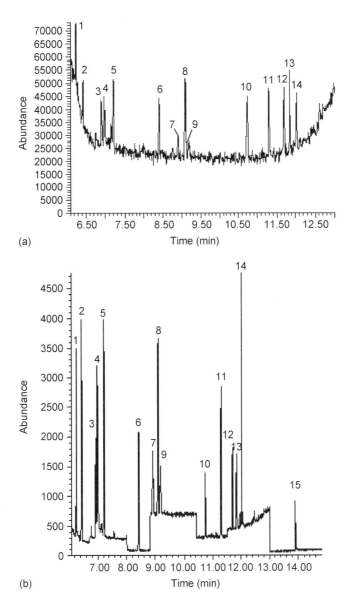

Figure 5.28 Full-scan mass chromatogram (a) of a mixture of organophosphorus pesticides (25 pg injected) and data on the same range of compounds collected in selected ion monitoring mode (b) (10 pg injected). Peak 15 (coumaphos) is not observed in the full-scan chromatogram. (Reproduced by permission of Agilent Technologies Limited, Palo Alto, CA, USA.)

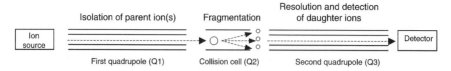

Figure 5.29 MS-MS process in a tandem quadrupole mass analyser.

ions, MS-MS is capable of exceptionally high sensitivity. Its selectivity, already high, can be enhanced still further through the use of selective ionisation. Figure 5.30 shows a mass chromatogram produced by negative CI-MS-MS of a drug metabolite, derivatised using fluorine-containing reagents. In this example, selection of fluorinated derivatisation reagents enhances sensitivity by improving negative ion stability.

Triple-quadrupole instruments may also be used to monitor all parent ions giving rise to a particular daughter ion by scanning Q1 while setting Q3 to transmit the fragment ion of interest. Another possibility is to monitor for the loss of a particular neutral fragment by scanning Q1 and Q3 at a fixed mass difference [1]. Ion trap instruments are not capable of operating in these two modes, since triple-quadrupole instruments can select parent ions with much greater flexibility.

Krahmer *et al.* [9] compared the qualitative and quantitative MS-MS performance of triple-quadrupole and ion trap mass analysers. They concluded that ion trap instruments were superior to tandem quadrupole analysers in terms of absolute sensitivity, while the precision of triple-quadrupole mass spectrometers was better than that of ion traps.

5.2.6   *Isotopically labelled surrogate standards*

The performance of an analytical method often depends more on the method employed for sample preparation than on the instrumental technique used for

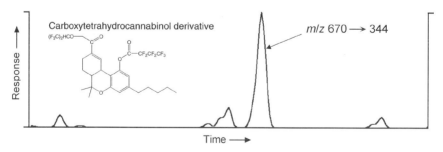

Figure 5.30 Negative chemical ionisation MS-MS chromatogram of 1 fg of a fluorinated derivative of a cannabis metabolite. (Reproduced by permission of ThermoFinnigan, Hemel Hempstead, UK.)

analyte measurement. In particular, analyte recoveries at low levels from challenging matrices are generally less than 100%, and may be quite variable. An ideal way to correct for these variations in recovery is to add a surrogate standard to each sample. A surrogate standard is a compound that behaves in the same way as the analyte during sample preparation, but may be separately quantified during instrumental analysis. The choice of a suitable surrogate may be problematical for conventional detectors. For example, if the analyte is ionisable, the surrogate should have an identical pK_a. It should also possess identical polarity to the analyte and not normally occur in samples.

The use of mass spectrometric detection allows the use of materials that fulfil all the above criteria but which are unsuitable for use with conventional detectors. These are analogues of the analyte itself that differ only in their mass. This difference is achieved by the replacement of one or more atoms in the molecule with heavier, stable isotopes. In practice, this usually means the replacement of ^1H or ^{12}C atoms by ^2H or ^{13}C, respectively. In all other important respects, analytes and their isotopically labelled surrogates are identical. Quantifiable amounts of surrogate standard are added to each sample prior to extraction. A poor recovery of the analyte from a particular sample will be accompanied by a correspondingly poor recovery for the surrogate standard, enabling a correction to be made. Figure 5.31 illustrates the use of a trideuterated analogue in the positive CI-GC-MS analysis of cotinine, a nicotine metabolite. The two species, although eluting close together, are perfectly resolved by the mass spectrometer.

5.2.7 Applications

A number of applications of GC-MS have already been used to illustrate various points. The purpose of this section is to present a number of typical applications in order to illustrate the capabilities of different GC-MS systems under normal operating conditions. It is not intended to be an exhaustive review of recent applications, and readers are encouraged to trawl the literature for material relevant to their own areas of interest.

5.2.7.1 Environmental applications

Natangelo *et al.* [10] examined the performance of a quadrupole (operating in SIM mode) and an ion trap (operating in GC-MS-MS mode) for the analysis of some relatively polar (in GC terms) pesticides in water using solid phase microextraction (SPME) for analyte pretreatment. Limits of detection were found to be roughly equivalent, ranging from $2 \, ng \, l^{-1}$ for propanil to $30 \, ng \, l^{-1}$ for myclobutanil. The ion trap provided better linearity and the quadrupole better precision on replicate samples. They concluded that both were capable of satisfactory performance for the analysis of low $ng \, l^{-1}$ levels of pesticides

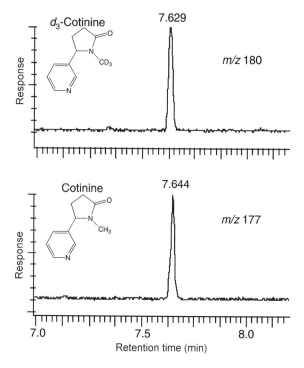

Figure 5.31 Mass chromatograms of cotinine (m/z 177) and d_3-cotinine internal standard (m/z 180). (Reproduced by permission of ThermoFinnigan, Hemel Hempstead, UK.)

in drinking water. Sarrión *et al.* [11] also used SPME for analyte isolation and preconcentration prior to injection into a GC-ion trap MS system. The analytes in this case were haloacetic acids formed in drinking water as a result of chlorine disinfection. The involatile acids were converted to their ethyl esters in a sealed vial and the SPME fibre was exposed in the headspace. The ion trap mass spectrometer was operated in EI-MS mode with AGC. The typical linear dynamic range for 30 ml sample volume and a headspace in the order of 5 ml was 0–125 $\mu g \, l^{-1}$ and limits of detection ranged from 10 to 200 $ng \, l^{-1}$. Ding and Chen used GC–ion trap MS to analyse carboxylated metabolites of alkylphenyl polyethoxylates following conversion in the GC injector to their butyl esters using tetrabutylammonium hydrogen sulfate [12]. Recoveries from spiked samples were 90–108%, with RSDs no higher than 8%.

Miermans *et al.* [13] evaluated the analysis of volatile organic compounds by purge-and-trap GC–ion trap MS, with d_8-toluene as internal standard. The limits of detection achieved with 25 ml sample volume ranged from 0.4 $ng \, l^{-1}$ to 30 $ng \, l^{-1}$. They noted differences between library and calibration spectra for some ethers and esters, and overcame this problem by compiling their own library of target compound spectra. Spaulding *et al.* [14] evaluated differences between EI and positive CI, using methane and pentafluorobenzyl alcohol (PFBOH) as reagent gases, for the analysis of derivatives of airborne carbonyl and hydroxycarbonyl moieties by GC–ion trap MS. Methane CI yielded $M+H^+$ as the dominant ion, while $M+H^+$ and $M+181^+$ species were observed at varying intensities in PFBOH CI spectra. The use of PFBOH as CI reagent gas offered significant advantages over the other ionisation modes, making the molecular weight of the detected species easier to deduce. Küchler and Brzezinski [15] evaluated GC–ion trap MS-MS with multiple reaction monitoring as an alternative to GC–high resolution mass spectrometry for the analysis of polychlorinated dibenzo-*p*-dioxins and furans in sewage effluents. They discussed the steps involved in optimising parameters for MS-MS, and concluded that sufficient selectivity and sensitivity were afforded by GC–ion trap MS-MS to enable individual congeners to be determined reliably at concentrations as low as 25 pg l^{-1} in sewage effluents.

Bowerbank *et al.* [16] capitalised on the fast scanning speed of a TOF mass analyser to analyse rapidly extremely low levels of atmospheric chemical warfare agents by solvating GC-MS. Solvating GC uses a mobile phase that is a supercritical fluid at the injector and becomes a gas at the column outlet owing to pressure reduction. Fast chromatography, combined with rapid, sensitive detection, was an improvement on portable ion mobility spectrometry and mass spectrometers without a chromatographic front-end. Polycyclic aromatic hydrocarbons were analysed by Davis *et al.* using GC-supersonic molecular beam HSI-MS [17]. HSI produced fewer fragment ions compared to EI, and provided the required method sensitivity for benzo[*a*]pyrene (1 ng l^{-1}) using just 5 ml sample volume and large-volume injection, compared to 1 l for HPLC-fluorescence methods. Vreuls *et al.* [18] evaluated the application of GC-TOF-MS to a variety of compounds of environmental importance, reporting on-column limits of detection in the region of 1–6 pg for organophosphorus insecticides, 4–60 pg for triazine herbicides and 0.3–6 pg for polycyclic aromatic hydrocarbons. Co-eluting peaks whose retention times differed by 0.3 or 0.15, at spectrum storage speeds of 10 and 20 Hz, respectively, were amenable to manipulation by a peak deconvolution algorithm, giving an interference-free mass spectrum. Better sensitivity was achieved using slower scanning speeds; signal:noise ratios of 1 and 0.1 were observed at scanning speeds of 10 and 200 scans s^{-1}, respectively.

Roach *et al.* [19] examined the application of three different ionisation modes to the quantification of underivatised patulin in apple juice. The mass analysers employed were a high-resolution magnetic sector instrument operated at 10 000 mass resolution, a quadrupole mass spectrometer and an ion trap system operating with external ionisation. Sensitivity in negative CI mode was superior to both EI and positive CI data, with on-column limits of quantitation of 1 ng, 4–10 ng and 10 ng, respectively using 5% phenylmethyl polysiloxane columns. These limits of quantification are unusually high for GC-MS work, and result from the patulin peak width at base exceeding 30 s as a consequence of adsorptive interactions between the polar, underivatised patulin and the nonpolar stationary phase. With a trifluoropropylmethyl stationary phase less adsorption was encountered and 8 pg of patulin on-column could be quantified using negative CI.

5.2.7.2   Biological and pharmaceutical applications

Niwa reviewed the use of mass spectrometry in the search for toxins in uraemic blood, including GC-MS for low-molecular-weight species such as organic acids, phenols and polyols [20]. Among the techniques employed were positive CI and EI with selected ion monitoring for the detection of 3-deoxyglucosone, and the use of $^{13}C_2$-oxalic acid as an internal standard for oxalic acid quantification. GC–ion trap MS using isobutane-moderated positive CI was employed by Teal et al. [21] for the detection and quantification of insect juvenile hormones. Some discrepancy was noted between previously reported isobutane CI spectra and those acquired under the authors' conditions. Pichini et al. [22] analysed trimethylsilyl derivatives of lorazepam (a member of the bezodiazepine family of drugs) in plasma and urine using GC-MS-MS with an ion trap detector. Calibration was linear over four orders of magnitude and the limit of detection was 0.1 ng ml^{-1} (1 ml sample volume; 5% of extract injected).

Muramic acid, a marker of bacterial peptidoglycan, was analysed in bacteria, house dust and urine as its acylated derivative by Bal and Larsson [23] using GC–ion trap MS-MS. Nonresonant excitation was employed, together with a broad parent ion isolation window in order to maximise sensitivity. Protonated molecular ions were observed for both the analyte and its ^{13}C-labelled analogue, presumably a result of in-trap ionisation and analyte basicity, despite operation in EI mode. With optimised MS-MS conditions, a signal:noise ratio of 400:1 was recorded for a 500 pg injection. Negative CI–ion trap MS-MS was the preferred detection method of Lee et al. [24] for fluorinated derivatives of clenbuterol extracted from urine. The instrument used for the study employed an external ion source, and concentrations as low as 50 fg ml^{-1} could be determined. LSD in urine at concentrations between 20 pg ml^{-1} and 2 ng ml^{-1} was quantified by Sklerov et al. [25] using internal ionisation ion trap MS-MS detection. A limit of detection was estimated at 14–20 pg ml^{-1}, and the correlation coefficient over the calibrated range was 0.9998. Within-day and between-day precision values did not exceed 5% and 12.1%, respectively. GC-TOF-MS was used by Taguchi et al. [26] to confirm the presence of myo- and chiro-inositol in glycosylphosphatidylinositol-anchored proteins from bovine erythrocyte membranes following hydrolysis and trimethylsilylation. Flavour volatiles in fruit were analysed rapidly by Song et al. [27] using solid phase microextraction–GC-EI-TOF-MS. Tomato and strawberry volatiles were analysed in under 4 and 3 min, respectively.

5.3 Gas chromatography–atomic emission detection

5.3.1 Theory and instrumentation

Element-specific chromatographic detection is now a well-established technique for the analysis of heteroatomic compounds [28]. Of the different types of atomic spectroscopy available, atomic emission detection (AED) is the most popular for gas chromatography owing to a combination of good sensitivity, selectivity and linear dynamic range. It also has the ability to cope with analysis in complex matrices and to monitor more than one element at once. Species with suitable chemical functionality may be derivatised with a reagent containing an appropriate element to improve the sensitivity and selectivity of detection [29]. In GC-AED, analytes eluting from the column are introduced into a plasma (a cloud of hot, ionised gas) where atomisation occurs. The constituent atoms of the compound gain sufficient thermal energy to excite some of their electrons to higher states. On returning to its ground state, the electron releases this energy as a quantum of light, the wavelength of which is characteristic of the element. Emissions of different wavelengths are resolved by a diffraction grating, and the intensity of monochromatic light characteristic of the analyte is measured by a photodiode.

5.3.1.1 Types of plasma source

Direct current plasmas (DCP) are formed as a result of an electrical discharge between two electrodes in an atmosphere of inert gas. The inductively coupled plasma (ICP) source creates a plasma discharge in an atmosphere of argon flowing through a quartz tube subjected to an r.f. field. Temperatures in an ICP source can exceed $9000°C$, resulting in extremely efficient sample atomisation. Microwave-induced plasma (MIP) is the source most commonly used in GC-AED. Plasma is formed in a stream of inert gas flowing through a microwave cavity. Power and gas consumption is much lower than in DCP or ICP instruments, and the plasma is normally at or slightly below atmospheric pressure. Atmospheric pressure plasmas are preferred, since these are more easily interfaced to GC and generally suffer less from post-column band broadening. Despite the fact that temperatures are lower in MIP than in ICP or DCP, high emission intensities are observed for most elements when helium plasma is used, owing to the high electron energies generated [29]. Reagent gases, such as hydrogen, oxygen or methane, may be used to suppress chemical interferences. Figure 5.32 shows a schematic diagram of a commercially available GC-AED instrument with a MIP source. Recently, Jerrell et al. reported a reduced pressure low-power ICP source for GC-AED that required only column flow to sustain the plasma [30]. The sensitivity of the prototype was inferior to that of MIP-AED systems, however, with limits of detection approximately three orders of magnitude higher in most cases.

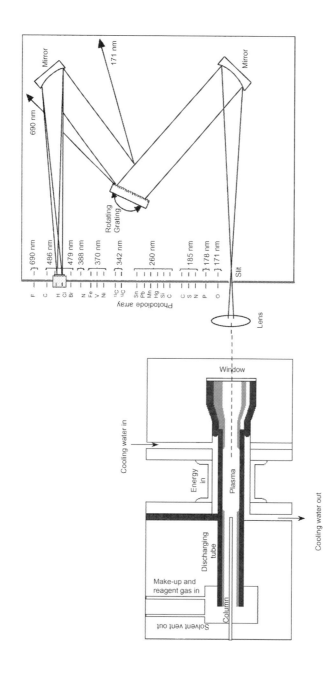

Figure 5.32 Schematic diagram of a MIP-AED system. (Reproduced by permission of Agilent Technologies Limited, Palo Alto, CA, USA.)

5.3.1.2   Sensitivity and selectivity

Sensitivity in GC-AED is a function of the intensity of light emitted by the element concerned, and varies widely [28]. For each element there are a number of emission lines that may be monitored, and the sensitivity of different emission wavelengths depends on the plasma used. Typical sensitivity data for selected elements are given in Table 5.2. Good selectivity depends upon the absence of spectral interferences from other elements at the measured wavelength. Selectivity varies with type of plasma, the resolving power of the instrument optics and operating conditions. Augusto and Valente investigated factors influencing the selectivity and sensitivity of fluorine detection by GC-MIP-AED [31], which included helium flow rates and microwave power settings. For a general discussion of selectivity in GC-AED, see the account of Sullivan and Quimby [32].

5.3.1.3   Simultaneous multielement detection

The simultaneous monitoring of more than one element may be achieved either by the rapid adjustment of the diffraction grating, allowing rapid sequential measurements of selected elements to be made, or by the positioning of a number of photodiodes to monitor several elements without scanning the monochromator. The main limitation of the first approach is that it is difficult to switch

Table 5.2 GC-MIP-AED sensitivity data

Element	Wavelength (nm)	Limit of detection (pg s^{-1})
^{1}H	656.3	7.5
^{2}H	656.1	7.4
B	249.8	3.6
C	193.1	1
N	174.2	30
O	777.2	75
F	685.6	40
Si	251.6	7.0
P	177.5	2
S	180.7	1.7
Cl	479.5	30
V	268.8	10
Mn	257.6	1.6
Fe	259.9	0.3
Ni	231.6	2.6
Ge	265.1	1.3
As	228.8	6.5
Br	478.6	60
Sn	284.0	1.6
Hg	253.7	0.1
Pb	283.3	0.17

Compiled from references 29 and 30, and manufacturer's specifications.

monochromator positions quickly enough to gather sufficient data points to characterise narrow GC peaks. This problem is magnified as the number of monitored elements increases. The second approach is the one adopted in a commercial GC-AED system that is able to monitor simultaneously up to four different elements. Its movable diode array covers a range of 25 nm, and this limits the combinations of elements that may be detected together (see Figure 5.32). Element-specific chromatograms, acquired using different wavelength settings during a series of repeat injections from the same sample, are shown in Figure 5.33.

5.3.1.4   Compound-independent calibration

The detector response in GC-AED is commonly assumed to be compound independent; 1 ng of carbon will give the same response whether it comes from a hydrocarbon or a heteroatomic molecule. The mole fraction of each element present in an unknown may therefore be calculated by reference to known compounds containing the elements in question. This approach assumes a detector response for each element that is independent of molecular structure, and permits the empirical formula of an unknown to be estimated. However, several authors have questioned this assumption. Yu *et al.* offered evidence that the relative responses for carbon and hydrogen were different in halogenated compounds [33]. Becker *et al.* [34] found a better correlation of element response factors for compounds having similar molecular weights as well as closely related elemental compositions. Bos and Barnett reported poor correlation of Se:C ratios when analyte concentration differed markedly from that of the reference compound [35]. Stevens and Borgerding [36] demonstrated a relationship between column flow rate and element response factor, and inferred that response factors could change during temperature-programmed GC with constant-pressure carrier gas control. For compounds containing 12 carbon atoms in addition to varying numbers of hydrogen, oxygen and sulfur atoms, however, element response factors varied by less than 5% from their target values [37].

5.3.1.5   Isotopic measurement

Characteristic emission lines of isotopes vary slightly, usually by a fraction of a nanometre, enabling the spectral resolution of, for example, ^{12}C and ^{13}C in GC-AED instruments. Thomas and Ramus [38] investigated the use of d_{10}-anthracene as a surrogate standard for the determination of anthracene. Deuterium was found to interfere with the hydrogen signal, although a corrected result could be obtained by subtracting the interference signal. The authors concluded that the use of labelled surrogate standards in GC-AED was indeed possible. Elbast *et al.* came to the same conclusion regarding the benefits of ^{13}C-labelled analyte molecules [39]. In their study, the pharmacokinetic behaviour of labelled and unlabelled caffeine was successfully investigated, comparing the oral route simultaneously with intravenous injection. More recently, the

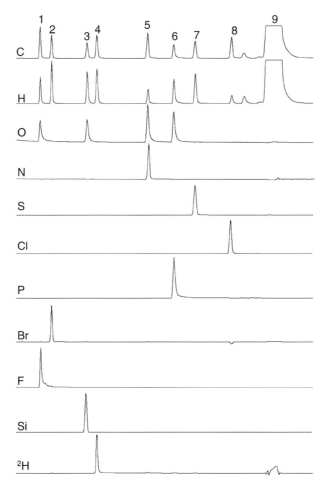

Figure 5.33 GC-MIP-AED chromatogram of a test mixture. Peaks identities: 1, 4-fluoroanisole; 2, 1-bromohexane; 3, tetraethylorthosilicate; 4, d_{22}-decane; 5, nitrobenzene; 6, triethylphosphate; 7, t-butyl disulfide; 8, 1,2,4-trichlorobenzene; 9, n-dodecane. (Reproduced by permission of Agilent Technologies Limited, Palo Alto, CA, USA.)

application of ^{15}N-labelled compounds to metabolite studies has been demonstrated [40].

5.3.2   Applications

5.3.2.1   Environmental applications

GC-AED is exceptionally well suited to the selective and sensitive detection, speciation and quantification of organometallic pollutants. These include tin-, lead- and mercury-containing species [41–43]. Orellana-Velado *et al.* [41] used a r.f. and d.c. glow discharge plasma source to measure alkylmercury compounds following extraction from marine biota. Limits of detection were similar to those obtained with MIP-AED instruments, and inorganic mercury could also be analysed following derivatisation with a Grignard reagent (butylmagnesium chloride). Organotin and alkyl lead compounds were determined simultaneously in sediment by Chau and Yang [42] using GC-MIP-AED following derivatisation with pentylmagnesium bromide. Calculated concentrations in reference materials were close to their certified values and limits of detection were 0.5 pg and 0.2 pg for tin and lead, respectively. Pereiro *et al.* [43] used a multicapillary GC column coupled to a MIP-AED system for the analysis of lead-, tin- and mercury-containing organometallic compounds in various sample types in under a minute. Most analytes were detectable at sub-picogram levels and calibration curves displayed excellent linearity.

Halogenated compounds of environmental concern have also been analysed by GC-AED [44,45]. Chlorinated insecticides were quantifiable at levels as low as 20 ppb in cod liver oil using compound-independent calibration. Results were in close agreement with data produced using GC-ECD [44]. Asp *et al.* constructed their own AED, using a 350 kHz r.f. discharge to generate a plasma inside the end of a 0.32 mm i.d. capillary column [45]. Their work showed no significant difference in the chlorine and bromine response factors in chlorinated and brominated biphenyls. Detection limits were inferior to those of both GC-MS operated in SIM mode and GC-ECD. Stuff *et al.* [46] used GC-MIP-AED to screen for the presence of chemical warfare agents in soil samples and a disused storage tank. Elements monitored included sulfur, phosphorus, chlorine, arsenic and fluorine. They used a combination of GC-AED and molecular weight data to assign possible molecular formulae to unidentified peaks. Hudak *et al.* evaluated the ruggedness of a GC-MIP-AED mounted in a mobile laboratory [47]. Intended applications included the mobile monitoring of areas contaminated with chemical warfare agents, chemical dump sites and polluted air. They concluded that there was no perceptible difference in performance between the mobile instrument and a laboratory-based system.

5.3.2.2   *Biological and pharmaceutical applications*
The drugs diazepam and chlorpromazine were analysed in human plasma using the on-column plasma approach referred to earlier (section 5.3.2.1, reference 45) by Bråthen *et al.* [48]. It was not possible to achieve high enough sensitivity for oxygen and nitrogen to enable therapeutic levels of these drugs to be measured. Monitoring of chlorine and sulfur channels, however, enabled detection of as little as 1.5 ng ml^{-1} and 3.8 ng ml^{-1} of diazepam and chlorpromazine, respectively, with minimal sample clean-up. Marhevka *et al.* [49] discussed a number of applications using GC-AED. These included the selective detection and quantification of a fluorocarbon in human plasma following accidental exposure,

and studies in rats using blood and urine samples to investigate the metabolism of fluorinated decanol. They also discussed applications involving derivatisation to enhance detectability, including formation of trifluoroethyl esters of fatty acids and the use of chlorodifluoroacetic anhydride to tag alcohols and amino acids. Pyrolysis-GC-AED was used by Voisin and co-workers to identify *Corynebacterium* species, a family of bacteria frequently involved in nonspecific infections [50]. Data collected using carbon, sulfur and nitrogen channels were broadly comparable with GC-FID results, although some anomalies were noted.

5.3.2.3 *Industrial applications*

GC-AED has been widely used for the identification and quantification of heteroatomic and organometallic compounds in petrochemical products. Pereiro and Lobinski [51] used multicapillary GC-AED to screen petrol for organolead compounds. Separation was achieved in under 18 s, compared to 6 min for conventional techniques, and the calculated value for a reference sample was in close agreement with the certified value. Limits of detection for tetramethyl lead and tetraethyl lead were 70 pg ml^{-1} and 180 pg ml^{-1} respectively, and the correlation coefficient (r^2) of both calibration curves exceeded 0.999. Changes in the concentration of nitrogen-containing compounds during the processing of diesel fuels were investigated by Wiwel *et al.* [52]. Silica solid-phase extraction cartridges were used selectively to extract nitrogen-containing species and other polar compounds from samples prior to analysis. The authors succeeded in identifying most of the higher-boiling nitrogen compounds present in diesel feedstock and finished product. Quimby *et al.* optimised selectivity for sulfur and nitrogen compounds in refinery liquids by adjustment of make-up and reagent gas flows [53]. Their conditions resulted in slightly poorer limits of detection, but this was more than compensated for by the improvement in selectivity achieved (10-fold for sulfur and more than 100-fold for nitrogen), which permitted more sample to be introduced.

5.4 Gas chromatography–Fourier transform infrared spectrometry

5.4.1 *Theory and instrumentation*

Infrared spectroscopy provides unique information regarding functional groups present in a molecule, and can usually distinguish between positional or conformational isomers. Absorbance by organic molecules in the infrared region of the spectrum arises as a result of bending and stretching of covalent bonds at different characteristic frequencies. The frequency of vibration is related to both bond order and the mass of the atom attached to the bond. Vibrational frequencies of multiple bonds tend to be higher than for single bonds, and the attachment of lighter elements to a particular functional group similarly results in absorbance at higher frequencies. For example, alkyl C–H bond stretching occurs at 3000–2840 cm^{-1} whilst alkyl C–Cl bonds vibrate more slowly (850–550 cm^{-1}) [54]. Older types of IR spectrometer used moving prisms or gratings to disperse polychromatic IR radiation into its component frequencies, passing each frequency in turn through the sample cell in order to obtain a full spectrum. This process would be much too slow for on-line data collection in GC. Instead, a Michelson interferometer (shown in part of Figure 5.34) is usually used to split light from the source into two out-of-phase beams, which are then combined to form an interferogram. The Michelson interferometer consists of a fixed and a moving mirror at right angles to each other. A partially reflecting beam splitter is placed so as to be at 45° to each mirror. Light from the source is split into two by the beam splitter, and each resulting beam is reflected by one of the mirrors. The two light beams are recombined at the beam splitter, but the beam from the moving mirror is now out of phase with the original IR beam, and forms an interference pattern. An IR spectrum may be derived if a Fourier transform (FT) of the interferogram is performed after the beam has travelled through the sample cell [55].

There are two common ways of interfacing GC with FTIR spectroscopy. The first involves deposition of solutes onto a cryogenically cooled surface as they leave the GC column. This technique is known variously as matrix isolation-FTIR (MI-FTIR) or direct deposition-FTIR (DD-FTIR). In MI-FTIR, analyte molecules are trapped in a band of frozen argon on the surface of a gold-plated drum; DD-FTIR typically involves analyte deposition on a transparent zinc selenide window at or below −193°C. In both techniques, FTIR detection may be carried out either concurrently or after the GC run and cryogenic deposition process has finished. This type of interface is sensitive to the build-up of injection solvent and ice on the deposition surface, but is capable of trapping in the region of 0.1–1 ng of analyte [56]. Experimental conditions can affect the appearance of IR spectra obtained in GC-MI-FTIR [57], although DD-FTIR spectra are generally similar in appearance to those of samples prepared as KBr discs [58].

An alternative approach is to direct the GC column effluent through a heated flow cell (commonly known as a light pipe) placed in the optical path of a FTIR spectrometer. The light pipe design most commonly used in GC-FTIR comprises a gold-plated tube closed at either end with potassium bromide windows. An instrument using this set-up is shown in Figure 5.34. Light pipe volume is an important variable in this type of instrument [55]. To achieve maximum sensitivity, the flow cell should be sufficiently large to accommodate most of a chromatographic peak. If the flow cell is too large, however, it is likely that the tail of one solute band will not have been swept clear of the detector's optical path before the next compound enters the cell, resulting in poor chromatographic resolution. Vapour phase IR bands have different widths and absorbance maxima from corresponding bands in solids or liquids [59]. This effect becomes more

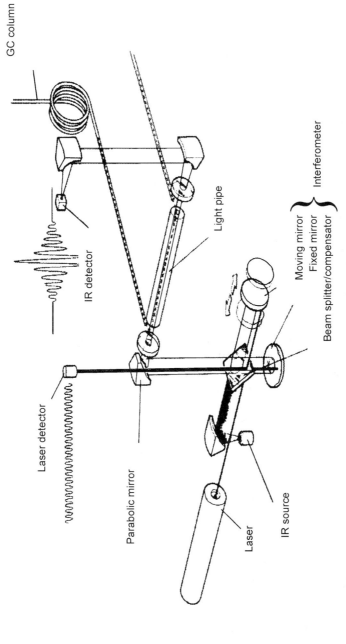

GC column

Laser detector

Parabolic mirror

IR detector

Light pipe

Laser

IR source

Moving mirror
Fixed mirror
Beam splitter/compensator

Interferometer

Figure 5.34 Schematic diagram of a GC–light pipe FTIR instrument. (Reproduced by permission of BioRad Limited, Veenendaal, The Netherlands.)

pronounced as the temperature of the gas phase molecules increases [60]. The greatest difference is observed for functional groups that can take part in intermolecular hydrogen-bonding interactions, whose liquid film IR spectra are characterised by broad absorption bands. The same bands appear as sharp peaks in vapour phase IR spectra, owing to the absence of intermolecular interactions [61]. In practice, this means that IR spectra acquired on a light-pipe instrument may not be directly comparable with libraries of IR data produced at room temperature. The sensitivity of light-pipe instruments is quite poor; tens of nanograms of analyte are generally required to produce a detectable response (Table 5.3) [62].

Despite the poor sensitivity of GC-FTIR systems, the qualitative spectral information provided can be invaluable for compound identification. There are many occasions when GC-MS is incapable of distinguishing between isomers of the same compound. A commonly encountered example is that of m- and p-xylene, which have virtually identical mass spectra. IR data are able to distinguish between isomers with different aromatic ring substitution positions (Figure 5.35).

5.4.2 Applications

Norton and Griffiths [59] compared the performance of direct deposition (DD) and light-pipe GC-FTIR instruments for the analysis of barbiturates. The DD instrument gave limits of detection and identification nearly two orders of magnitude lower than those of the light-pipe system. The spectra produced by each technique for the same compounds were markedly different, and chromatographic resolution was similar. Budzinski et al. [62] commented on the poor sensitivity of commercially available light-pipe instruments, but commented on the complementary nature of GC-FTIR and GC-MS. GC-FTIR succeeded in distinguishing between a number of different isomers with similar EI mass spectra. Similarly, Visser et al. [63] favoured GC-DD-FTIR for the separation and identification of substituted polycyclic aromatic hydrocarbons following flash vacuum pyrolysis of ethynylated aromatics. Zhang et al. [64] used

Table 5.3 GC-FTIR detection limits

Compound	Limit of detection (ng)
Phenol	25
2,4,6-trichlorophenol	30
2,4-dinitrophenol	75
p,p'-DDE	50
Dieldrin	60
Endrin	80
3-Methylphenanthrene	95

From reference 62.

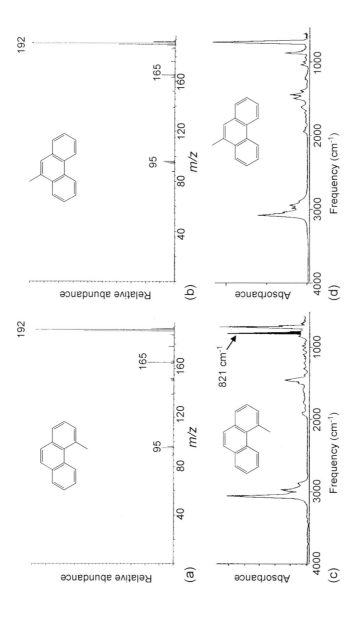

Figure 5.35 Electron impact mass spectra (a,b) and IR spectra (c,d) of 4-methylphenanthrene (a,c) and 9-methylphenanthrene (b,d). Note the characteristic strong absorption band at 821 cm^{-1} in the IR spectrum of 4-methylphenanthrene. (Reproduced from [8] by permission of Elsevier Limited.)

GC–light pipe FTIR to determine the compositions of light oil fractions in high-temperature coal tar distillates on the basis of retention index and spectral data. Dirinck *et al.* [65], while acknowledging the greater sensitivity and ease of spectral comparison with DD- and MI-FTIR, favoured light pipe FTIR for its greater simplicity and robustness. They created their own library of vapour phase IR spectra in order to identify and quantify different amphetamine analogues in illegal drug formulations. Large-volume injection GC-DD-FTIR was used by Hankemeier *et al.* [66] to detect organic contaminants in raw and potable waters at concentrations between 0.1 and 1 $\mu g \, l^{-1}$ from as little as 20 ml sample volume. Acquired IR spectra were successfully searched against libraries of data obtained using KBr discs.

5.5   Applications of combined techniques

5.5.1   GC-MS and GC-AED

5.5.1.1   Environmental applications
Hooker and De Zwaan reviewed applications of combined GC-MS and GC-AED, including the identification of impurities, analysis of reaction mixtures and selectivity improvement in complex matrices [67]. Unknown compounds in wastewater samples pre- and post-treatment were identified by van Stee *et al.* [68] using GC with a post-column split to AED and MS detectors. Compounds detected included plasticisers, fragrances and fatty acids, and the authors were able to draw valuable conclusions regarding the efficiency of the water treatment process with respect to various compound types. Vegetable and water samples were analysed for pesticides and other heteroatomic pollutants using a similar instrumental set-up, with the addition of large-volume injection. Elemental detection limits for AED using this split-flow system were typically 30–50 pg for sulfur, phosphorus and carbon, 400–800 pg for chlorine, bromine, nitrogen and hydrogen, and 2–4 ng for fluorine and oxygen. On-line sample preparation was coupled to separate GC-MS and GC-AED systems for the screening of water samples for organic pollutants [69]. A temperature-programmed retention index system was developed to enable data comparison between the two systems. This was particularly useful to correct for differences between AED runs, since the reagent gas conditions required to optimise sensitivity for different elements resulted in slight variations in retention time. The authors succeeded in identifying several unknown compounds in clean and dirty water samples. Frischenschlager *et al.* [70] used parallel GC-AED-MS to detect and identify pesticides and solvents in water samples taken at various points along the course of a river. GC-AED data were susceptible to matrix interference, and the authors concluded that a revised sample clean-up strategy would be necessary in order

to achieve required limits of detection. GC-AED was compared with GC-ECD, GC–ion trap MS and GC–high resolution MS for the analysis of haloacetic acids in water by Williams *et al.* [71]. High-resolution MS and ECD gave the best detection limits (less than 10 pg for each compound), with ion trap mass spectrometry detecting less than 100 pg. GC-AED had limits of detection 10-fold higher than the ion trap. In evaluating concentrations of haloacetic acids in environmental samples using these techniques, the poorer sensitivity of the AED system was found to be a major limitation.

5.5.1.2 Industrial applications

Clarkson and Cooke used thermal desorption GC-MS and headspace GC-AED to identify an unknown volatile compound in processed tobacco [72]. Final identification was achieved through a combination of both techniques, and confirmed using a reference standard. Pyrolysis GC-AED combined with GC-MS was capable of identifying brominated flame retardants in polymers [73]. The major advantage of AED was increased selectivity, and the type of flame retardant present could be identified by peak pattern recognition.

5.5.2 GC-MS and GC-FTIR

Auger *et al.* [74] succeeded in identifying an unknown male insect pheromone using GC-MS data for molecular weight and fragmentation information, and GC-DD-FTIR data for identification of functional groups. GC-MS and GC-DD-FTIR were shown to be complementary techniques for the analysis of dimethylphenanthrenes in crude oil and extracts of sedimentary rocks [75]. Results of FTIR screening were confirmed by fractionation using molecular sieves and further analysis by GC-FTIR and GC-MS.

Pyrolysis products of valine and leucine were identified using GC–light pipe FTIR–MS by Basiuk [76]. He used these data to establish the mechanisms of thermal decomposition, which included dehydrogenation, dealkylation and loss of HNCO. Compounds causing contact dermatitis were investigated by Tomlinson and Wilkins [77] using multidimensional (heartcutting) GC–light pipe FTIR–MS. Matching of spectra against commercially available vapour phase IR and mass spectral libraries succeeded in identifying potential irritants in 8 out of 13 cases. The same authors, in partnership with Sasaki, reviewed multi-dimensional GC and GC-FTIR applications in the fields of organic chemistry, biology, polymer and petroleum chemistry, and environmental protection [78]. Thermal desorption GC–light pipe FTIR–MS was used by Krüsemann and Jansen [79] to identify unknown polymers. Positive CI-GC-MS using ammonia reagent gas provided molecular weight information, while EI-GC-MS and FTIR data were used to deduce compound structures. Jackson [80] used

GC-MI-FTIR and GC-MS to identify and quantify caffeine and chlorinated pesticides. Recognisable caffeine IR spectra were acquired from as little as 40 pg of material: three orders of magnitude greater sensitivity than light pipe instruments. The use of GC-MS detection reduced the risk of missing smaller GC peaks with the less-sensitive IR instrument, as well as providing complementary data. Krock *et al.* [81] described a combined GC-MS and GC-FTIR instrument in which the carrier gas flow was split post-column, 10% to MS and 90% to FTIR. Following crude separation on a precolumn, fractions were collected on a series of cryotraps prior to sequential desorption and separation on the main analytical column. Unleaded petrol was analysed to evaluate this instrument, and the authors endorsed its use in the analysis of complex mixtures. Krock and Wilkins later extended the application of this system to mixtures including the Grob test mix, to evaluate its performance with compounds of varying polarity and functionality [82]. The same authors used multidimensional GC-FTIR-MS to identify compounds present in paint thinners [83].

Listemann *et al.* [61] commented on the application of GC–light pipe FTIR–MS to analysis of reaction products following organic synthesis. GC-MS could not be relied upon to provide useful diagnostic information in every case, while the absence of hydrogen bonding in gas phase IR spectra gave rise to sharp absorption bands for N–H and O–H vibrations.

5.6   Future directions and developments

One of the main challenges facing analytical chemists over the past 25 years or so has been to achieve ever lower limits of detection for hazardous or environmentally damaging compounds in complex matrices. Techniques such as liquid–liquid and solid phase extraction have been used to great effect to preconcentrate analytes and, to a varying extent, remove interfering matrix material. The law of diminishing returns applies to sample clean-up, however. This is particularly true where matrix and analyte polarity are closely matched. In addition, different samples in ostensibly the same matrix can pose very different sample preparation problems. For example, the term 'soil' covers a multitude of sample types, including petrochemical sludges from disused gas works and clean sand! Many of the recent innovations mentioned in this chapter, such as low-cost MS-MS, selective and sensitive ionisation techniques for GC-MS (HSI and FI) and the fast scanning, medium mass resolution capabilities of GC-TOF-MS, should result in better method performance for these difficult analyses, through improved selectivity and sensitivity. There will still, of course, be a need to remove matrix compounds that would compromise chromatographic robustness.

By comparison, GC-AED, and especially GC-FTIR, are relatively insensitive techniques. They occupy an important niche in the organic analyst's inventory, however, because of the unique and complementary information that they provide. Compared to GC-MS they are not widely used techniques. Because instrument sales of IR and AED detectors are not as buoyant as for GC-MS, costs tend to be higher.

References

1. J.R. Chapman, *Practical Organic Mass Spectrometry: A Guide for Chemical and Biochemical Analysis*, 2nd edn, Wiley, Chichester, 1993.
2. S.G. Lias, J.E. Bartmess, J.F. Liebmann, J.L. Holmes, R.D. Levin and W.G. Mallard, *J. Phys. Chem. Ref. Data*, **1988**, *17* (supplement 1), 1.
3. A. Danon and A. Amirav, *J. Chem. Phys.*, **1987**, *86*, 4708.
4. S. Dagan and A. Amirav, *Int. J. Mass Spectrom. Ion Processes*, **1994**, *133*, 187.
5. S. Dagan, A. Amirav and T. Fujü, *Int. J. Mass Spectrom. Ion Processes*, **1995**, *151*, 159.
6. R.E. March and J.F.J. Todd, *Practical Aspects of Ion Trap Mass Spectrometry*, Volume 1– *Fundamentals of Ion Trap Mass Spectrometry*, CRC Press, Boca Raton, FL, 1995.
7. C. Andalò, G.C. Galletti and P. Bocchini, *Rapid Commun. Mass Spectrom.*, **1998**, *12*, 1777.
8. S.C. Davis, A.A. Makarov and J.D. Hughes, *Rapid Commun. Mass Spectrom.*, **1999**, *13*, 237.
9. M. Krahmer, K. Fox and A. Fox, *Int. J. Mass Spectrom.*, **1999**, *190/191*, 321.
10. M. Natangelo, S. Tavazzi, R. Fanelli and E. Benfenati, *J. Chromatogr. A*, **1999**, *859*, 193.
11. M.N. Sarrión, F.J. Santos and M.T. Galceran, *J. Chromatogr. A*, **1999**, *859*, 159.
12. W.-S. Ding and C.-T. Chen, *J. Chromatogr. A*, **1999**, *862*, 113.
13. C.J.H. Miermans, L.E. van der Velde and P.C.M. Frintrop, *Chemosphere*, **2000**, *40*, 39.
14. R.S. Spaulding, P. Frazey, X. Rao and M.J. Charles, *Anal. Chem.*, **1999**, *71*, 3420.
15. T. Küchler and H. Brzezinski, *Chemosphere*, **2000**, *40*, 213.
16. C.R. Bowerbank, P.A. Smith, D.B. Drown, W. Alexander, W.J. Jederberg, K.R. Still and M.L. Lee, *Drug Chem. Toxicol.*, **1999**, *22*, 57.
17. S.C. Davis, A.A. Makarov and J.D. Hughes, *Rapid Commun. Mass Spectrom.*, **1999**, *13*, 247.
18. R.J.J. Vreuls, J. Dallüge and U.A.Th. Brinkman, *J. Microcolumn Sep.*, **1999**, *11*, 663.
19. J.A.G. Roach, K.D. White, M.W. Trucksess and F.S. Thomas, *J. AOAC Int.*, **2000**, *83*, 104.
20. T. Niwa, *Mass Spectrom. Rev.*, **1997**, *16*, 307.
21. P.E.A. Teal, A.T. Proveaux and R.R. Heath, *Anal. Biochem.*, **2000**, *277*, 206.
22. S. Pichini, R. Pacifici, I. Altieri, A. Palmeri, M. Pellegrini and P. Zuccaro, *J. Chromatogr. B*, **1999**, *732*, 509.
23. K. Bal and L. Larsson, *J. Chromatogr. B*, **2000**, *738*, 57.
24. J. Lee, D. Lho, M. Kim, W. Lee and Y. Kim, *Rapid Commun. Mass Spectrom.*, **1998**, *12*, 1366.
25. J.H. Sklerov, K.S. Kalasinsky and C.A. Ehorn, *J. Anal. Toxicol.*, **1999**, *23*, 474.
26. R. Taguchi, J. Yamazaki, M. Takahashi, A. Hirano and H. Ikezawa, *Arch. Biochem. Biophys.*, **1999**, *363*, 60.
27. J. Song, L. Fan and R.M. Beaudry, *J. Agric. Food Chem.*, **1998**, *46*, 3721.
28. P.C. Uden, in *Element-Specific Chromatographic Detection by Atomic Emission Spectroscopy*, ACS Symposium Series 479, (ed. P.C. Uden), American Chemical Society, Washington DC, 1992, p. 1.
29. P.C. Uden, in *Detectors for Capillary Chromatography* (eds. H.H. Hill and D.G. McMinn), Chemical Analysis, Vol. 121, Wiley, New York, 1992, p. 219.
30. L.J. Jerrell, M.R. Dunn, J.E. Anderson and H.B. Fannin, *Appl. Spectrosc.*, **1999**, *53*, 245.
31. F. Augusto and A.L.P. Valente, *J. Microcolumn Sep.*, **1999**, *11*, 23.
32. J.J. Sullivan and B.D. Quimby, in *Element-Specific Chromatographic Detection by Atomic Emission Spectroscopy*, ACS Symposium Series 479 (ed. P.C. Uden), American Chemical Society, Washington DC, 1992, p. 62.

33. W. Yu, Y. Huang and Q. Ou, in *Element-Specific Chromatographic Detection by Atomic Emission Spectroscopy*, ACS Symposium Series 479 (ed. P.C. Uden), American Chemical Society, Washington DC, 1992, p. 44.
34. G. Becker, A. Colsjö and C. Östman, *Anal. Chim. Acta*, **1997**, *340*, 181.
35. R. Bos and N.W. Barnett, *J. Anal. Atom. Spectrosc.*, **1997**, *12*, 733.
36. N.A. Stevens and M.F. Borgerding, *Anal. Chem.*, **1998**, *70*, 4223.
37. W. Elbast, M.S. Caubet, D. Deruaz and J.L. Brazier, *Anal. Lett.*, **1999**, *32*, 3111.
38. L.C. Thomas and T.L. Ramus, *J. Chromatogr.*, **1991**, *586*, 309.
39. W. Elbast, F. Besacier, D. Deruaz and J.L. Brazier, *J. Chromatogr. B*, **1997**, *690*, 115.
40. W. Elbast, M.S. Caubet, D. Deruaz and J.L. Brazier, *Anal. Lett.*, **1999**, *32*, 1627.
41. N.G. Orellana-Velado, R. Pereiro and A. Sanz-Medel, *J. Anal. Atom. Spectrosc.*, **1998**, *13*, 905.
42. Y.K. Chau and F. Yang, *Appl. Organometallic Chem.*, **1997**, *11*, 851.
43. I.R. Pereiro, A. Wasik and R. Lobinski, *J. Chromatogr. A*, **1998**, *795*, 359.
44. S.I. Semb, E.M. Brevik and S. Pedersen-Bjergaard, *Chemosphere*, **1998**, *36*, 213.
45. T.N. Asp, S. Pedersen-Bjergaard and T. Greibrokk, *J. High Resolut. Chromatogr.*, **1997**, *20*, 201.
46. J.R. Stuff, W.R. Creasey, A.A. Rodriguez and H.D. Durst, *J. Microcolumn Sep.*, **1999**, *11*, 644.
47. G.J. Hudak, P.S. Demond and L. Russel, *J. Hazardous Mater.*, **1995**, *43*, 155.
48. V. Bråthen, E. Lundanes and S. Pedersen-Bjergaard, *J. High Resolut. Chromatogr.*, **1999**, *22*, 123.
49. J.S. Marhevka, D.F. Hagen and J.W. Miller, in *Element-Specific Chromatographic Detection by Atomic Emission Spectroscopy*, ACS Symposium Series 479 (ed. P.C. Uden), American Chemical Society, Washington DC, 1992, p. 117.
50. S. Voisin, F.N.R. Renaud, J. Freney, M. de Montclos, R. Boulieu and D. Deruaz, *J. Chromatogr. A*, **1999**, *863*, 243.
51. I.R. Pereiro and R. Lobinski, *J. Anal. Atom. Spectrosc.*, **1997**, *12*, 1381.
52. P. Wiwel, K. Knudsen, P. Zeuthen and D. Whitehurst, *Ind. Eng. Chem. Res.*, **2000**, *39*, 533.
53. B.D. Quimby, D.A. Grudoski and V. Giarrocco, *J. Chromatogr. Sci.*, **1998**, *36*, 435.
54. R.M. Silverstein, G.C. Bassler and T.C. Morrill, *Spectrometric Identification of Organic Compounds*, 5th edn, Wiley, New York, 1991.
55. D.F. Gurka, in *Detectors for Capillary Chromatography*, (eds. H.H. Hill and D.G. McMinn), Chemical Analysis Monographs, Wiley, New York, 1992, p. 251.
56. T. Visser, M.J. Vredenbregt and A.P.J.M. de Jong, *J. Chromatogr. A*, **1994**, *687*, 303.
57. W.M. Coleman and B.M. Gordon, Analysis of natural products by gas chromatography/matrix isolation/infrared spectrometry, in *Advances in Chromatography*, Volume 34, Marcel Dekker, New York, 1994, p. 57.
58. S. Bourne, A.M. Haefner, K.L. Norton and P.R. Griffiths, *Anal. Chem.*, **1990**, *62*, 2448.
59. K.L. Norton and P.R. Griffiths, *J. Chromatogr. A*, **1995**, *703*, 383.
60. C. Joblin, P. Boissel, A. Léger, L. d'Hendecourt and D. Défourneau, *Astron. Astrophys.*, **1995**, *299*, 835.
61. M.L. Listemann, F.J. Waller and F.L. Herman, *Spectroscopy*, **1993**, *8*, 41.
62. H. Budzinski, Y. Hermange, C. Pierard, P. Garrigues and J. Bellocq, *Analusis*, **1992**, *20*, 155.
63. T. Visser, M. Sarobe, L.W. Jenneskens and J.W. Wesseling, *Fuel*, **1998**, *77*, 913.
64. M.J. Zhang, S.D. Li and B.J. Chen, *Chromatographia*, **1992**, *33*, 138.
65. I. Dirinck, E. Meyer, J. Van Bocxlaer, W. Lambert and A. De Leenheer, *J. Chromatogr. A*, **1998**, *819*, 155
66. T. Hankemeier, E. Hooijschuur, R.J.J. Vreuls, U.A.Th. Brinkman and T. Visser, *J. High Resolut. Chromatogr.*, **1998**, *21*, 341.
67. D.B. Hooker and J. De Zwaan, Analytical problem solving with simultaneous atomic emission–mass spectrometric detection for gas chromatography, in *Element-Specific Chromatographic Detection by Atomic Emission Spectroscopy* ACS Symposium Series 479, (ed. P. C. Uden), American Chemical Society, Washington DC, 1992, p. 132.
68. L.L.P. van Stee, P.E.G. Leonards, R.J.J. Vreuls and U.A.Th. Brinkman, *Analyst*, **1999**, *124*, 1547.

69. T. Hankemeier, J. Rozenbrand, M. Abhadur, J.J. Vreuls and U.A.Th. Brinkman, *Chromatographia*, **1998**, *48*, 273.
70. H. Frischenschlager, C. Mittermayr, M. Peck, E. Rosenberg and M. Grasserbauer, *Fresenius J. Anal. Chem.*, **1997**, *359*, 213.
71. D.T. Williams, F.M. Benoit and G.L. Le Bel, *Proc. 1992 Water Quality Technology Conf.*, Toronto, American Water Works Association, 1993, p. 1297.
72. P. Clarkson and M. Cooke, *Anal. Chim. Acta*, **1996**, *335*, 253.
73. F.C.-Y. Wang, *Anal. Chem.*, **1999**, *71*, 2039.
74. J. Auger, S. Rousset, E. Thibout and B. Jaillais, *J. Chromatogr. A*, **1998**, *819*, 45.
75. S.J. Fisher, R. Alexander, L. Ellis and R.I. Kagi, *Polycyclic Aromatic Compounds*, **1996**, *9*, 257.
76. V.A. Basiuk, *J. Anal. Appl. Pyrolysis*, **1998**, *47*, 127.
77. M.J. Tomlinson and C.L. Wilkins, *J. High Resolut. Chromatogr.*, **1998**, *21*, 347.
78. M.J. Tomlinson, T.A. Sasaki and C.L. Wilkins, *Mass Spectrom. Rev.*, **1996**, *15*, 1.
79. P.V.E. Krüsemann and J.A.J. Kansen, *J. Chromatogr. A*, **1998**, *819*, 243.
80. P. Jackson, *Anal. Proc.*, **1993**, *30*, 394.
81. K.A. Krock, N. Ragunathan and C.L. Wilkins, *J. Chromatogr.*, **1993**, *645*, 153.
82. K.A. Krock and C.L. Wilkins, *Anal. Chim. Acta*, **1993**, *277*, 381.
83. K.A. Krock and C.L. Wilkins, *J. Chromatogr. A*, **1994**, *678*, 265.

6 Method validation in gas chromatography

R. D. McDowall

6.1 Introduction

The concept of validation has gained considerable popularity within the last 15 years, but it is hardly a new idea. The Public Analyst service and Official Agricultural Analysts were set up in the latter half of the nineteenth century and bodies such as the Institute of Petroleum in the UK and the American Society for Testing Materials (ASTM) have been standardising analytical methods for over 50 years.

The purpose of this chapter is to discuss qualification validation of analytical methods using GC. Part of the discussion is the difference between qualification and validation.

Equipment qualification will be covered first, as the gas chromatograph needs to be fit for the purpose envisaged. When funds are limited, there is a temptation to use equipment for purposes other than that for which it was designed. No one should expect a GC designed for gas analysis to be able to analyse high-boiling compounds such as triglycerides and, although this may be an exaggerated example, it must be stressed that the equipment purchased should be demonstrated by the manufacturer as being suitable for the proposed work.

6.2 Definition of key terms

It is important to define the main terms used in this chapter so that the following discussion is built on firm foundations. The three terms defined here are calibration, equipment qualification and method validation.

Equipment calibration. This is a service function, where the term calibration implies that adjustments can be made to bring a system into a state of proper function. Such adjustments are generally best left to trained service engineers.

Calibration is inextricably linked to preventative maintenance and then to equipment qualification. Whenever calibration involves adjustments of the type described above, it is important to document the activity and, where appropriate, the user needs to re-qualify the instrument concerned. It is important to realise that this term can be confused with equipment qualification.

Equipment qualification. Equipment qualification is the demonstration that an item of equipment—the GC—is fit for the intended purpose. This implies that all the parameters utilised by the methods run on that instrument are within acceptable limits.

Method validation. This is the demonstration that an analytical method is fit for the intended purpose. It implies that the analyst knows why the method is required and the acceptable values of key method validation parameters.

Building on these last two definitions, the differences and interactions between the two will now be discussed.

6.3 Equipment qualification and method validation

Qualification of analytical instruments cannot be achieved through method validation. Equipment qualification provides the foundation to develop, validate and run analytical methods within the operating range measured. If a method is developed using parameters outside of this range, the analytical instrument needs to be re-qualified before method validation can proceed.

It is important to recognise the difference between equipment qualification and method validation. In some analytical scientists' minds these are the same and therefore, by validating a method, the equipment is considered qualified. This is wrong.

It should be realised that equipment qualification assesses the performance of modules or the system over the complete operating range of the instrument that the laboratory anticipates using. For instance, if the carrier gas flow for a gas chromatograph is qualified over the range 5–20 ml min^{-1} for packed column analysis, then any method utilising flows within this range can be run. However, if the analyst now wants to use a capillary column with a flow rate of 0.5 ml min^{-1}, this flow rate is outside the range and the instrument is unqualified. Clearly, the parameters to qualify and their various ranges must be specified before purchasing an instrument.

6.4 Equipment qualification (EQ)

6.4.1 The overall EQ process

The background and approach discussed in this chapter have been developed primarily from within the pharmaceutical industry [1]. Equipment qualification is the way to ensure that the GC is fit for purpose and consists of four basic elements and four questions that must be answered for successful completion of the process [1–3].

1. Design qualification: Defining the quality parameters that are required of the equipment. This has been discussed above.
2. Installation qualification: Assurance that the intended equipment is received as designed and specified. Does the instrument work to the manufacturer's specification? This is normally achieved on installation by the manufacturer's engineers.
3. Operational qualification: Does the instrument work the way the analytical scientist wants it to?
4. Performance qualification: Confirmation that the equipment consistently continues to perform as required. Does it continue at all times to operate within the parameters monitored?

This model is known as the 4Qs model, based on the four stages of qualification.

The questions posed above need to be addressed adequately at each key stage before progressing to the next one in the process. If you fail to do this, you will have unresolved problems carried forward to the next stages.

6.4.2   Equipment qualification to ensure efficient method transfer

One compelling reason for equipment qualification, is the need to transfer methods between laboratories [1]. Why are so many collaborative trials a failure? The answer lies in the fact that the key analytical variables have not been identified and controlled through specification and/or procedural practice. These may lie within the method but more often are due to the operating parameters of the equipment or system, some of which may not be recognised at first or may be regarded as trivial.

The need to standardise on a specific manufacturer's GC column (and sometimes individual production batch) is well known and will not be elaborated further here. However, the same principle holds for equipment: if temperature is a key factor, how can it be specified if there is no assurance that instrument A's temperature readout is the same as instrument B's? Without an adequate instrument specification there can be no reliable control. Under these circumstances, methods will lack the robustness and reliability needed for fitness for their intended purpose.

6.4.3   Design qualification (DQ)

Design qualification is totally absent in many laboratories, so that money may be wasted in purchasing inappropriate instrumentation. However, the lack of DQ has ramifications throughout the whole of the 4Qs approach. Without a specification that acts as a guide, laboratory management should not be overly surprised if equipment does not work as expected or anticipated.

The best approach to design qualification is summarised as writing down what the equipment is required to do. This can cover several areas, dependent on the scope and complexity of the equipment or system to be purchased. The following areas are typically to be considered in the purchase of a gas chromatograph [1].

1. *Technical requirements.* Detail the operating specification of the chromatograph. Consideration must be given to both the upper and lower ranges of each operating parameter as this provides an input to the operational qualification tests. Is the instrument wanted for a single task that will not change (e.g. isothermal analysis using packed columns and a flame ionisation detector) or will the instrument be required for several different tasks of varying complexity (e.g. isothermal and temperature programming, capillary and packed columns, range of injection modes and several detectors)? The first is relatively easy to specify; the second is much more difficult, to the point where it may be decided that a single instrument will not suffice.

2. *Environmental considerations.* Where is the instrument to be placed and are there any considerations that the manufacturer should know about? For example, an instrument for process analysis should be very robust for nonanalytical staff to operate compared with an instrument used for research purposes (equipment may often be used beyond its design specifications in research). Such considerations will have impacts on robustness, reliability and serviceability of the instrument purchased. Other factors may need to be considered: for example, the amount of space available for the instrument.

3. *Sample presentation.* How will samples be introduced to the instrument? If an autosampler is to be used, this will also require a technical specification under item (1).

4. *Data acquisition needs.* Will the system have its own data system or must it be linked to an existing data system? What constraints are placed upon the equipment?

5. *Operability factors.* What training is required to use the system and what are the service requirements?

6. *Health and safety issues.* Are specific requirements needed regarding electrical supply, radio interference, exhaustion of excess sample and solvent, etc.?

7. *Integration* with other equipment or interfacing with computer applications should be specified. A typical example may be the transfer of data to a spreadsheet or integration of the system with a Laboratory Information Management System (LIMS).

8. *Cost–benefit analysis.* This is the overall financial justification of the equipment based on balancing all the costs of the system against the benefits once it is operational. Cost–benefit analysis is often difficult to predict. Sometimes an apparatus acquired as a speculative purchase may justify the purchase as the result of unforeseen circumstances such as

a plant problem. On other occasions the cost–benefit may prove to be over-optimistic.

All these needs should be formalised in a design specification and instruments from possible vendors should be evaluated against these criteria. It may be that in the light of equipment available the criteria may have to be made less severe.

6.4.4   Installation qualification (IQ)

In the majority of cases an engineer from the vendor will arrive to install the instrument; the engineer will work to a plan and will record that the installation has been done correctly.

A certification body such as UKAS (United Kingdom Accreditation Service) may require evidence that the installation engineer has been trained to execute this work. Commonly, at the end of the IQ a simple test of the chromatograph should be performed with a standard mixture to show that instrument works to the manufacturer's specification. It is sometimes found that the instrument may work satisfactorily with the manufacturer's standard mixture but not with the customer's, which may be more demanding.

6.4.5   Operational qualification (OQ)

If the design qualification has been specified in detail, installation by the supplier's qualified engineer should be sufficient to verify the various test parameters such as the accuracy of the oven temperature control and the accuracy of the temperature readout. The DQ test parameters must be checked using equipment that is calibrated to internationally traceable standards.

6.4.6   Modular and holistic qualification

Furman et al. [4], discussing the validation of computerised liquid chromatographic systems, presented the concepts of modular and holistic qualification. Modular validation is the qualification of the individual components of a system such as pump, autosampler, column heater and detector of an HPLC. The authors state that "Calibration of each module may be useful for trouble shooting purposes, such tests alone cannot guarantee the accuracy and precision of analytical results." Therefore, the authors introduced the concept of holistic validation, in which the whole chromatographic system was also qualified to evaluate the performance of the system. The same approach must be applied to a GC. Where individual units such as the oven, detector and autosampler may be within limits, the GC as a whole may fail a holistic test owing to the additive effects of errors with the component parts.

6.4.7 Performance qualification (PQ)

Performance qualification must be maintained throughout the working life of the instrument by monitoring and recording key operational parameters each time the instrument is run. Additionally, a standard sample may be run at various intervals ranging from one standard in ten to one to one for critical analyses. If the results for the standard do not satisfy the expected results then the cause(s) must be found immediately.

When the instrument is serviced either routinely or after a breakdown, all or part of the qualification should be repeated and the results recorded before the instrument is returned to service.

6.5 Method validation

This section is intended to introduce the scientific and regulatory requirements for analytical method validation. While the emphasis is on European requirements [5, 6], many of these are now common to the United States and Japan owing to the International Conference on Harmonisation (ICH) [7, 8]. All four of these references are the basis for the discussion that follows.

6.5.1 Documenting the method

Validation studies are an essential part of all methods for demonstrating fitness for use in many industries; any analytical procedure must be described in sufficient practical detail including the following:

- sample (aliquot size) and sampling instructions
- storage conditions
- reference standards used including any internal standard
- reagents and their quality and how they are prepared
- chromatographic equipment with set up and equilibration details
- calibration curve preparation and the calibration model used
- example chromatograms including a reagent blank, matrix blank and typical samples correctly labelled
- any necessary calculations and their application
- all relevant data should be detailed together with any necessary formulae to show how the results were calculated
- the accuracy and precision to be expected.

6.5.2 Defining the validation parameters to measure

The purpose of validation is to demonstrate that the proposed analytical procedure is suitable for its intended purpose. It is possible to measure appropriate validation parameters such as precision and accuracy simultaneously and thus to provide a sound overall knowledge of the capabilities of the procedure.

In the case of pharmaceuticals validation, data may have to be submitted as part of an application for a marketing authorisation. Carr and Warlich [9] suggest which parameters need to be considered when designing validation studies for particular situations.

6.5.3   Analytical reference substances

As gas chromatography is not, in general, an absolute analytical technique, reference substances for validation must be available. Standards must be well characterised with documented purity levels (acceptable levels being dependent on the intended use of the material). This includes any internal standard and degradation materials or impurities. Where appropriate, there must be an on-going program for characterising these materials by independent analytical techniques (differential scanning calorimetry, mass spectrometry, IR spectroscopy, NMR spectroscopy, etc.).

6.5.4   Scope of validation

Analytical validation is required for all methods where results are used to make decisions, for example in product release.

1. *Identification tests*: intended to ensure the identity of an analyte in a sample by comparing an appropriate property—e.g. spectrum, chromatographic behaviour, or chemical reactivity—of the sample to that of a reference standard.
2. *Qualitative or limit test impurity controls*: intended to reflect accurately the purity of a sample.
3. *Quantitative analyses* of the active moiety in active substances or finished products and other selected components in finished products (to measure the analyte present in a sample).
4. *Revalidation* may be required where there have been changes to the production process or in analytical procedure. The degree of revalidation required will be dependent on the nature of the changes.

The types of validation studies required depend on the type of test being considered, as indicated in Table. 6.1.

6.5.5   Specificity versus selectivity

Gas chromatography is a specific and absolute analytical technique if it is used with high-resolution mass spectrometry, but in general it is a relative and selective method. The validation procedure should confirm the ability of the

Table 6.1 Validation parameters recommended for different analytical methods

Validation parameter	Identification	Related substances, quantitative[a]	Related substances, limit test[a]	Assays
Accuracy		✓		
Precision				
repeatability		✓		
intermediate precision		✓[b]		✓[b]
Specificity[c]	✓	✓	✓	✓
Limit of detection (LOD)		✓	✓	
Limit of quantification (LOQ)		✓		
Linearity		✓		
Range		✓		

[a]Includes related substances, heavy metals and residual solvents as well as impurities and degradation products.
[b]Or reproducibility, if undertaken.
[c]Lack of specificity (incomplete discrimination) in one method may be compensated for by the use of one or more supporting analytical procedures, which, together, give the necessary level of discrimination.

procedure to assess unequivocally the presence of the analyte in the presence of other components that may be expected to be present—impurities, degradation products and matrix components. The validation studies needed depend on the use to which the method is to be put.

Individual methods taken in isolation may not demonstrate complete discrimination. In such cases, it may be necessary to use a combination of two or more orthogonal methods that can be shown, in combination, to show an adequate level of discrimination.

For identification tests, the method should be shown to be capable of ensuring the identity of an analyte and discriminating between structurally closely related compounds. A known reference sample should give no false positives or false negatives. It should be confirmed by experiment that structurally closely related compounds do not result in a positive result, but the choice of potential interfering compounds should be based on a sensible scientific judgement as to which interferences are likely.

For impurity tests and assays, chromatographic procedures should be shown to have a suitable level of specificity and representative chromatograms provided with the labels for individual components. For critical GC separations, the resolutions of two components eluting close to each other should be investigated to a suitable level.

Discrimination between the analyte and impurities should be demonstrated. Sample spiking studies or comparisons with results from other validated methods should be used where possible.

6.5.6   Accuracy

Accuracy is the closeness of agreement between the value that is accepted as the true value and the value that is found. It should be established over the prescribed range over which a procedure may be used. In terms of the well-known archery target analogue, accuracy means that all the analytical results lie within the 'bull's-eye'. Accuracy is usually established by the analysis of reference samples of known composition made up from pure components or by the analysis of certified reference materials. The true value should, whenever possible, be established by means of an independent analytical method such as IR or UV spectroscopy or mass spectrometry. For example, the accuracy of the determination of a metal by atomic absorption spectroscopy can be verified by means of a 'wet chemical' gravimetric analysis.

Accuracy may be determined by the process of 'standard addition' in which known amounts of a pure component of a mixture are added to the mixture and the original amount is found by interpolation to zero addition. This technique is required when it is impossible to obtain a 'pure' sample, for example for the determination of phenol (conjugates) in urine.

6.5.6.1   Precision

Precision is the closeness of agreement (degree of scatter) between a series of measurements obtained from multiple samples of the same homogeneous sample under prescribed conditions. Again in terms of the archery target, a high precision means that all the analytical results are close together on the target but not necessarily within the bull's-eye. A set of results with high precision but poor accuracy implies that there are constant systematic errors in the determination. The source(s) of these errors should be found and eliminated, but where this proves impossible it may be permissible to introduce an experimentally determined correction factor to convert the high-precision results to highly accurate values (again in target shooting terms, equivalent to 'aiming off').

Repeatability (short-term precision).   This is the precision of the method under the same operating conditions over a short interval of time (daily). The best way to describe these is

- same laboratory
- same analyst
- same reagents and solutions
- same instrument

Repeatability should be estimated using a minimum of nine determinations over the prescribed range; for example, using three replicates at three concentrations, or a minimum of six determinations at the 100% test concentration.

Long-term precision. This takes into account intralaboratory variations such as different days and different reagents. These are best expressed as

- same laboratory
- different analysts
- different reagents and solution
- different days

The need for and extent of this type of data depend on the conditions under which the method will be used. Where it is investigated, the use of an experimental matrix design is encouraged.

Reproducibility. This relates to interlaboratory variations from collaborative studies and is usually applied as part of the standardisation of the method (such as for the inclusion of the method into a set of methods such as ASTM, Institute of Petroleum or a pharmacopoeia. For each type of precision, the following data are required:

- standard deviation (SD)
- relative standard deviation (RSD) (coefficient of variation)
- confidence intervals (if required).

6.5.7   Limit of detection (LOD)

This is the lowest amount of analyte that can be detected (but not necessarily quantified) in a sample by an analytical procedure. It is usually validated by the analysis of samples containing known concentrations of the analyte. The minimum level at which the analyte can be reliably detected should be established.

In the case of GC methods, the signal:noise ratio approach may be used, in which signals from samples containing known low concentrations of the analyte are compared with those from samples containing no analyte. A signal:noise ratio of 2–3:1 is generally accepted for estimating the detection limit. The method for determining the detection limit should be clearly stated. If the detection limit is determined from visual evaluation or based on signal:noise ratio, relevant chromatograms will be accepted as justification. Validation studies to determine the limit of detection may be required for trace analysis methods.

6.5.8   Limit of quantification (LOQ)

This is the lowest amount of analyte that can be quantitatively determined in a sample with suitable precision and accuracy by a given analytical procedure. The value can be determined by the analysis of samples containing known concentrations of the analyte and this information used to establish the minimum level at which the analyte can be quantified with acceptable accuracy and precision.

The signal:noise ratio approach may be used where the signals obtained from samples with known low concentrations of the analyte are compared with those from blanks. A signal:noise ratio of 10:1 is commonly accepted. The minimum concentration is then calculated at which the analyte can be reliably quantified.

The residual standard deviation of the regression line or the standard deviation of the y-intercept of the calibration curve may also be used. The magnitude of the analytical background response to blank samples may also be used to calculate the standard deviation.

The data required are the limit of quantification and the method for its determination. Where the value is estimated by calculation or extrapolation, the value should subsequently be validated by the analysis of a suitable number of samples at or near to the estimated quantification limit.

6.5.9   Linearity

Linearity is the ability of the method to obtain results that are directly proportional to the concentration or amount of analyte in the sample over a defined concentration range. It should be demonstrated using dilutions of the substance to be determined within a given range, using dilutions of the substance and/or separate weighings of a synthetic mixture of components present in the samples. The plot of the signals obtained as a function of concentration of the analyte should be subjected to an appropriate statistical method (e.g. regression using the least-squares technique).

Data from the regression line may help to provide mathematical estimates of the degree of linearity. A minimum of five concentrations is recommended to be used in the construction of the curve and the data provided to support the validation of the method, including the correlation coefficient, the y-intercept, the slope of the regression line and the residual sum of the squares.

For some GC detectors, such as the flame photometric detector in the sulfur mode, a linear result may not be possible. In such cases, the analytical response should be described by an appropriate function of the concentration of the analyte and confirmed by the appropriate experiments.

6.5.10   Range

This is the interval between the upper and lower concentration or amount of analyte in the sample for which it has been demonstrated that the analytical procedure has a suitable level of precision and accuracy (and, ideally, linearity). It is normal to derive this information from the linearity studies. The acceptable linearity, accuracy and precision will depend on the intended use of the procedure. In general, all these properties will be poorer for the determination of trace components than for percentage quantities. During method development, the range around a probable specified limit should be considered.

6.5.11 Robustness

This is a measure of the ability of the analytical procedure to remain unaffected by small variations in the method parameters. It provides a measure of the reliability of the method during normal use. The flame ionisation detector is extremely robust, but some of the selective detectors used in GC are notoriously lacking in robustness.

Ideally, robustness validation should be undertaken during the development phase and the type and the amount of data required depend on the intended use of the method. Factors may include the effect of the analytical solution stability and the effect of different equipment and of extraction time on results obtained. Where measurements show susceptibility to the effects of these types of variation in analytical conditions, the conditions used in the test procedure should be suitably controlled and/or a precautionary statement should be included in the procedure. In GC, examples of variables that should be considered include different column lots or manufacturers; column temperature; carrier gas flow rate; sample size; method of sample introduction; and detector variables.

System suitability tests using parameters established for a given procedure (using a total system of equipment, electronics, analytical operations and samples to be analysed) should be an integral part of the analytical procedures.

6.5.12 Analyte stability

It is possible for the instrument to show instability over time and it is also possible for the sample to show instability (due to volatility, light instability, hydrolysis, oxidation, pH effects, interactions with other materials in the solution, interaction with the container, etc.). In both cases this instability will result in poor accuracy and precision. For example, loss of low-boiling compounds from a sample will result in the overestimation of higher-boiling components, especially if normalisation is used for quantification. All analytes should be shown to be stable for at least the period required for the completion of the analysis—up to 6 hours for a manual method, up to 15 hours for an automated method, or longer to allow for repeat analysis.

Stored samples should be analysed and the RSD of the assays determined. An RSD < 2% should be shown for the assay of the major component of the material under investigation. In the case of GC it should be confirmed that no unexpected additional peaks form between the first application to the column and the last one.

6.5.13 Revalidation of a method

Any change to an analytical method should trigger consideration of the need to revalidate the method. This need will depend on the nature of the changes to the method or its application introduced. Instances in which such additional work may be required include

- transfer of a product from development to production (especially for nonroutine tests)
- where there has been a significant change to the manufacturing methods (change of the source of a component, starting materials or composition of the product)
- change in the synthetic route or process
- modification of a chromatographic procedure (such as sample preparation, or system components)

GC analytical methods used to analyse biological materials may require at least partial revalidation if a different matrix or a different species is used, and if new staff are introduced to the use of a procedure or a lapsed assay is reintroduced.

6.5.14   Internal or external standard method?

GC methods can use either external or internal standards. In general, internal standards give higher accuracy except for the analysis of gaseous mixtures. However, the use of internal markers will result in an excessive amount of work if many components are to be determined. For GC-MS work the use of stable isotopes as internal markers is highly desirable.

6.5.14.1   Internal standardisation

Internal standardisation uses a compound that is added in a fixed amount to all standards and unknown samples in the analytical run. It is used as a reference peak as means of compensation for small variations in sample size. It can also be used for a different purpose to monitor the efficiency of extraction from a matrix. A marker compound is added at the start of the sample preparation and is carried through the processing; therefore it should have similar physico-chemical properties to the analyte and should elute after the peak of interest on the resulting chromatogram so that it does not hide any degradation products or metabolites. Again, stable isotopes are ideal for GC-MS work if they are available.

Calculation is achieved by measuring the peak areas of the analyte and the internal standard and calculating the analyte:internal standard ratio. This latter value is fed into the calibration used in the method and a value for the amount or concentration is calculated.

6.5.14.2   External standardisation

External standardisation uses standards with no additional compound added. The peak areas of the standards are used to construct the appropriate calibration model and peak areas of the analyte in the unknowns are input to calculate the final amount or concentration.

External standardisation assumes that there are no sample preparation losses and that recovery is consistent in all samples and standards. It is best used for gas analysis or for methods where the sample preparation is relatively simple.

6.5.15 Calibration methods

Although this is part of method development, the selection of the correct calibration method is essential for any GC analytical method. Options for overall calibration of GC methods (as opposed to detector calibration) include

- response factor
- linear regression with the following combinations:

  forced through zero
  not forced through zero
  unweighted
  weighted

- nonlinear; these methods need to be justified carefully

Linear regression is used widely as a calibration method. The assumption inherent in this is that all the error is contained in the response (y-axis) rather than the amount or concentration (x-axis). Furthermore, variance should be similar over the concentration range measured (homoscadastic); this is found by plotting the residuals of the observed points against concentration. Where the variance rises at the lower concentrations, the calibration model does not hold, as the data are now heteroscadastic.

However, if this situation occurs, the problem is usually due to the highest concentration values having a bigger influence than the lower concentration points. Thus, small errors in the highest standards will distort the calibration curve and cause problems with measuring low concentrations. This problem can be overcome by weighting the curve, usually with factors of x^{-1} or x^{-2}. This reduces the influence of the higher value standards and is better for wide dynamic range assays. Confirming that the model holds should be undertaken during method development and is done by plotting the residuals to show that the scatter is constant over the dynamic range of the method.

References

1. C. Burgess, D.G. Jones and R.D. McDowall, *Analyst*, **1998**, *123*, 1979.
2. P. Bedson and M. Sargent, *Accreditation and Quality Assurance*, **1996**, *1*, 265.
3. M. Freeman, M. Leng, D. Morrison and R.P. Munden, *Pharm. Technol. Eur.*, **1995**, 7(10), 40.

4. W.B. Furman, T.P. Layloff and R.T. Tetzlaff, *JAOAC Int.*, **1994**, *77*, 1314.
5. Analytical validation (CPMP guideline), in *The Rules Governing Medicinal Products in the European Community*, Vol. III, Addendum, July 1990. Commission of the European Communities, Office for the Official Publications of the European Communities, Luxembourg, 1990.
6. *Guidelines on the Quality Safety and Efficacy of Medicinal Products for Human Use*, Commission of the European Communities, Office for Official Publications of the European Communities, Luxembourg, 1990.
7. ICH guideline: Validation of analytical procedures: definitions and terminology, in *The Rules Governing Medicinal Products in the European Community*, Vol. III, Part 1, 1996, *Guidelines on the Quality Safety and Efficacy of Medicinal Products for Human Use*, Commission of the European Communities, Office for the Official Publications of the European Communities, Luxembourg, 1996.
8. ICH guideline: Validation of analytical procedures: methodology (adopted December 1996), Commission of the European Communities, Office for the Official Publications of the European Communities, Luxembourg, 1996.
9. G.P.R. Carr and J.C. Warlich, Analytical validation, in *International Pharmaceutical Product Registration Aspects of Quality Safety and Efficacy* (eds. A.C. Cartwright and B.R. Matthews), Ellis Horwood, London, 1994.

7 Faster gas chromatography

Carl A. Cramers, J.G.M. Janssen,
M.M. van Deursen and J. Beens

7.1 Introduction and scope

Minimal-time operation in gas chromatography (GC) has been a research topic ever since the introduction of open tubular columns more than 40 years ago. Giddings [1] proved theoretically and Desty *et al.* [2] demonstrated experimentally that fast separations can be achieved by vacuum-outlet operation and by applying narrow-bore columns, respectively. The lack of adequate instrumentation hindered the application of high-speed separations in daily practice in the following years.

Recently, a revival of interest can be observed that is driven by such applications as process control, high-throughput analysis or simply increasing the speed of routine analyses. In the meantime the instrumentation has improved and now offers the necessary small time constants for detection and data acquisition and allows for higher temperature programming rates. At the same time, capillary column technology has become mature.

7.2 Options for increased speed of separation

Speeding up the separation, the main subject of this chapter, can be carried out in various ways. The first step should always be to see whether trading-in resolution for time is allowed; analysis time is very strongly dependent on resolution. This concept is treated extensively in the section dealing with guidelines for retention time reduction.

If this approach is impossible, the concept of resolution normalised conditions comes into play. It includes comparison of capillary, multicapillary and packed columns, based on reduced (dimensionless) parameters. This treatment is described in full detail in a recent review article by the authors of this chapter [3].

After this theoretical introduction, several state-of-the-art options for fast GC will be described in more detail:

- narrow-bore columns, and sample capacity-related problems
- multicapillary columns
- vacuum outlet gas chromatography

- fast temperature programming
- comprehensive two-dimensional gas chromatography

7.2.1 Resolution normalised conditions

In order to compare the kinetic conditions for the optimum speed attainable with different column types, it is essential to start with the concept of resolution normalised conditions. This means that in the comparison the required number of theoretical plates N_{req} (equation (7.9) below) for a given separation of a critical pair is considered to be constant.

Relative band broadening in chromatography is expressed in terms of the height H of a theoretical plate. The apparent plate height is the measured value, given operationally as

$$H = L\frac{\sigma^2}{t_R^2} \tag{7.1}$$

where L is the column length, t_R is the retention time of a solute, and σ^2 is the variance of the eluting peak. The quotient $(t_R/\sigma)^2$ is called the plate number: $N = L/H$.

Defining d as the inner diameter d_c of open-tubular columns, or the particle size d_p in packed columns, the dimensionless, reduced parameters in Table 7.1 can be defined, where u_o and $D_{m,o}$ represent the linear carrier gas velocity and the binary solute/carrier gas diffusion coefficient at column outlet pressure p_o; p_i is the column inlet pressure; K is the distribution coefficient; and V_s and V_m are the volumes of the stationary and mobile (gas) phase, respectively. The values of the reduced parameters at optimum chromatographic conditions are given in Table 7.2.

The column hold-up time is

$$t_m = \frac{L}{u} = \frac{NH}{u} = \frac{NH}{fu_o}$$

Table 7.1 Definition of reduced parameters

Reduced plate height	$h = \frac{H}{d}$
Reduced linear carrier gas velocity	$v = \frac{u_o d}{D_{m,o}}$
Reduced pressure	$P = \frac{p_i}{p_o}$
Retention factor	$k = K\frac{V_s}{V_m}$
Column resistance factor	Φ

Table 7.2 Approximate values of reduced parameters at optimum chromatographic conditions (minimum plate height or maximum plate number)

Parameter	Packed column	Capillary column
Reduced plate height, h_{min}	2	0.8
Reduced gas velocity, v_{opt}	3	5
Column resistance factor, Φ	~ 1000	32

where u is the average linear carrier gas velocity and f is the James–Martin gas compressibility correction factor

$$f = \frac{3(P^2 - 1)}{2(P^3 - 1)} \tag{7.2}$$

The retention time of a solute, $t_R = t_m(1 + k)$, can be expressed in reduced parameters as follows:

$$t_R = \frac{hNd^2}{vfD_{m,o}}(1 + k) \tag{7.3}$$

For laminar viscous flows of ideal gases the Hagen–Poiseuille equation reads

$$v = \frac{d^3 p_o}{2\Phi \eta L D_{m,o}}(P^2 - 1) = \frac{d^2 p_o}{2\Phi \eta N h D_{m,o}}(P^2 - 1) \tag{7.4}$$

in which η is the dynamic carrier gas viscosity. For high-pressure drop columns ($P \gg 1$), as normally encountered in high-speed GC-MS or vacuum-outlet GC in general, equation (7.4) can be rewritten as

$$v = \frac{d^2 p_i^2}{2\Phi \eta N h p_o D_{m,o}} \tag{7.5}$$

Making use of the fact that f approaches $3/2P$ for high pressure drops, substitution of equation (7.5) into (7.3) yields

$$t_R = \frac{4N^2 h^2 \eta \Phi}{3 p_i}(1 + k) \tag{7.6}$$

The product $h^2\Phi$ is called the separation impedance.
    Solving for p_i from equation (7.5),

$$p_i = \frac{(2Nhv\Phi \eta p_o D_{m,o})^{1/2}}{d} \tag{7.7}$$

and substitution in equation (7.6) finally gives

$$t_R = \left[\frac{8}{9} N^3 \frac{h^3 \Phi}{v} \frac{\eta}{p_o D_{m,o}} \right]^{1/2} d(1 + k) \qquad (7.8)$$

A similar equation for packed columns was derived by Knox and Saleem [4].

The number of theoretical plates required for a given separation problem can be calculated from the well-known resolution equation:

$$N_{req} = \left[4 \frac{k+1}{k} \frac{\alpha}{\alpha - 1} R_s \right]^2 \qquad (7.9)$$

where $R_s = (t_{R,2} - t_{R,1})/4\sigma$ is the resolution between two subsequently eluting peaks, σ is the average standard deviation of the two peaks, k is the retention factor of the last eluting compound, and $\alpha = k_2/k_1$ is the relative retention of the two solutes.

Equations (7.8) and (7.9) clearly show all factors affecting the speed of analysis. For multicomponent mixtures, as encountered in practice, equation (7.9) is still largely valid, since the 'critical pair' of peaks to be separated is similar to the two-component mixture to be resolved.

Equations (7.8) and (7.9) explicitly show all parameters affecting analysis time for both packed and (multi) capillary columns for high pressure-drop conditions as almost always encountered in fast chromatography.

Conclusions (from equations (7.8) and (7.9))
Carrier gas.

$$t_R \propto \left[\frac{\eta}{p_o D_{m,o}} \right]^{1/2}$$

This gas phase term has a minimum value for hydrogen as carrier gas, with helium as second best (about 60% slower). Other gases are at least two times slower.

Plate number, retention factor and relative retention. Applying equations (7.8) and (7.9) it appears that the analysis time is very dependent on the required plate number ($N_{req}^{3/2}$). Extremely fast analysis can only be obtained if N_{req} is small. Minimisation of N_{req} is therefore always a primary task; this is done by proper stationary phase and column temperature selection ($k_{opt} \approx 2$). Maximising the relative retention α by a proper choice of the stationary phase is a primary task for obtaining short analysis times.

Particle size, column diameter. Decreasing the particle size in packed columns or the column diameter of open-tubular capillaries is an effective way to speed up the separation process. Because the optimum reduced plate heights h_{min} have fixed values (Table 7.2), a decrease of d results in a proportionally decreased value of $H_{\min}(H = hd)$. Therefore, the column length $L = NH$ can be decreased by the same factor in order to yield the same plate number N. For example, a 5 m × 0.05 mm i.d. capillary column has about the same theoretical plate number as a 25 m × 0.25 mm i.d. column.

Vacuum outlet. Although not explicitly shown by the equations presented, faster separations can also be achieved by applying vacuum outlet to the column [5]. As mentioned before, $f \to 3/2P$ for high pressure-drop columns. Substitution of this expression in equation (7.3) yields $t_R \propto p_i$. Therefore, for any given column operated under optimum chromatographic (minimum H) conditions, vacuum-outlet operation is always the ultimate optimum, because this corresponds with the lowest possible optimum inlet pressure. The gain in speed of analysis by vacuum-outlet operation of short wide-bore columns is quite substantial and can be used to advantage for nondemanding separations. However, the gain in speed by vacuum-outlet operation is rapidly decreased by application of higher pressure-drop (narrower bore) columns.

Comparison of packed and open tubular columns. Equations (7.6), (7.7) and (7.8) allow the comparison of packed and open-tubular columns for a given separation problem under normalised conditions, that is for the same plate number, solute and carrier gas (N, k, η, p_o, $D_{m,o}$ and T constant). Using the (optimum) parameter values from Table 7.2, the reduced term $(h^3 \Phi / V)^{1/2}$ in equation (7.8) amounts to 52 for packed and 1.8 for capillary columns, both employed under optimum chromatographic conditions (Table 7.2). This means that equal analysis times can be obtained by capillary and packed columns whenever $d_c = 29d_p$. Introducing these values (d_p and $29d_p$) in equation (7.6) directly shows about a 200-fold increased inlet pressure for packed columns compared to open-tubular ones for equal analysis times. This factor, the ratio of the separation impedances $h^2 \Phi$, for packed columns ≈ 4000 and for capillary columns ≈ 20 (Table 7.2).

Temperature-programmed conditions. In the foregoing treatment analysis time minimisation was discussed for isothermal analysis. In an extensive treatment Schutjes *et al.* [6] record that under resolution normalised conditions the temperature programming rate $r = \partial T / \partial t$ is proportional to $1/t_{m,iso}$.

Consequently, in linear temperature-programmed analysis the programming rate has to be increased significantly when using narrow-bore (or small-particle) columns as compared to standard ones, requiring instrumentation offering these high programming rates (e.g., by resistive heating).

Sample capacity. Another important factor is the effect of characteristic diameter reduction on the sample capacity of fast columns. This is not a problem for packed columns; the speed of analysis is related to d_p, but the sample capacity can be varied at will by increasing the column diameter, a large advantage in this respect for packed columns. In open tubular columns the speed as well as the sample capacity is related to the inside column diameter d_c, as will be discussed later.

7.2.2   Guidelines for retention time reduction

In the theory section, equations were derived that describe the influence of various parameters on the analysis time. From the equations (7.8) and (7.9) it is evident that there are basically three ways to reduce the analysis time of a given GC separation. First, one should carefully minimise the resolution. Only those peaks that are really important should be separated. In doing so, one should not target a higher resolution than is strictly needed. A resolution of 1.0 might be sufficient for a quantitative analysis if a not too high degree of accuracy is needed. Resolution of 1.5 is sufficient for all analyses, even those requiring the utmost accuracy. When optimising resolution, the target should be 'the lowest possible resolution for the lowest possible number of peaks'. The second option for speeding up a separation is through maximising the selectivity of the stationary phase. Select the stationary phase, or combination of stationary phases, that is best capable of distinguishing between the various analytes in the sample. In addition to stationary phase selection, the selection of the temperature (program) is also important as this also affects selectivity. Finally, the kinetics of the separation process can be optimised to result in faster radial equilibration. The use of columns with reduced inner diameters, vacuum outlet operation or the use of hydrogen as the carrier gas are examples of this route.

This section focuses on the practical implementations of the theoretical equations derived earlier. Guidelines will be given that assist the analyst in answering the question of which approach for faster analysis should be selected for a given separation. Two steps can be distinguished in this process. In the first step the resolution is brought back to the minimum required level. Simple methods for achieving this are the use of a shorter column, a gas velocity above optimum or, for example, faster temperature programming. Significant time savings are possible in this way. If a further reduction of the analysis time is required, a method that minimises time at constant resolution should be implemented. Especially for the latter situation, the equations derived earlier can be of great help.

In the following sections we address the various options for reducing the analysis time in GC in a more qualitative way. An exhaustive overview of the various approaches for reduction of analysis time in GC is given. Advantages and disadvantages of the various options are compared. By combining this overview of options for faster analysis with a system for categorising chromatograms, we will be able to provide general guidelines on which 'enhanced speed option' is preferable for a given separation problem. The discussion presented below centres around a table listing the various methods for faster analysis, and a method for the classification of chromatograms recently developed in our laboratories. A strategy for selecting the preferred method for speed optimisation for a given separation is obtained by combining the table with options for increased speed with the chromatogram classification system.

Before we start to discuss the various options in more detail, one final remark is appropriate. In the entire analytical procedure the actual separation of the sample is only one aspect. Significant time savings are also possible in the sample preparation part of the protocol, for example, or in the interpretation of the data. Discussion of improvements in these areas is outside the scope of the present chapter; we restrict ourselves to the separation part. Finally, a chromatographer should always bear in mind that other, non-chromatography-based, methods might be better suited to resolve the particular problem at hand.

7.2.2.1 Methods for speeding up separations
In recent years numerous methods for speeding-up GC separations have been proposed, illustrating the great attention this subject has attracted. Potential methods to reduce the time of separation include

- use of shorter columns
- use of open columns with reduced inner diameter
- use of faster temperature programming
- use of pressure/flow programming
- use of (micro) packed columns with small particles
- application of more selective stationary phases
- use of coupled columns or two-dimensional chromatography
- use of hydrogen as the carrier gas
- operation of the column at above-optimum linear velocities

The wide variety of potential approaches for faster GC seriously complicates the situation for the analyst interested in reducing the analysis time of the application. The question of which approach to select is clearly not trivial. This part of the chapter tries to assist the analyst in selecting the best approach. The discussion of options is limited to open-tubular columns.

Unfortunately, there are no (or almost no) universal solutions to the enhanced speed problem; that is no single method will result in a significant time reduction for all applications. The use of columns with reduced inner diameter is coming close to being a universal solution, but in many cases there are other options for faster separation that are to be preferred over the use of narrow-bore columns. In particular this is the case for separations that are over-resolved. Very often options that are excellent for speeding up a particular GC application can lead to no gain or an unacceptable resolution loss in other situations. Potential methods for reducing the analysis time in GC are listed in Table 7.3.

Table 7.3 Options for speeding up a capillary GC separation

Solution number	Option	Effect on resolution (all other parameters constant unless mentioned otherwise)	Effect on elution time	Remarks	Selected references
1	Shorter column length	Loss of resolution	I-GC[a]: reduced proportionally to column length TP-GC[b]: slightly reduced	Reduce inlet pressure	[7]
2	Higher carrier gas velocity (above Golay optimum)	Loss of resolution	I-GC: reduced inversely proportionally to u TP-GC: slightly reduced	Keep split ratio constant	[8]
3A	Higher isothermal temperature (I-GC)	Loss of resolution	t_R reduced by factor 2 for ca. every 15°C T increase	–	
3B	Higher initial temperature (TP-GC)	Loss of resolution	Gain generally small (typically a few minutes)	–	
3C	Higher final temperature (TP-GC)	Loss of resolution (in later part of chromatogram only)	Gain typically a few minutes, unless peaks elute late in the final isothermal stage	Do not exceed the maximum operating temperature	
4	Change temperature to obtain a retention factor of 1.7–2 (I-GC)	Constant resolution	Adjust temperature to elute critical pairs at a retention factor around 1.7–2. Practical applicability limited to narrow boiling range samples	–	[9,10]
5A	Faster temperature programming	Usually loss of resolution. Ocasionally a gain possible for two distinctly diffifrent analytes	Significant gains possible, especially if analytes elute over a wide temperature range	Retention time repeatability might be adversely affected	
5B	Convert an I-GC separation to a TP-GC method	(Generally) resolution is lost	Very significant time reductions possible	–	[11]–[14]
6	Pressure/flow programming	Usually slight loss of resolution	Some time gain possible due to faster elution of highest-boiling analytes	–	
7	Lower film thickness	Loss of resolution (T constant)	Real gain only if going from a thick film to a normal film	Adjust injected amount/split ratio to	[15,16]

			column. Otherwise small gain only (similar to option IV)	avoid overloading	[17]
8	Use of hydrogen as the carrier gas	Resolution constant (adjust velocity and T programming rate accordingly)	Gain of factor 1.5–2	For safety reasons use a hydrogen monitor detector	[18]
9	Reduce column inner diameter	Resolution constant (adjust velocity, column length and temperature programming rate)	Time savings at least proportional to diameter reduction. Even higher gains possible	Adjust inlet pressure, length and injected amounts. Minimise dead volumes/extra column peak broadening	[6]
10	Use more selective stationary phase or apply coupled columns.	Significant resolution improvements possible	Significant gains possible	–	[19,20]
11	Apply vacuum outlet conditions	Resolution constant	Time savings up to a factor 5 possible for short wide-bore columns Effect negligible for standard columns	Ideal in case of MS detection	[21]
12	Apply back flush	Resolution constant	Gain only if sample contains very high-boiling analytes that are not important	Requires special instrumentation	
13	Use selective detection	Dramatic improvements possible, but applicability very 'problem dependent'	Large reductions possible	–	
14	Apply mass spectrometric detection	Use of deconvolution software or selected ion chromatograms can result in large improvements	Large reductions possible	–	[22]
15	Use (conventional) two-dimensional GC	Significant improvements possible for a limited number of peak pairs	Overall time generally (slightly) longer	Requires dedicated instruments	[23]
16	Apply turbulent flow conditions	Practical applicability limited Not a relevant option	Very fast separations possible, but only for weakly retained species	Pressure drop extremely high	[24,25]

a I-GC, isothermal GC.

b TP-GC, temperature-programmed GC.

Comprehensive two-dimensional GC is a new mode of two-dimensional chromatography with an extremely high resolving power. It is discussed in more detail later in this chapter and in Chapter 8.

Table 7.3 provides a more or less exhaustive overview of the various options for reducing the analysis time in capillary GC. More detailed (theoretical) backgrounds of the various options can be found in the earlier sections of this chapter or in the references cited in the last column of the table. The actual instrumental consequences of speeding up a separation will be discussed under 'Practical consequences of fast GC' later. Irrespective of which route is selected, the detector, injector and data system will have to perform to certain criteria in order to benefit fully from the potentials of a given enhanced-speed method. The consequences of changes in the method intended to reduce the analysis time on other parameters such as detection limits and loadability will also be briefly addressed.

The most important question to be answered when considering the various methods for speeding up a separation is whether the separation is performed at optimal resolution. If a resolution loss can be tolerated, selecting a system that provides less (but still sufficient) resolution in a shorter time is the first step on the route towards faster analysis. If the analysis time in the minimum acceptable resolution situation still exceeds the desired or permitted time, other options that reduce the analysis time at constant resolution can be exploited. The strategy for analysis time reduction in GC hence should be as follows.

1. Minimise the resolution to a value that is just sufficient.
    If a further reduction is necessary:
2. Implement a method that reduces analysis time at constant resolution.

Excellent options for reducing the analysis time at the expense of resolution are the use of a shorter column (solution 1 in Table 7.3), operating the column at a higher carrier gas velocity (solution 2) or the use of a higher isothermal temperature (in isothermal GC) or a higher initial and/or final temperature in temperature-programmed separations (solutions 3A–C). The exact gains of these alterations in the chromatographic conditions are not always easy to predict. As an example, a reduction of the column length in isothermal GC will result in a proportional reduction of the analysis time (and a concomitant square-root loss of resolution). In a temperature-programmed separation in which a wide temperature range is covered, however, a length reduction will only marginally affect the elution time of the last-eluting compound. Here the total analysis time is determined by the time it takes for the oven to reach a temperature sufficiently high to impart a sufficient vapour pressure to the last component. In the latter case faster temperature programming is a rewarding option for reducing the analysis time. Column length clearly is not. Faster programming (solution 5A) will virtually always result in a loss of resolution. This is because the same number of peaks is now forced into a narrower time window. The same is the case when an isothermal separation is transformed into a temperature-programmed method (solution 5B). This can also result in a significant time saving, but again it can only be applied to 'over-resolved' chromatograms. An additional disadvantage of fast programming is the sub-stantially higher elution temperatures of the analytes [26]. Two other methods that can be used to reduce the analysis time of chromatograms that are 'over-resolved' are flow/pressure programming (solution 6) and the use of a column with a lower film thickness (solution 7). Again the benefits of flow/pressure programming are restricted to isothermal separations (or isothermal final parts of a temperature-programmed run). In temperature-programmed analyses the effect of flow is generally negligible relative to the effect of temperature.

When looking to the influence of film thickness on analysis time, two dis-tinctly different situations can be distinguished. If thick-film columns (column diameter divided by four times the film thickness < 100) are used, potential speed and resolving power are lost. This does not mean that there are never good reasons for using columns coated with thick films of a stationary phase. Reasons to opt for the use of thick-film columns could be to provide sufficient retention at ambient temperature or to improve the sample capacity of the system. An enhanced sample capacity might be desired when limited-sensitivity detectors, such as the Fourier transform infrared (FT-IR) detector, are used. If a thick-film column is replaced by a normal film column, resolution improves while simultaneously the analysis time is reduced. If a normal/thin-film column is already in use, a further reduction of the film thickness will reduce the analysis time, but generally only at the expense of resolution. Basically this variation in experimental conditions is very much comparable to operating the column at a higher temperature.

The different methods described above share one aspect: they all result in a loss of resolution. In more positive terms, time is gained by minimising excess resolution. This is clearly not a problem for situations in which each peak is surrounded by baseline. However, if analysis time has to be reduced without sacrificing resolution, other options have to be exploited. Two generally appli-cable options for reducing the analysis time at constant resolution are the use of hydrogen as the carrier gas (solution 8) or the use of columns with a reduced inner diameter (solution 9). The physical backgrounds for the faster analyses obtained with such systems were discussed in detail in the earlier sections: radial equilibration is enhanced owing to the higher diffusion coefficients or the reduced diffusion distances. An alternative method for enhancing radial equilibration is the use of vacuum outlet conditions (solution 11). As previously shown, this option is only valid for columns with a low resistance to flow, that is short wide-bore columns (e.g. length 10 m, inner diameter 0.53 mm). A final method that is based on enhanced radial transfer is turbulent flow GC (solution 16). This method unfortunately is of little or no use for daily practice. In case the chromatogram contains only one region with a critical pair/group of peaks, it is recommended to adjust temperature to elute this critical region at a retention factor of around 1.5–2 (solution 4). A very important option for faster analysis that has been somewhat neglected since the advent of (fused-silica) capillary columns is the use of more selective stationary phases or series-coupled columns of different polarity (solution 10). In recent decades column manufacturers (and users!) have tried to solve all chromatographic problems according to a 'one type suits all' approach. Indeed, long and narrow nonpolar columns provide a wealth of plates, but the use of columns of a carefully tuned selectivity might be a faster route to realise the desired resolution.

Series-coupled columns with control of the mid-point pressure are the ultimate system as they allow continuous control of the column selectivity [27]. (Conventional) Two-dimensional chromatography is a very sophisticated and very powerful form of coupled column separations (solution 15) and is an excellent method for speeding up very complex separations. Unresolved peaks can be transferred to a second column for further separation on a different stationary phase. Significant time savings are possible, but the instrumentation required is expensive and more difficult to operate than standard GC equipment. A very sophisticated form of two-dimensional GC is a recently introduced technique known as comprehensive two-dimensional GC (GC × GC) [28]. This technique is discussed in more detail in later sections and is not included in the present discussion. A final example of an instrumentally more demanding way to reduce the analysis time is the back flush technique (solution 12). This method is attractive for samples that contain large amounts of unimportant high-boiling material. In standard chromatographic separations the elution of this material would require lengthy conditioning of the column at the final temperature of the temperature program. Using column back flush, this problem can be circumvented.

In the discussion of the various methods for faster analysis presented above, it has been assumed that all peaks in the chromatogram are equally important. Very often this will not be the case. As an example, in the analysis of pesticides in an environmental extract most of the compounds often represent matrix interferences. If a universal, nonselective detector is applied to such samples, each of the compounds of interest has to be resolved from the other pesticides and from the co-extracted interferents. In such cases the required resolution can be significantly reduced by using a selective detector (e.g. a nitrogen phosphorus detector or a sulfur chemiluminescence detector). When using such a detector, the pesticides of interest have only to be separated from each other. A full separation from the matrix compounds is no longer necessary, unless the matrix compounds exhibit some response on the selective detector or interfere with the detection principle. The use of selective detectors can hence be an attractive option for faster separation (solution 13). Analogous to a selective (element) detector, a mass spectrometric detector can also provide the required additional selectivity and hence reduce analysis time (solution 14). Partially resolved peaks can be 'separated' through mass spectral deconvolution, the only requirement being that the co-eluting peaks contain one or more unique mass fragments. When considering these options for faster analysis, it is important to bear in mind that most of the selective detectors have limited selectivity.

Moreover, detectors such as the flame photometric detector are notorious for their quenching problems.

7.2.2.2 Chromatogram classification system

Capillary gas chromatography is routinely applied to an extremely wide range of analytical problems in an equally diverse range of application areas. This results in an almost infinite number of different chromatograms, so that it is impossible to discuss how to speed up each of these individual separations. Fortunately, when taking a closer look at the full range of GC applications, it can be seen that the various chromatograms tend to cluster and groups of more or less similar chromatograms can be distinguished. This observation inspired us to design a system for the classification of chromatograms. Chromatograms that exhibit similar characteristics are clustered into specific classes. Next, for each cluster, guidelines can be given on how to speed up that specific group of separations.

In our chromatogram classification system the two most important degrees of freedom are the number of peaks and the difference in boiling point between the first- and the last-eluting peaks. The latter parameter is nicely reflected in the temperature range that has to be spanned for full elution of the sample. Having defined two axes for the chromatogram classification (number of peaks and temperature range), we can identify four different 'extreme chromatograms'. This is illustrated schematically in Figure 7.1.

The lower left corner represents a chromatogram with only a few peaks of compounds that have boiling points more or less in the same range. A typical example might be the separation of benzene, toluene and xylene (BTX). A second extreme would be a small number of peaks with large differences in boiling point (lower right corner). An example of this is given in Figure 7.2, which shows the GC separation of four ethylene glycol oligomers. A third extreme is a sample in which there are a large number of peaks with only minor differences in boiling point. Here a typical example would be a narrow distillate fraction from a mineral oil. Finally, the fourth extreme would be a chromatogram showing a very large number of peaks that differ widely in boiling point. Typical examples here are essential oils, flavour extracts, complex environmental extracts and (total) mineral oil samples. The ultimate example of an upper right corner chromatogram is the chromatogram of a simulated distillation. An infinite number of components covering an elution range as high as 450°C are eluted to determine the boiling range of the heavy oil product.

The four-chromatogram classification system described above clearly is too simple to allow accurate classification of the enormous range of GC applications. A further refinement is necessary. Normal chromatograms are often combinations of 'high peak density' areas with elution ranges in which the peaks are more homogeneously distributed over time. A second consideration is that in real chromatograms not all peaks may be equally important. An example of this

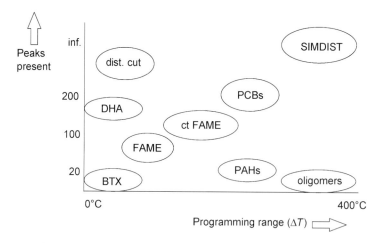

Figure 7.1 Schematic representation of the overall application range of gas chromatography. Abbreviations: dist. cut, narrow distillation cut from a mineral oil; SIMDIST, simulated distillation (thousands of compounds are eluted to characterize the boiling range of an oil); DHA, detailed hydrocarbon analysis (separation of some 200 C_1 to C_7 hydrocarbons); PCBs, polychlorinated biphenyls (209 isomers containing 1–8 chlorine atoms); FAME, fatty acid methyl esters (routine analysis to characterise edible oils and fats); ctFAME, *cis/trans* FAME separation (as FAME; for each double bond it is determined whether the bond is *cis* or *trans* configured); PAH, polyaromatic hydrocarbons (16 compounds are measured in most routine analyses); BTX, benzene, toluene, xylene (routinely measured as indicators for air pollution); oligomers, mono to (e.g.) hexamers of polystyrene, ethylene glycol, etc.

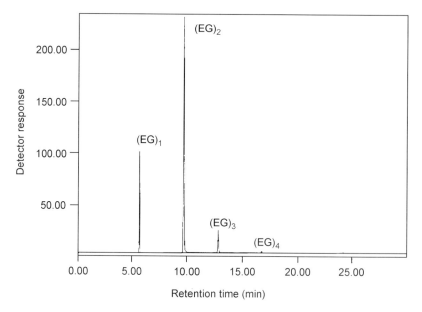

Figure 7.2 Example of one of the chromatogram classes. Separation of four ethylene glycol (EG) oligomers. Temperature program 40°C to 320°C at 15°C min^{-1}.

is the analysis of a few target compounds in a complex extract. On the basis of these considerations we identified nine basic classes of chromatograms. Two of these classes are divided into three subclasses. The resulting 13 basic GC chromatogram classes are shown schematically in Table 7.4. A brief description of the different classes is given below.

Type I   Experimental chromatograms are assigned a class I classification if there is only a limited number of peaks, well separated in an isothermal separation. The main characteristic of a class I chromatogram is that each peak is surrounded by large segments of baseline.

Type II   Class II chromatograms show a low number of well-resolved peaks next to one or more critical pairs/groups. Here also the separation is performed under isothermal conditions. The general type II class is subdivided into three subclasses, depending on whether the critical pair/group is positioned in the early part (IIA), the final part (IIC) or at multiple positions (IIB) in the chromatogram.

Type III Chromatograms of class III are crowded with peaks. There is hardly any baseline at all (isothermal conditions).

Type IV   Type IV chromatograms contain only a limited number of peaks and large segments of baseline. In this respect the chromatograms are similar to class I chromatograms, the difference being that the separation is now temperature programmed. The temperature range is relatively narrow ($\sim$100°C).

Type V   As type IV, but the temperature range now is large (e.g. 300°C).

Type VI   Comparable to type II chromatograms, but now the separation is temperature programmed.

Type VII Temperature-programmed analogue of type III.

Type VIII Similar to chromatogram classes III and VII. The chromatogram represents a crowded chromatogram with hardly any free baseline and lots of peaks. A distinct difference with the two other classes is that now only a limited number of peaks are relevant (target compound analysis in a complex matrix). In the present example the relevant compounds are the peaks A, B, C and D.

Type IX   In this class all the important information is in the first part of the chromatogram. After the important peaks have eluted, long conditioning at a high temperature is required to remove high-boiling compounds from the chromatographic pathway.

7.2.2.3   Selecting the optimum method for minimum-time operation
Now that we have discussed both the various options for speeding up GC separations and the system for the classification of chromatograms, we can

Table 7.4 Chromatogram classification system

Chromatogram type

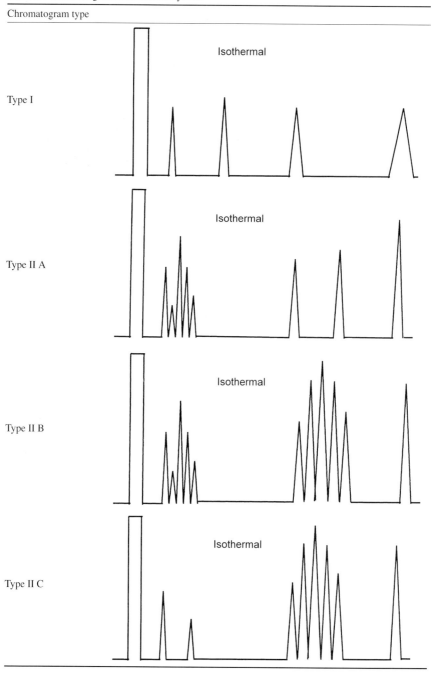

Table 7.4 (continued)

Chromatogram type

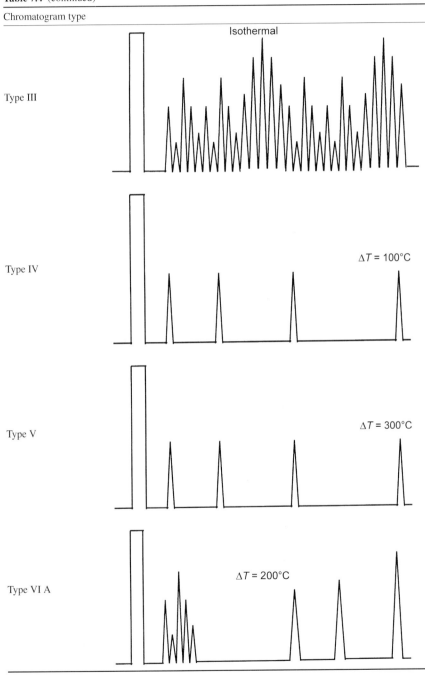

Table 7.4 (continued)

Chromatogram type

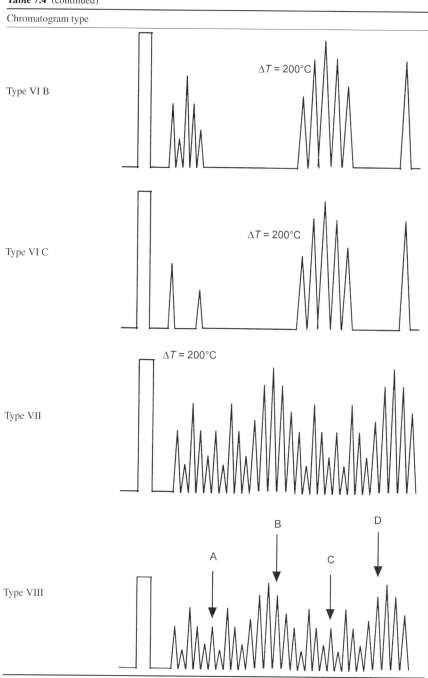

Table 7.4 (continued)

Chromatogram type

Type IX

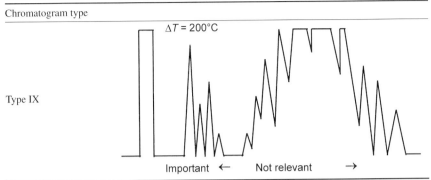

systematically judge the applicability of each of the options for each of the types of chromatograms. This process will be illustrated for two types of chromatograms selected more or less arbitrarily from Table 7.4. The results for the other chromatogram classes are summarised in Table 7.5.

Type I chromatograms contain a limited number of peaks, well separated under isothermal conditions. Each peak is surrounded by large empty segments of baseline. The type I chromatogram is clearly over-resolved. The first step in reducing the analysis time starts with trading-in resolution for time. Good options for achieving this are shortening the column length (solution 1), or operating the column at a much higher velocity (solution 2). Both options are viable but also have their limits. Very short columns would require very low inlet pressures. Eventually the required pressure might become too low, resulting in problems with pressure stability. The use of high carrier gas velocities, on the other hand, is limited by the maximum inlet pressure of the system. Other ways of speeding up a type I separation are increasing the isothermal temperature of the column (solution 3A), transforming the method into a temperature programmed separation (solution 5B), or using a thinner-film column at constant temperature (solution 7). Pressure or flow programming (solution 6) could be used as it also results in an increased gas velocity.

By incorporating one or more of the solutions discussed, the resolution can be brought back to the minimum and a significant time reduction is possible. Most of the other options in Table 7.4 would also result in a shorter analysis time for this type I separation. They are, however, clearly more difficult to implement and therefore should not be the first option considered for speeding up class I chromatograms. Once the various options for speeding up this type I separation have been fully exploited, the final chromatogram is basically no longer a type I chromatogram. It now belongs to class II, III, VI or VII depending on the way the resolution reduction was realised and the exact effects of the route selected on the elution times of the different peaks.

Table 7.5 Overview of preferred options to speed up a given GC separation

Chromatogram type	Increased speed option number																		
	s 1	s 2	s 3A	s 3B	s 3C	s 4	s 5A	s 5B	s 6	s 7	s 8	s 9	s 10	s 11	s 12	s 13	s 14	s 15	s 16
I	++	++	++	NA	NA	−	NA	++	+	+	+	+/−	−	+	−	−	−	−	−
IIA	−	−	−	NA	NA	+	NA	++	+	−	+	+	+	+/−	−	−	+ if …	+	−
IIB	−	−	−	NA	NA	+	NA	−	−	−	+	+	+	+/−	−	−	+ if …	+	−
IIC	−	−	−	NA	NA	+	NA	−	−	−	+	+	+	+/−	−	−	+ if …	+	−
III	−	−	−	NA	NA	−	NA	−	−	−	+	+	−	−	−	−	+ if …	−	−
IV	+	+	NA	+	+	NA	++	NA	+	+	+	+	−	+	−	−	−	−	−
V	−	−	NA	+	+	NA	++	NA	−	−	+	+	−	+	−	−	−	−	−
VIA	−	−	NA	−	+	NA	++	NA	−	−	+	+	+	+/−	−	−	+ if …	+	−
VIB	−	−	NA	−	+−	NA	−	NA	−	−	+	+	+	+/−	−	−	+ if …	+	−
VIC	−	+	+	+	−	NA	−	NA	−	−	+	+	+	+/−	−	−	+ if …	+	−
VII	−	−	−	−	−	−	−	NA	−	−	+	+	−	−	−	−	+ if …	−	−
VIII	−	−	−	−	−	−	−	NA	−	−	+	+	−	+/−	−	+	+ if …	−	−
IX	−	−	NA	−	+	NA	+	NA	−	+	+	+	−	−	+	−	−	−	−

++, Significant gain or very easy to achieve reasonable gain.
+, Reasonable gain or significant gain slightly more difficult to realise.
−, Not a viable option.
+/−, Only a significant option for low-plate number separations (medium-resolution GC) or in combination with MS detection for selected ion monitoring.
+ if …, Only a valid option if neighbouring peaks contain unique ions.
NA, not applicable.

If chromatograms of type III have to be made faster, the options 1 to 6 are generally of no use. They would all result in an unacceptable loss of resolution. The elution time of the last peak would be reduced, but clearly there is a difference in elution time and analysis time, a difference that is often overlooked in the literature. A reduction of the analysis time of a type III chromatogram can only be obtained through the implementation of one of the 'constant resolution' methods from Table 7.4. Valid options are the use of columns with a reduced inner diameter (solution 9), or using hydrogen as the carrier gas (solution 10). The use of a more selective stationary phase is probably of little benefit. The chromatogram has almost no empty baseline so that the use of another stationary phase would probably result in other co-elutions. Applying vacuum outlet conditions (option 11) could be attractive, but only if the chromatogram requires a relatively low plate number (<20 000) so that a short, wide-bore column can be used. Options 12 and 13 (back flush or selective detection) are of no use. Solution 14, on the other hand, might be an attractive option. If all neighbouring peaks contain unique mass fragments, one could seriously consider applying one of the resolution loss options, such as reducing the column length and compensating the resolution loss though spectral deconvolution. Two-dimensional chromatography (option 15) is only suited if the sample contains one region or a few narrow regions with insufficient resolution. Finally, turbulent flow (option 16) can only be applied for chromatograms in which a limited number of peaks (typically less than three or four) elute at low capacity factors [25]. In Table 7.5 the applicability of each of the options for each of the classes is summarised.

7.3   Practical consequences of fast GC

In the previous sections, guidelines were presented on how to reduce the analysis time of given GC separations. The factors basically comprise column parameters (length, diameter, etc.) and operational settings (temperatures, flow, etc.). When implementing these methods a number of issues not directly related to the column might require special attention:

- pressure drop
- column loadability
- detection limits
- injection band width
- extracolumn band broadening
- time constant for detection
- elution temperature/maximum column temperature
- coupling to mass spectrometry

7.3.1   Pressure drop

Pressure might become a limiting factor for a number of options for faster GC. As an example, the use of shorter columns is limited by the minimum inlet pressure required for stable operation of the carrier gas system. In the case of long narrow-bore columns or turbulent flow operation, problems might be encountered at the high-pressure end.

7.3.2   Column loadability

Especially in the case of narrow-bore columns, column loadability or sample capacity is an important factor to consider. In an extensive theoretical and practical study on the sample capacity of open-tubular columns Ghijsen *et al.* [30] concluded that: the maximum sample capacity, C_{max}, for columns with an equal phase ratio leading to maximally 10% peak broadening is given by

$$C_{max} \propto \beta d_c^3 \qquad (7.10)$$

In this equation β is a proportionality factor: $0.05 < \beta < 1.8$; $\beta \approx 1.8$ for solutes and stationary phases with similar properties and $\beta \approx 0.05$ for solutes and stationary phases of dissimilar structures (e.g. alcohols as solutes on a nonpolar poly(dimethyl siloxane phase). According to equation (7.10) for narrow-bore columns, the sample capacity is drastically reduced. Overloading, manifested by leading peaks, will occur even at minute sample quantities.

7.3.3   Detection limits

The minimum detectable amount (Q_0), the lowest quantity of solute that can be distinguished from the noise, is given by

$$Q_0 = \sqrt{2\pi} \frac{4R_n}{S} \sigma_t \qquad (7.11)$$

where R_n is the detector noise, S is the detector sensitivity and σ_t is the total band width. The minimum detectable amount is favoured by a reduction of the inner diameter ($Q_0 \propto d$ [31]) and basically by any other method that results in faster analysis. This is caused by the decrease of σ_t when the analysis time is reduced. The minimum detectable concentration, on the other hand, is not always improved when methods for faster chromatography are implemented. This because the injection volume might have to be reduced to avoid an excessive contribution of the injection to the overall peak width. Whereas this effect is marginal for most of the options for faster analysis summarised in Table 7.3, the minimum detectable concentration increases dramatically when working with columns with a reduced inner diameter. Narrow-bore columns therefore are not suitable for trace analyses. Also, detectors have to be very sensitive to detect the low quantities eluting from a narrow-bore column.

7.3.4   Injection band width

To minimise the contribution of the input band width to total band broadening, the injected sample plug has to be narrow in comparison to the total chromatographic band broadening. For faster analyses the residence time of the components in the column is reduced. As discussed above, this results in a reduced widening of the chromatographic zone. Injection therefore becomes more critical. This is especially true for isothermal analyses. In temperature-programmed separations, zone focusing will occur in the column inlet.

The theoretical value for the contribution of a plug injection to band broadening (isothermal column operation) is given by

$$\sigma_i = \frac{w}{\sqrt{12}} \tag{7.12}$$

where w is the width of the injected plug. Typical injection band widths (σ_i) required for very fast analyses on short narrow-bore columns are approximately 1–3 ms [32]. Especially for the narrow-bore option for fast GC injection band-width is critical. It is for this reason that the development of injection systems compatible with narrow-bore GC has received considerable attention in the literature.

7.3.5   Extracolumn band broadening

The overall band broadening is determined by the sum of the contributions of the column, injection, detection and other (extra)column effects. Because peak widths in fast GC are narrow, extracolumn contributions of, for example, press-fits or other column connections have to be minimised. This puts stringent demands on the coupling of columns, or the coupling to detection and injection devices. Again, especially in the case of narrow-bore columns operated under isothermal conditions, dead-volume-free operation is critical. Because of the low column flow in narrow columns, even tiny dead volumes can result in a significant zone broadening. Evidently, the dead volume problem is less significant when considering columns at high flow rates or in cases of temperature programmed operation.

7.3.6   Time constant for detection

To be able to describe a peak accurately in fast GC, fast detection is required. For peaks that are approximately 1 s wide at baseline level, a data acquisition rate of 10 points s^{-1} is required. Also, the overall time constant of the detector should be at least ten times smaller than the width of the peak. If this requirement is met, the peak height is reduced by less than 0.5%. When considering detector time constants it is important to realise that there are basically two contributions that together make up the overall time constant of the detector. The first is the volumetric time constant. This is more or less the time required to fill the sensing volume of the detector and the volume between the column outlet and the actual location of measurement. In addition, there is an electronic time constant, a delay time associated with the (parasitic) capacity of the electronics. The first contribution can be minimised by the addition of make-up gas. When implementing a method for faster GC it might be necessary to adjust the make-up flow. If one is working with a concentration-sensitive detector this might adversely affect the detection limits. The second contribution cannot be affected by the instrument user. The user should make sure that the particular instrument is suited for fast GC.

7.3.7   Elution temperature/maximum column temperature

The temperature at which a solute elutes from a temperature-programmed GC column is determined by the temperature programming rate among other factors. A rule of thumb is that the optimum temperature programming rate is $10°C$/void time [33]. At higher rates, not only will the resolution decrease but the elution temperature will increase. This means that when analysing samples with high final boiling points, where the maximum allowable temperature for the stationary phase is necessary to elute the highest-boiling analytes, increasing the temperature programming rate is not an option. A second drawback of rapid programming is that the increased elution temperatures may affect samples that contain thermally labile analytes.

7.3.8   Coupling to mass spectrometry

An important trend in GC is the ever-increasing need for positive identification and the need for more flexible systems that allow the analysis of a wide variety of samples on one system. These last two trends clearly result in a strong requirement for mass spectrometric detection. Combination of fast GC with mass spectrometric detection is by no means trivial. Several types of mass spectrometers are available for coupling to GC systems. These systems differ in the way that the ion fragments, formed from the molecules eluting from the GC column, are separated according to their mass. Important mass analysers are the ion trap, the sector instrument, the quadrupole and the time-of-flight mass spectrometer. The performances of each show differences in terms of acquisition rates, detection limits, mass spectrometric resolution and quality of the mass spectra obtained. The choice of the most suitable mass analyser is very much dependent on the composition of the sample, the detection limits and the speed of separation.

For an accurate description of a chromatographic peak in a chromatogram, at least 15–20 data points across a peak are required [34]. Typical acquisition rates of scanning mass spectrometers like the ion trap, the quadrupole and the sector instrument range from 10 to 20 spectra per second in the full-scan mode.

Table 7.6 Comparison of calculated [71] plate number, retention time and peak width at baseline level for a standard, fast, very fast and ultrafast separation. Conditions for these calculations are $\beta = 62.5$, compound = hexane, $T = 330$ K, carrier gas = helium, peak width at half-height (2.354σ)

Type of analysis	Column diameter (mm)	Length (m)	Plate number	Retention time (s)	Peak width at half-height (s)
$P_{out} = 100$ kPA					
Standard	0.32	25	90 000	160	1
Fast	0.05	10	260 000	60	0.2
Very fast	0.05	1	25 000	2.0	0.03
Ultrafast	0.05	0.3	7 000	0.40	0.01
$P_{out} = 0$ kPA					
Standard	0.32	25	75 000	100	0.7
Fast	0.050	10	230 000	60	0.2
Very fast	0.050	1	22 000	2.0	0.03
Ultrafast	0.050	0.3	6 300	0.30	0.01

Reprinted from [41] with permission of Elsevier Science.

Hence, only chromatographic peaks with a width at half-height of 0.5 s or more can be accurately represented.

Peak widths obtained from narrow-bore columns are usually slightly above 1 s. Scanning mass spectrometers are hence capable of offering the speed required for mass spectrometric detection.

In Table 7.6 peak widths are calculated for analyses on a standard column, a narrow-bore column and two extremely short columns. Here, the influence on band broadening of only the column itself is taken into account. From the table it is clear that for very fast separations on short columns the spectral acquisition rate of scanning mass spectrometers is too low. Time-of-flight mass spectrometers, however, can provide up to 500 full spectra per second.

7.4   Practical implementations of selected options for fast GC

7.4.1   Narrow-bore columns

It was understood in the early days of GC that miniaturisation is the obvious route towards faster separations. This holds both for packed columns, where smaller particles yield faster separations, and for open-tubular columns, where the analysis time can be reduced significantly by reducing the inner diameter of the capillary column. The use of narrow-bore open-tubular columns as an effective method for reducing the analysis time under constant resolution conditions has been described by numerous authors [2, 6, 35–41, 78, 79]. Numerous applications are described in these and other articles. Good results were reported, for example, in (petro-)chemical analysis, environmental monitoring, process control, flavour research and fat and oil analysis. A special area of application of narrow-bore columns, comprehensive two-dimensional chromatography, will be described below.

A wide range of narrow-bore columns have been used. The stationary phases used were mostly nonpolar phases, because great difficulties were encountered in coating highly polar phases into narrow-bore columns. Commercial instrumentation for the use of narrow-bore columns has become available only in recent years. Modern GC instruments allow operation at inlet pressures as high as 10 bar. Maximum programming rates for the oven are around $100°C\,min^{-1}$. Moreover, suitable injection and detection devices are now available for all but the fastest separations. Split, splitless, programmed temperature vaporisation and even on-column injectors have been modified to match the performance of fast narrow-bore columns [42]. Similar progress has also been made in the area of detection. Fast detectors incorporating fast electronics have been described by various authors. The range of detection devices studied includes flame ionisation detection [43], thermal conductivity detection [24, 25], electron capture detection [44], photoionisation detection [45] and various mass spectrometric detection devices [44]. A number of representative separations performed on narrow-bore columns are shown below.

Figure 7.3 shows the separation of a mixture of chlorinated organic solvents. The separation was performed on a 5 m, 0.05 mm column coated with $0.2\,\mu m$ of a CP-Sil 5 stationary phase. Detection was performed using an ECD operated at a very high make-up gas flow of $900\,ml\,min^{-1}$. The inlet pressure was 20 bar.

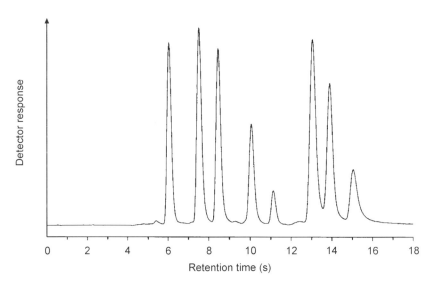

Figure 7.3 Analysis of a mixture of halogenated hydrocarbons. Column: 5 m × 0.05 mm i.d., coated with $0.2\,\mu m$ CP-Sil 5. Experimental conditions: $T = 110°C$; inlet-pressure = 20 bar; split-flow = $1000\,ml\,min^{-1}$; injection $20\,\mu l$ headspace; carrier gas hydrogen; detection ECD; make-up flow ECD = $900\,ml\,min^{-1}$. Components in order of elution: chloroform; 1-iodopropane; 1,1,2-trichloroethane; 1-iodobutane; 1,1,2-trichloropropane; tribromomethane; 1,1,2,2-tetrachloroethane; di-iodomethane [78].

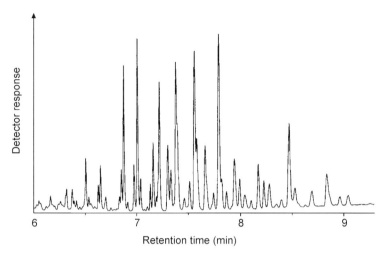

Figure 7.4 Analysis of a PCB extract (Arochlor 1260) of transformer oil. Column: 5 m × 0.05 mm i.d., coated with 0.2 μm CP-Sil 5. Experimental conditions: $T = 50°C$ (3 min) ballistically heated to 280°C; carrier gas hydrogen; detection ECD; make-up flow ECD = 400 ml min^{-1}; inlet-pressure = 12 bar, splitless time = 3 min, injection: 0.3 μl [78].

The carrier gas was hydrogen. The same column was used for the separation of a PCB extract (see Figure 7.4). The analysis time for this separation was 9 min compared to around 45 min on a standard-bore column.

Figure 7.5 shows the high-speed narrow-bore separation of a diesel oil sample. A prerequisite in the development of the fast capillary GC method for this application was the baseline separation of the biomarkers pristane and phytane from the preceding normal alkanes. The detector used for this analysis was a flame ionisation detector (FID). A second example of a fast separation with FID detection is given in Figure 7.6, which shows the analysis of a mixture of alkylaromatics on a 8.6 m, 0.046 mm SE-30 column.

An example of a fast separation using thermal conductivity detection (TCD) is shown in Figure 7.7. Here a mixture of eight organic solvents is separated in less than 2 min. A critical concern in the coupling of narrow-bore columns with TCD is the fairly large cell volume of this detector. The standard solution to this problem would be the use of make-up gas. Owing to the extremely narrow band widths of the peaks eluting from narrow-bore columns, very high make-up flows would be required, but since the TCD is a concentration-sensitive detector, this would result in an unacceptable reduction of the peak intensity. To avoid this problem, the linear velocity was enhanced by operating the detector at vacuum conditions.

A detector of rapidly growing importance is the mass spectrometer (MS). As mentioned above, various MS detectors have been coupled to narrow-bore GC. A parameter to consider here is the scan rate of the MS. Typically one would

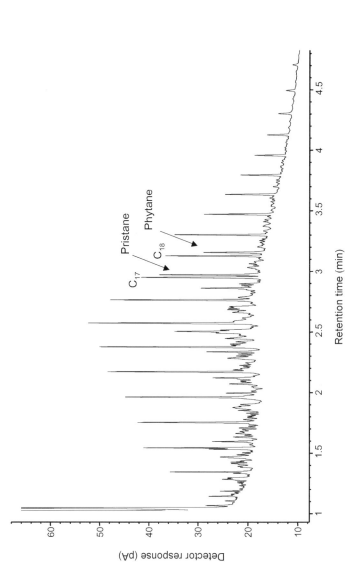

Figure 7.5 Analysis of diesel oil in hexane. Column: 10 m × 0.05 mm i.d., coated with a 0.1 μm SE-30 stationary phase. GC: HP6890 (Agilent Technologies, Wilmington, DE, USA). Experimental conditions: $T = 40°C$ ballistically heated to 320°C; carrier gas helium; split flow = 200 ml min^{-1}; inlet-pressure = 10 bar; injection volume = 1 μl (500 ppm).

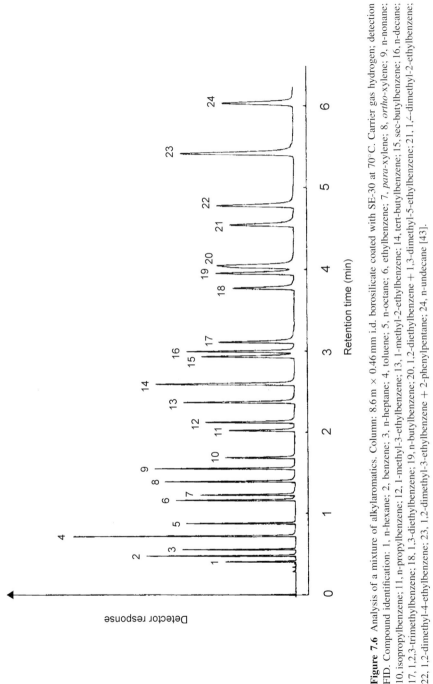

Figure 7.6 Analysis of a mixture of alkylaromatics. Column: 8.6 m × 0.46 mm i.d. borosilicate coated with SE-30 at 70°C. Carrier gas hydrogen; detection FID. Compound identification: 1, n-hexane; 2, benzene; 3, n-heptane; 4, toluene; 5, n-octane; 6, ethylbenzene; 7, *para*-xylene; 8, *ortho*-xylene; 9, n-nonane; 10, isopropylbenzene; 11, n-propylbenzene; 12, 1-methyl-3-ethylbenzene; 13, 1-methyl-2-ethylbenzene; 14, tert-butylbenzene; 15, sec-butylbenzene; 16, n-decane; 17, 1,2,3-trimethylbenzene; 18, 1,3-diethylbenzene; 19, n-butylbenzene; 20, 1,2-diethylbenzene + 1,3-dimethyl-5-ethylbenzene; 21, 1,4–dimethyl-2-ethylbenzene; 22, 1,2-dimethyl-4-ethylbenzene; 23, 1,2-dimethyl-3-ethylbenzene + 2-phenylpentane; 24, n-undecane [43].

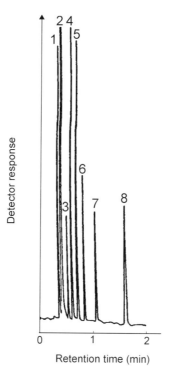

Figure 7.7 Analysis of solvents. Column: 9 m × 0.50 mm i.d., coated 0.2 μm SE-54 at 60°C. Detection TCD (140 μl cell volume) at reduced pressure (4 mmHg). Inlet pressure 20 bar. Compound identification: 1, acetone; 2, dichloromethane; 3, butanal; 4, ethylacetate; 5, dichloroethane; 6, cyclohexane; 7, n-heptane; 8, toluene [53].

like to collect some 5–10 spectra over the peak. For quadrupole and ion trap MS instruments this means a peak width at baseline level of around 1 s. Rapid scanning instruments are necessary for faster analyses. Fast GC-MS has been reviewed by Leclercq and Cramers [31]. Figure 7.8 shows one of the fastest separations ever obtained. A mixture of hydrocarbons was separated on a short narrow-bore column (30 cm, 0.05 mm). Detection was performed by time-of-flight MS. The spectral acquisition rate was 500 spectra s^{-1}. In fairness it must be said that this is not a separation that can be performed in a routine environment. Dedicated instrumentation not yet commercially available was used for injection and control of the equipment [32, 41, 76]. Given the speed of development in this area it is to be expected that reliable equipment for separations on this time scale will become available in the not too distant future.

One of the major factors obstructing the use of fast narrow-bore columns at this moment is the limited loadability of the columns. Typically, the working range of a narrow-bore column—the region between detection limit and

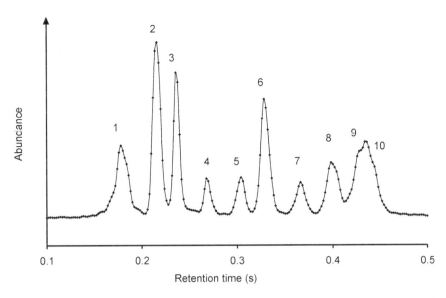

Figure 7.8 Ultrafast analysis of 10 compounds within 500 ms on a 0.3 m × 0.05 mm column with a 0.17 μm thickness nonpolar OV-1 phase, using a cryogenic focusing inlet system. Inlet pressure = 450 kPa; carrier gas helium; split flow = 400 ml min^{-1}; $T_{injector}$ = 250°C; T_{oven} = 75°C. Injection and sampling: 1 μl, headspace (≈1 ng/compound); detection, time-of-flight mass spectrometer (LECO, St Joseph, MI, USA); $T_{transferline}$ = 275°C; T_{source} = 200°C, scan rate = 500 spectra s^{-1}, from mass 40 to 200. Compounds: 1, pentane; 2, 2,3-dimethylbutane; 3, hexane; 4, benzene; 5, heptane; 6, methylcyclohexane; 7, toluene; 8, *trans*-1,4-dimethylcyclohexane; 9, octane; 10, *cis*-1,4-dimethylcyclohexane. Reprinted from [41] with permission of Elsevier Science.

maximum amount where overloading starts to occur—varies from 0.05 ng (detection limit) to a few nanograms (maximum loadability). Either more sensitive detectors or columns with an increased sample capacity are necessary. The first alternative is clearly something the instrument manufacturers should address. Bundled multicapillary columns, the subject of the next section, are a way of increasing the loadability.

7.4.2 Multicapillary columns

The multicapillary column, which consists of a bundle of short narrow-bore capillaries, benefits from the speed of a narrow-bore column and the high sample loadability of a wide-bore column. This type of column became commercially available very recently. It is made by combining some 900 capillaries with inner diameter of 0.04 mm into a bundle. The advantages of the multicapillary column in comparison to a single narrow-bore column are the higher sample capacity, the higher flow rate and, consequently, the lower minimum detectable concentration of solute. In practice the multicapillary column is very easy to use and can easily be installed in any GC. Detection normally takes place by FID [77]. The optimum inlet pressure of 375 kPa results in a column outlet flow of 200 ml min^{-1}. For an optimal performance of the FID the air/hydrogen flows of the detector have to be increased from 400/40 to 400/100 ml min^{-1}. An example of a BTEX (benzene, toluene, ethyl benzene, xylene) analysis within 1 min performed on a multicapillary column is shown in Figure 7.9. The experimentally obtained plate number was approximately 12 500. A plate number of 25 000 was predicted theoretically. The multicapillary column therefore is especially suitable for relatively simple mixtures, requiring only low plate numbers.

To obtain the maximum performance, the column has to meet very stringent requirements. For example, each of the capillaries should have exactly the same diameter. If the 900 capillaries differed significantly in diameter, this would cause serious band broadening. Other important parameters that should remain constant are the length and the film thickness of the capillaries.

If each of the 900 individual capillaries in the multicapillary column were identical, the chromatogram obtained from this column could not be distinguished from that of a single capillary with identical dimensions. However, if the capillaries are not exactly identical, 900 different chromatograms will be obtained. The final chromatogram now will be the sum of the 900 individual chromatograms.

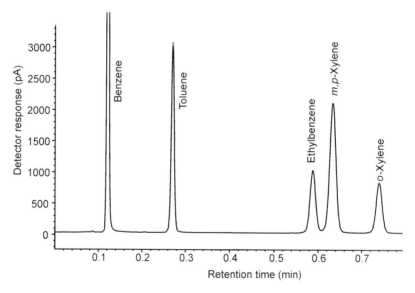

Figure 7.9 Headspace analysis of BTEX on a multicapillary column (Alltech Associates Inc., Deerfield, IL, USA) within 1 min. Column length 1 m, capillary diameter 0.04 mm, number of capillaries 900. Inlet pressure 375 kPa; injection 20 µl vapour; temperature program, 40°C (0.5 min) to 200°C at 25°C min^{-1}; split-flow 800 ml min^{-1}; column outlet flow 200 ml min^{-1}; detector FID.

In previous work we have studied the influence of variations in the individual capillaries on the final chromatogram [47]. In Figure 7.10 the effect of a 0.001 mm variation in capillary diameter of some of the capillaries is illustrated. The curve for 10x shows the situation for a total of 900 capillaries, of which 10 capillaries have a diameter that is 0.001 mm larger than the nominal diameter of 0.04 mm, 10 capillaries have a diameter that is 0.001 mm smaller than the nominal diameter, and the remaining 880 capillaries have the actual 'correct' diameter. For each point in time t, $h(t)$ is calculated for all three diameters. Next, the relation (7.13) is used to calculate the overall height at time t:

$$h(t)_{\text{multicap}} = 10h(t)_{39\,\mu m} + 880h(t)_{40\,\mu m} + 10h(t)_{41\,\mu m} \qquad (7.13)$$

Three peaks appear in the first (predicted) chromatogram. The first peak shows the elution from the 10 capillaries with a diameter of 0.041 mm; the large second peak corresponds to the 880 'correct' capillaries; and the third peak results from the 10 capillaries with a diameter of 0.039 mm. From Figure 7.10 it can be concluded that with nonuniform diameters band broadening occurs. The simulation is an ideal case in which only one variation (0.001 mm) is

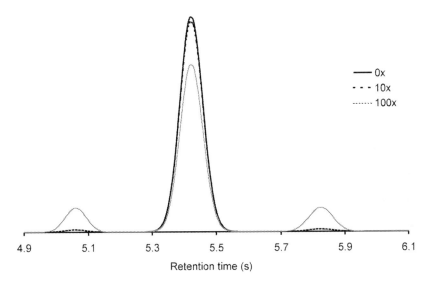

Figure 7.10 Effect of variations in diameter on the peak shape. 0X (thick line): the ideal situation when all capillaries have exactly the same diameter of 0.04 mm. 10X (broken line): the situation when for a total of 900 capillaries, 10 capillaries have diameters that are 0.001 mm larger than the nominal diameter of 0.04 mm, 10 capillaries have diameters that are 0.001 mm smaller than the nominal diameter, and the remaining 880 capillaries have the actual 'correct' diameter. 100X (dotted line): the situation that for a total of 900 capillaries, 100 capillaries have diameters that are 0.001 mm larger than the nominal diameter of 0.04 mm, 100 capillaries have diameters that are 0.001 mm smaller than the nominal diameter, and the remaining 700 capillaries have the actual 'correct' diameter. Reprinted from [47] with permission of Wiley-VCH.

considered. Although the assumption of, say, a Gaussian distribution of diameter makes the models somewhat more complicated, the equations derived above can also be used to simulate such a situation. The influence of variations in film thickness and length of the capillaries on the peak shape can be determined in a similar way. From these calculations it can be concluded that very tight manufacturing specifications must be imposed on the precision of the capillary diameter and the film thickness to obtain an optimal separation efficiency.

7.4.3 Fast temperature programming

An attractive solution for reducing the analysis time of an over-resolved chromatogram is the use of a higher temperature programming rate [74] (solution 4A from Table 7.3) or, if the separation is performed at constant temperature, converting the method into a temperature-programmed-separation (solution 4B). Since the first temperature-programmable GC became commercially available in the late 1950s, the heating capacity of the ovens has been significantly improved. In the current generation of GC instruments the maximum temperature-programming rate is limited to approximately 1 to $2°C\,s^{-1}$. With the standard oven and heater designs it is difficult to obtain higher rates owing to temperature stability problems and long cooling times. To enable the use of higher heating rates while still maintaining sufficient temperature stability and acceptable cooling times, a number of systems based on resistive heating have been developed. To adopt such a technique, a capillary column can be coated with a conductive material, a coil can be wrapped around or run parallel to the column, or the capillary column can be put into a metal tubing.

Jain and Phillips [48] successfully applied a conductive coating to the capillary column, and obtained very fast analyses in the seconds range. Ehrmann et al. [49] tested two concepts of resistive heating. One was based on the use of a metal tube as a heater (coaxial at-column heater) and the second used a metal wire that ran parallel to the column as the heater (colinear at-column heater). It was found that a colinear at-column heater resulted in a more reliable system than the coaxial heater and also enabled rates of $10°C\,s^{-1}$ to be achieved. The latter suffered from differences in expansion coefficients of the metal and fused silica, causing damage of the fused-silica column.

In the resistive heating technique, large amounts of heat are produced in systems that have very low thermal masses. Small changes in heat production can hence result in major temperature fluctuations, which in turn can result in a poor retention time stability. For their system, Ehrmann et al. [49] reported a relative standard deviation (RSD) of retention times of better than 1%. MacDonald and Wheeler [12, 13] reported an even better retention time RSD of lower than 0.2% for the commercially available EZ-Flash system. The latter value is comparable to that found using a conventional GC oven operated at high heating rates. The retention time repeatability (RSD) was better than 0.3%. The typical average peak width at half-height was about 0.4 s at a heating rate of $4°C s^{-1}$.

An example of a fast analysis of $C_{10}-C_{42}$ at a temperature programming rate of $4°C s^{-1}$, is shown in Figure 7.11 [46]. The elution time of C_{42} here is less than 1.5 min. Elution of the same compound in a normal GC with a maximum heating rate of $40°C min^{-1}$ roughly takes 10 min. The detection limit and the sample capacity are 10 pg and 200 ng, respectively. The resulting working range is extremely large. The good results for the working range obtained here are the result of the wide-diameter column that is used.

From Figure 7.11 it can be seen that some band broadening occurs for the high-boiling compounds C_{40} and C_{42}. This is most likely caused by the presence of cold spots between the injector and the resistively heated column or between the column and the detector. Similar effects were also observed by Ehrmann *et al.* [49]. A drawback of rapid programming is the significantly increased elution temperature of the later-eluting compounds. This could result in problems, for example, with samples that contain thermally labile analytes.

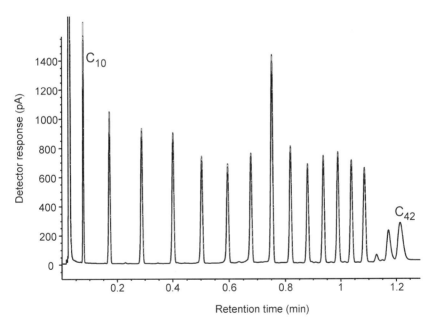

Figure 7.11 Fast analysis of a normal alkane mixture ($C_{10}-C_{42}$) within 1.5 min, using the EZ-Flash system (Thermedics Detection Inc., Chelmsford, MA, USA). Temperature program EZ-Flash 80–375°C at $4°C s^{-1}$; inlet-pressure = 300 kPa; split flow = 500 ml min^{-1}; temperature program GC-oven 80–140°C at $30°C min^{-1}$; injector temperature program 350–550°C at $16°C s^{-1}$. Reprinted from [46] with permission of Wiley-VCH.

As already discussed, the typical application area for fast heating techniques is that of samples consisting of a low number of compounds with widely differing boiling points (chromatogram type V). An application investigated was the separation of an industrial glycol mixture. This mixture consisted of four oligomers with large differences in boiling range, MEG (monoethylene glycol; $C_2H_6O_2$), DEG (diethylene glycol; $C_4H_{10}O_3$), TEG (triethylene glycol; $C_6H_{14}O_4$) and TTEG (tetraethylene glycol; $C_8H_{18}O_5$), dissolved in methanol. The boiling points of these compounds are 198, 244, 285 and 328°C, respectively. An analysis of this sample with a standard GC on a standard column temperature-programmed from 40 to 320°C at 15°C min^{-1}, took about 9 min. Using a fast temperature program with the EZ-Flash, the analysis time was reduced to less than 0.5 min. An example of a chromatogram is shown in Figure 7.12.

7.4.4   Vacuum outlet operation

A number of techniques for faster analysis are based on enhancing the rate of radial equilibration in the column. Techniques that belong to this group include the use of columns with a reduced inner diameter, use of hydrogen as the carrier gas, turbulent flow GC and vacuum outlet operation [73, 75] (solutions 8, 7, 15

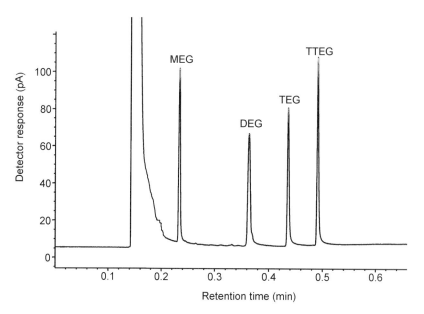

Figure 7.12 Fast analysis of (MEG) monoethylene glycol ($C_2H_6O_2$), (DEG) diethylene glycol ($C_4H_{10}O_3$), (TEG) triethylene glycol ($C_6H_{14}O_4$) and (TTEG) tetraethylene glycol ($C_8H_{18}O_5$). Inlet pressure: 45 kPa; split flow 300 ml min^{-1}; $T_{EZ-Flash}$ 40–300°C at 15°C s^{-1}; T_{oven} 40–100°C at 120°C min^{-1}. Reprinted from [46] with permission of Wiley-VCH.

and 10). As has already been stated, the benefits of vacuum outlet operation can only be fully exploited for short wide-bore columns (inner diameter of 0.53 mm). High analysis speeds can be obtained as a result of the sub-ambient pressure conditions present in the column, and the consequently higher diffusion coefficient of the solute in the mobile phase. The advantage of using a wide-bore column is the high sample capacity. A flow restrictor is required for vacuum operation by coupling a wide-bore column to a mass spectrometer. Three types of restrictors have been proposed [40]. The first method uses a narrow bore pre-column (60 cm length × 0.10 mm i.d.); the second method relies on the use of an SFC-type restriction; whereas the third method uses of a micro-injection valve. The flow restrictions are necessary to decrease the column outlet flow in order to keep the vacuum level in the MS system at an optimum level. From the results of a comparative study it was concluded that the performance of a short narrow-bore column as the restriction is comparable to that of an SFC-type restriction. A fast rotating micro-injection valve was used to enable direct injection of the sample onto the column at sub-atmospheric pressures. The best performance was obtained using the pre-column restrictors.

Despite the attractive speed and loadability characteristics of wide-bore columns operated under vacuum outlet conditions, this approach to faster analyses has so far received little attention. Most likely this is a result of the experimental difficulties associated with the use of wide-bore columns operated under vacuum outlet conditions. Vacuum outlet conditions are most readily obtained using a mass spectrometer as the detection device. Direct coupling of a short and/or wide-bore column to a standard benchtop MS, however, will result in operational problems. First, the carrier gas inlet and the injection system have to be operated at sub-ambient pressures. Second, the high column outlet flow might increase the pressure in the ion source of the mass spectrometer to a level exceeding the tolerable limit. Typical (optimum) column outlet flow rates of a wide-bore column with an inner diameter of 0.53 mm, and a length of 10 m are approximately 7–10 ml min^{-1}. In general, the maximum pumping capacity of an MS is already reached at a column flow rate of 5 ml min^{-1}. On the other hand, if the pumping system of the MS does have sufficient capacity to maintain pressure at an acceptable level, a complication will be that the inlet pressure in the injector will decrease to sub-ambient values. Although advantageous in terms of speed, this might cause additional practical problems.

If short wide-bore columns are to be used in conjunction with standard benchtop MS instruments, a restriction can be coupled to the column inlet. The flow is now restricted to an acceptable level and the injection system can operate at above-atmospheric pressures and low-pressure conditions still prevail throughout the entire column. Such a column inlet restriction can readily be obtained by coupling a narrow-bore column (e.g. 0.10 mm × 60 cm) at the inlet position of the analytical column, using a zero-dead-volume connector [50]. A disadvantage of such a coupling is the possible occurrence of dead volumes. An alternative solution is the use of an SFC integral tapered restriction prepared from the column inlet. In the latter case the restriction is an integrated part of the column, which implies that no coupling piece is needed [51]. This restriction system can act as a benchmark in terms of maximum achievable column efficiencies.

When the capacity of the pumps is high enough, one can also consider to coupling a wide-bore column directly to the MS, without using a pre- or post-column flow restriction. Sample introduction now has to take place at sub-atmospheric inlet pressures. In this section the possibilities and limitations of wide-bore columns under vacuum conditions as a means to fast GC will be demonstrated.

In 1981 Cramers et al. [52] derived an equation describing the gain in speed of analysis using vacuum outlet as compared to atmospheric outlet operation:

$$G = \frac{\overline{u}_{opt,vac}}{\overline{u}_{opt,atm}} = \frac{\overline{P}_{opt,atm}}{\overline{P}_{opt,vac}} = \frac{(P^3_{i,opt,atm} - P^3_{atm})}{(P^2_{i,opt,atm} - P^2_{atm})^{3/2}} \tag{7.14}$$

where $\overline{u}_{opt,vac}$ and $\overline{u}_{opt,atm}$ are the average optimum gas velocities through a capillary column at vacuum outlet and atmospheric outlet conditions, respectively; $\overline{P}_{opt,vac}$ and $\overline{P}_{opt,atm}$ are the average optimum column pressures at vacuum outlet and atmospheric outlet conditions, respectively; $P_{i,opt,atm}$ is the optimum inlet pressure at atmospheric outlet conditions; and P_{atm} is the atmospheric pressure.

From equation (7.14) it is clear that for short wide-bore columns the gain in speed is much higher than for long narrow-bore columns. For the short wide-bore column the pressure inside the column is low over its entire length. The resulting high diffusion coefficients result in a high speed of analysis.

As a contrast, if long narrow-bore columns are operated under vacuum outlet conditions, only a fraction of the column length is operated at sub-ambient pressures. This means that the gain in speed is negligible.

7.4.4.1 Sample loadability and film thickness

As already mentioned, three types of restrictions can be used in the coupling of wide-bore columns to a mass spectrometer: a narrow-bore pre-column, an SFC restriction and a micro-injection valve. In Figure 7.13 the narrow-bore pre-column and the SFC restriction are shown schematically.

For the SFC restriction no coupling is needed since this restriction is fabricated directly on the column. This means that leaks cannot occur and dead-volume problems are absent. The disadvantage of this type of restrictor is that producing it is a rather time-consuming task. The retention time repeatability obtained with the SFC restriction was better than 0.5% RSD. The area repeatability varied between 5% and 10% (measured for hexachlorobenzene and undecane). A high efficiency can be obtained only by minimising band

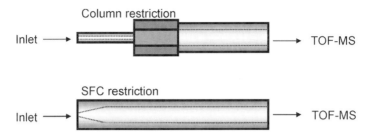

Figure 7.13 Schematic drawing of a narrow-bore pre-column (Varian-Chrompack, Middelburg, The Netherlands) and an SFC restriction (produced in house).

broadening caused by the injection system. This means that sample introduction has to be very fast. For the two restriction methods using the narrow-bore pre-column and the SFC inlet, sample introduction was performed in the split mode at high split flows (approximately 10–200 ml min^{-1}). The split ratio is around 2 to 100. In the third (valve) method, sample introduction was achieved by electrically actuating the sample valve. The performance of this valve for use in high-speed separations on narrow-bore columns has been studied previously [53]. Owing to the extremely small loop volume and the fast valve rotation time, injection pulses are in the millisecond range. Because the valve can be operated at temperatures up to 275°C, it can be used as an injection device for fast GC. However, only gaseous samples can be injected, which means that the range of compounds that can be analysed is limited. The performance of these three restriction methods is shown in Figure 7.14. The average carrier gas velocities observed are approximately 200 cm s^{-1}.

As can be seen from Figure 7.14, the performances of the narrow-bore pre-column set-up and the SFC restrictor are very comparable. With the valve system, a significantly poorer performance is obtained. On the basis of these results it can be concluded that the best restriction method is the use of a narrow-bore pre-column connected to the analytical column using a low-dead-volume column connector. For compounds that might exhibit interaction with the connector material, the use of an SFC integral tapered restrictor is a good alternative. The plate number obtained ranged from 17 000 to 20 000 plates at average carrier gas velocities of 150 to 180 cm s^{-1}. Analyses times were in the seconds range. The use of a wide-bore column coupled to a mass spectrometer is therefore very suitable for speeding up the analysis time of simple mixtures. An important advantage of this route to faster analyses is the high sample capacity of the wide-bore column.

Most of the options for faster GC summarised here result in a reduction of the sample capacity of the columns. This is particularly the case for the narrow-bore option. An advantage of wide-bore columns is the high sample loadability

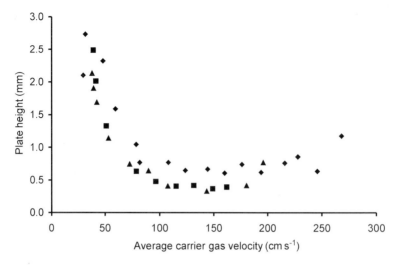

Figure 7.14 Comparison of the HU-curves of a thin-film wide-bore column coupled to a TOF mass spectrometer using the three restriction devices at the column inlet side. ▲, Narrow-bore pre-column-restriction; ■, SFC restriction; ◆, micro-injection valve. Compound, hexane; $k = 0.3$; temperature 60°C; column 9 m × 0.53 mm × 0.25 μm CP Sil 8 CB.

of such columns. The maximum sample capacity of a GC column, Q_s, can be described by [30]

$$Q_s = \frac{5\pi}{2} \beta'' \frac{(1 + k_0)^2}{k_0^2} p_s d_f d_c H \qquad (7.15)$$

where β'' is a solute–liquid phase specific factor, k_0 is the capacity factor at infinite dilution and p_s is the density of the stationary phase. From this equation it follows that the sample capacity is proportional to d_c^3 (as already described in a previous section). The sample capacity is thus drastically decreased when using narrow-bore columns. It might be interesting to increase the sample loadability by increasing the film thickness. However, the disadvantage of an increasing film thickness is the increasing C_s-term in the plate height equation. This means that at a certain film thickness, the influence of slow diffusion in the stationary phase on the plate height can no longer be neglected. In other words, at too high a film thickness the separation becomes less efficient and separation time increases. This is the case for narrow-bore columns but also for wide-bore columns operated at vacuum outlet conditions. If a wide-bore column is operated at vacuum outlet conditions, the stationary phase starts to contribute significantly to band broadening at much lower thicknesses than in case of atmospheric outlet operation.

The influence of the film thickness on the efficiency of a GC separation can be described using the following relations [54].

$$\frac{H_{\min}}{u_{o,\mathrm{opt}}} = 2C_{m,o}f_1 + C_s(2f_2 + y_2) \tag{7.16}$$

In equation (7.16), $H_{\min}$ is the minimum plate height at optimum conditions, $u_{o,\mathrm{opt}}$ is the optimum carrier gas velocity at outlet conditions, $C_{m,o}$ is the resistance to mass transfer in the mobile phase at column outlet conditions, C_s is the resistance to mass transfer in the stationary phase and y_2 is a function of P. At vacuum outlet: $P \to \infty$, $f_1 = 9/8$, $f_2 = 3/(2P)$ and $y_2 = -\frac{1}{2}f_2$. With $f_2 u_{o,\mathrm{opt}} = \overline{u}$ this leads to

$$\frac{H_{\min}}{\overline{u}} = \frac{18C_{m,o}}{8f_2} + \frac{3C_s}{2} \tag{7.17}$$

Equation (7.17) can be rewritten as

$$\frac{H_{\min}}{\overline{u}} = \frac{C_{m,o}}{f_2}\left(\frac{18}{8} + \frac{3}{2}\frac{C_s f_2}{C_{m,o}}\right) \tag{7.18}$$

With the following equations for $C_{m,o}$ and C_s [55],

$$C_{m,o} = \frac{11k^2 + 6k + 1}{24(k+1)^2}\frac{r^2}{D_{m,o}} \tag{7.19}$$

$$C_s = \frac{2k}{3(k+1)^2}\frac{d_f^2}{D_s} \tag{7.20}$$

and together with equations (7.21) and (7.22) [52],

$$f_2 D_{m,o} = \overline{D_m} \tag{7.21}$$

$$\overline{p} = \frac{2p_i}{3} \tag{7.22}$$

where r is the column radius, equations (7.17) to (7.20) can be rearranged to

$$\frac{H_{\min}}{\overline{u}_{\mathrm{opt}}} = \left(\frac{3}{2}\frac{11k^2 + 6k + 1}{24(k+1)^2}\frac{r^2 p_i}{P_1 D_{m,1}}\right) + \left(\frac{k}{(k+1)^2}\frac{d_f^2}{D_s}\right) \tag{7.23}$$

Together with equation (7.24),

$$t_R = \frac{NH_{\min}}{\overline{u}_{\mathrm{opt}}}(1 + k) \tag{7.24}$$

the total analysis time can be calculated. For a narrow-bore column (e.g. 0.05 mm), at a film thickness of only 0.1–0.15 µm the stationary phase starts to contribute to the overall peak width and hence the analysis time. With this equation it can be determined at which film-thickness the negative influence of slow mass transfer in the stationary phase is larger than the positive influence of operating a wide-bore column at lower inlet pressures. This is shown in Figure 7.15. From the figure it can be concluded that, to take full advantage of the gain in speed when operating a wide-bore column (10 m × 0.530 mm) at vacuum conditions, the film thickness should not exceed approximately 1.5 µm.

In Table 7.7 a comparison is made between the analysis speed on a wide-bore column used at vacuum outlet conditions and two narrow-bore columns operated at ambient outlet pressures. It can be concluded that a 0.05 mm i.d. column with a length of only 80 cm, which has the same plate number and phase ratio, in combination with a flame ionisation detector, would be faster than a wide-bore column at vacuum outlet conditions. This short 0.50 mm column however, is not easy to use in practice. Hence, a comparison with a 0.05 mm column with a length of 5 m would be more appropriate. From a comparison of the results shown in Table 7.7, it is evident that the efficiency of the wide-bore

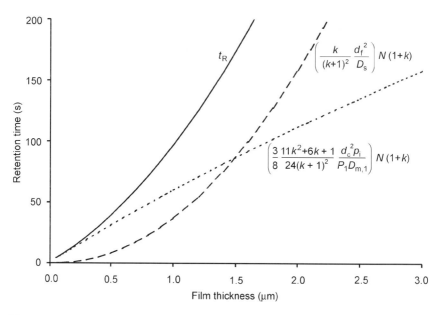

Figure 7.15 Influence of the film thickness in a wide-bore column (10 m × 0.53 mm) on the analysis time (calculated using equations (16) and (17)). Solid line, total retention time; broken line, influence of stationary phase on analysis time; dotted line, influence of diffusion in mobile phase on analysis time. Plate number: = 20000; compound, nonane; $T = 60°C$; detector, mass spectrometer. The inlet pressure P_i was calculated using the equation described in work of Leclercq [71].

Table 7.7 Comparison between the calculated analysis speed, retention time, plate number and sample capacity of a wide-bore column operated at vacuum conditions and two 0.05 mm columns at ambient outlet pressures. Compound is nonane; $k = 1.6$ for column 1 and 3, $k = 7$ for column 2; temperature 60°C

Column	P_{out} (kPa)	N	$\bar{u}$ (cm s^{-1})	t_R (s)	Q (ng)
(1) 9 m × 0.53 mm × 0.25 μm	0	20 000	120	20	500
(2) 5 m × 0.05 mm × 0.1 μm	100	100 000	40	90	5
(3) 0.8 m × 0.05 mm × 0.02 μm	100	20 000	110	2	0.5

column is less than that of the narrow-bore column. For very complex mixtures the narrow-bore column is hence clearly more suitable.

The analysis speed of a wide-bore column at vacuum conditions is approximately 2–3 times higher than that of a wide-bore column operated at atmospheric conditions. A distinct disadvantage of wide-bore columns is that the plate number is not very high. Wide-bore columns operated at vacuum conditions, therefore, are not very suitable for complex separations. For mixtures with a low number of compounds and a wide boiling point range, however, wide-bore columns operated at vacuum outlet conditions can yield a significantly faster separation.

7.4.4.2 Resolving peaks by mass spectrometry

The combination of a chromatographic separation technique with mass spectrometric detection is a very powerful tool for the study and identification of organic compounds in complex samples. When the identity of a compound is known, it can also be advantageous to use MS detection. With MS detection target peaks can readily be identified in crowded chromatograms. By using extracted ion traces, nonseparated peaks can even be quantified.

Nonseparated peaks can be 'separated' on the basis of mass spectral differences. Mass spectrometry hence offers an alternative route towards faster analysis. Spectral deconvolution of the Leco Pegasus II software offers the possibility to deconvolute and identify overlapping peaks. However, some chromatographic separation is required for the deconvolution algorithm to recognise the presence of two or more compounds in one peak [14]. Figure 7.16 shows the identification of the co-eluting compounds octane and *cis*-1,4-dimethylcyclohexane (see Figure 7.8). The separation here amounts to 10 ms or five spectra (rectangle). Deconvolution is certainly a helpful tool in the separation of complex samples and for the identification of co-eluting peaks. Spectral deconvolution methods should be applied with some care, however.

7.4.5 Comprehensive two-dimensional gas chromatography

One of the most interesting applications of very fast GC is in comprehensive two-dimensional GC (GC × GC). This technique can formally be regarded as equivalent to planar bed separation schemes such as thin-layer chromatography

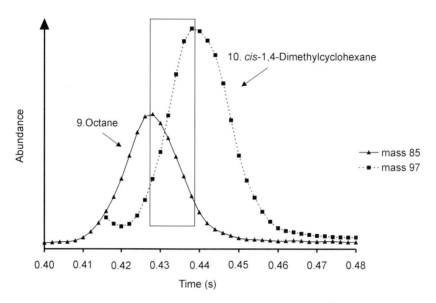

Figure 7.16 Deconvolution of octane and *cis*-1,4-dimethylcyclohexane, peaks numbers 9 and 10 of the ultrafast separation shown in Figure 7.8. In this graph the unique masses 85 and 97 are plotted, which are specific for the two compounds. Reprinted from [41] with permission of Elsevier Science.

(TLC), except that both dimensions of separation are gas chromatographic [56, 57]. Comparably it provides a two-dimensional separation in which sample substances are distributed over a retention plane formed by the operation of two independent columns. Separation and analysis in two dimensions is much more powerful than in one. A retention plane has much more peak capacity than a retention line and so can accommodate much more complex mixtures [56, 57]. Component identification is potentially more reliable because each substance has two identifying retention measures rather than one. Separations are likely to be more structured in two dimensions, leading to recognisable patterns characteristic of the mixture source [56].

In GC × GC two independent GC separations are applied to an entire sample. The sample is first separated on a normal-bore high-resolution capillary GC column in the programmed-temperature mode. All of the effluent of this first column is then focused in very many, very narrow fractions at regular, short intervals and subsequently injected onto a second capillary column, which is short and narrow to allow for very rapid separations. The resulting chromatogram has two time axes (retention on each of the two columns) and a signal intensity corresponding to the peak height.

Retention in GC is basically governed by two factors, the volatility of the analyte (pure component vapour pressure, p_i^0) and its interaction with the stationary phase (activity coefficient at infinite dilution, γ_i^∞). By selection of a nonpolar column, the separation in the first dimension is mainly based on the volatility of the analytes ('boiling-point separation'). The analytes will be eluted from the first column at different temperatures, but with very similar volatilities at the time (temperature) of elution. The second separation, which is so fast as to be essentially isothermal at the temperature of elution from the first dimension, is completely determined by the activity coefficients of the solutes. Both enthalpic (energy of interaction) and entropic factors contribute to γ_i^∞, so that the separation on the second column can be based on polarity, molecular geometry, size, etc. Thus the separations in the two dimensions can be completely independent, or orthogonal.

Since every peak eluting from the first dimension column is cut into five or six fractions, the separation of each of these fractions on the second dimension column must be finished within a few seconds. This puts stringent requirements on the re-injection and the separation for the second column and on the detection device. The band width of the pulse at which the focused fraction is re-injected onto the second column must be very narrow ($<$ 10 ms). This focusing and re-injection are performed with a modulation system. Three types of modulators are in use at present. One uses a thick-film modulation capillary for retaining the fraction and a heating device for focusing and re-injection [58]. The second type applies cryogenics for retaining and focusing the fraction in the front end of the second column and the heat capacity of the oven air to re-inject the fraction [59]. A recently introduced system [60] with no moving parts provides the advantages of both the other modulators. Small pulses of cooled and/or heated nitrogen cool and heat very short sections of the modulation capillary in turn, in order to focus and re-inject the fractions in narrow pulses. The dimensions of the second column (length about 1 m, i.d. 0.10 mm and a thin-film stationary phase) are chosen such that it provides the necessary speed. This means that it finishes the separation of all the constituents of the fraction within a few seconds and before the next fraction is re-injected. Finally, the detector must have a small rise time and permit a high sampling rate ($\geq$ 100 Hz). Most modern FID detectors fulfil these requirements. When using a mass spectrometer, only the fast scanning time-of-flight (TOF) spectrometers with scanning frequencies of up to 500 scans s^{-1} are suitable [40, 41, 60].

Figure 7.17 represents a schematic of the GC $\times$ GC process. A recent review of the technique of GC $\times$ GC is given in [28].

Analytes belonging to the same chemical group type will have approximately the same activity coefficients and hence exhibit about the same second-dimension retention times. This will result in ordering in the chromatogram, where compounds of the same group type are arranged in bands along the 2D-plane [62]. Identification is thus not only facilitated but is also very reliable. An example of such an ordered chromatogram, representing the GC $\times$ GC separation of a kerosene, is given in Figure 7.18.

The abundant crowded band of spots at the lower end of the figure represents the saturates, where each high intensity n-alkane can be seen as a white spot at

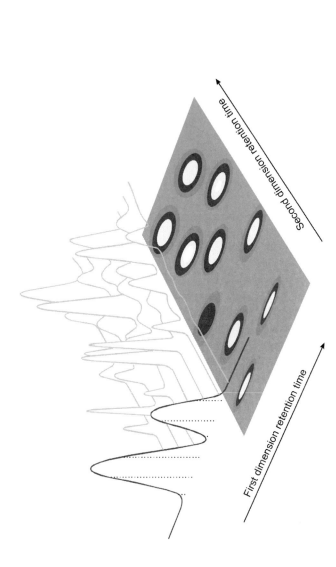

Figure 7.17 Schematic of the GC × GC process. The resulting data file is presented as a contour plot, which is the projection of the individual second-dimension chromatograms.

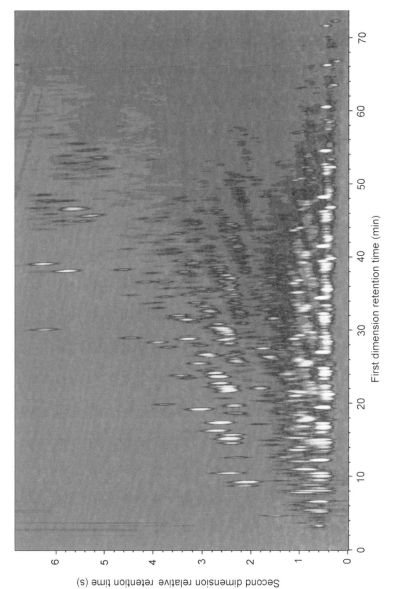

Figure 7.18 Contour plot of a GC × GC-separation of a kerosene.

regular time intervals. Between the n-alkanes are the branched alkanes, from alkanes with a low degree of branching just in front of the n-alkane, up to the highly branched ones with ever lower retention times. Above the alkanes, 'roof-tiles' of cyclic alkanes are present, each tile representing different isomers with the same number of carbon atoms, again with the lowest branching on the right side of the tile (highest retention time) and the ever higher branched ones more to the left side of the tile. About in the middle of the plane the different groups of aromatics with one ring (monoaromatics) can be found, from toluene at a retention time of about 5 min, through the xylenes (three spots) at around 10 min and the monoaromatics with nine C-atoms from 15 through 20 min. The rest of the monoaromatics again are ordered in 'roof-tiles' with the same number of C-atoms from 10 (20–30 min) through 15 C-atoms (50–60 min). Again, within these tiles the different branching of the isomers are arranged in accordance with the arrangement in the cyclic alkanes tiles—the lowest branching on the right side of the tile (highest retention time) and the ever higher branched ones more to the left side of the tile.

Finally, in the top of the plane the two-ring aromatics can be clearly seen. From naphthalene at a retention time of 30 min, through the two methylnaphthalenes (around 40 min) up to a cloud of C_3-naphthalenes (13 C-atoms) at 55 min.

The different group types can simply be extracted from the underlying data file. A representation (a reconstruction) of the group of diaromatics from such a data file is presented in Figure 7.19.

Likewise, in Figure 7.20 the reconstructed one-dimensional chromatogram of the monoaromatics is represented. Finally in Figure 7.21 the reconstruction of the one-dimensional separation of the total sample, showing the complexity of the sample and the inability of one-dimensional GC to separate all of the constituents properly.

GC × GC separations offer extremely high peak capacities (approximately equal to the product of the peak capacity of the first and the second column). Because of the orthogonality of the two separation mechanisms, a very substantial fraction of this theoretical peak capacity may actually be used in practical separations.

Calibration and measurement of substance quantity in two dimensions is not fundamentally different from that in one dimension. A given substance passes through the series of two chromatographic columns and eventually reaches the detector, which produces a proportional signal. Either the peak signal or the integral over all the time the substance passes through the detector can be measured and related to substance quantity. A GC × GC peak exists on a data plane and its integral must be taken over both dimensions to give a peak volume, which is directly proportional to the quantity of substance forming the peak. The peak on the plane is composed of several second-dimension peaks as the same substance reaches the detector in several sequential secondary chromatograms. The integral is the sum of the integrals of each of these individual peaks. Peak

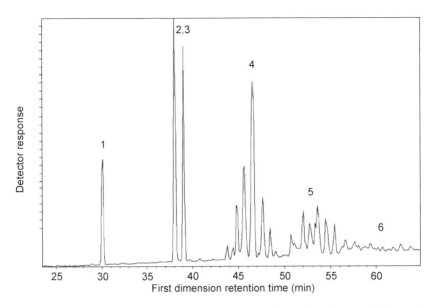

Figure 7.19 Reconstructed one-dimensional chromatogram of the selected group of diaromatics: 1, naphthalene; 2, 2-methylnaphthalene; 3, 1-methylnaphthalene; 4, C_2-naphthalenes; 5, C_3-naphthalenes; 6, C_4- and higher naphthalenes.

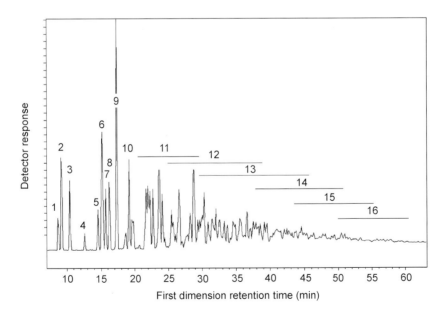

Figure 7.20 Reconstructed one-dimensional chromatogram of the monoaromatics: 1, ethylbenzene; 2, *para-* + *meta*-xylene; 3, *ortho*-xylene; 4, isopropylbenzene; 5, n-propylbenzene; 6, 1-methyl-3- + 1-methyl-4-ethylbenzene; 7, 1,3,5-trimethylbenzene; 8, 1-methyl-2-ethylbenzene; 9, 1,2,4-trimethylbenzene; 10, 1,2,3-trimethylbenzene; 11, C_{10}-monoaromatics; 12, C_{11}-monoaromatics; 13, C_{12}-monoaromatics; 14, C_{13}-monoaromatics; 15, C_{14}-monoaromatics; 16, C_{15}-monoaromatics.

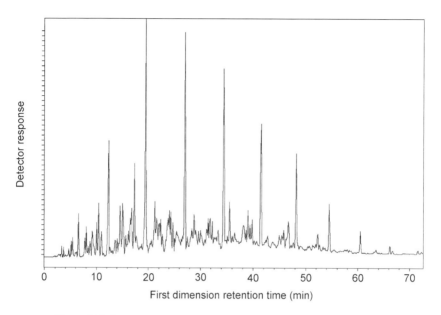

Figure 7.21 Reconstructed one-dimensional chromatogram of the total sample.

integration in GC × GC has been the subject of a study [63–65] and it has been demonstrated that this peak integration provides high-quality quantification.

A number of GC × GC applications have been studied and their practicality has been demonstrated, but the potential applications are far more widespread. The number of components in a sample for which GC × GC becomes the method of choice depends on the specific nature of the sample and analytes, but in most cases it is probably substantially less than one hundred and may often be less than about thirty.

Since petroleum products are readily available complex mixtures of well-controlled composition, they have been the first choice for samples to demonstrate the separation and ordering capabilities of GC × GC [62–66].

The environment, both natural and human-affected, contains exceedingly complex mixtures. Most environmental chemistry to date has focused on individual substances rather than the mixtures that contain the substances. A number of those complex environmental mixtures have been submitted to GC × GC with satisfactory results [66–70].

For a detailed account of two dimensional GC see Chapter 8.

References

1. J.C. Giddings, *Anal. Chem.*, **1962**, *34*, 314.
2. D.H. Desty, A. Goldup and W.T. Swanton, in *Gas Chromatography* (eds. N. Brenner, J.E. Callen and M.D. Weiss), Academic Press, New York, 1962, pp. 105–138.

3. C.A. Cramers, H.-G. Janssen, M.M. Van Deursen and P.A. Leclercq, *J. Chromatogr. A*, **1999**, *856*, 315.
4. J.H. Knox and M. Saleem, *J. Chromatogr. Sci.*, **1969**, *7*, 614.
5. C.A. Cramers and P.A. Leclercq, *CRC Crit. Rev. Anal. Chem.*, **1988**, *20*, 117.
6. C.P.M. Schutjes, E.A. Vermeer, J.A. Rijks and C.A. Cramers, *J. Chromatogr.*, **1982**, *253*, 1.
7. S. Dagan and A. Amirav, *J. Am. Soc. Mass Spectrom.*, **1996**, *7*, 737.
8. R.P.W. Scott and G.S.F. Hazeldean, in *Gas Chromatography 1960* (ed. R.P.W. Scott), Butterworths, London, 1960, p. 144.
9. G. Guiochon, *Anal. Chem.*, **1966**, *38*, 1020.
10. E. Grushka and G. Guiochon, *J. Chromatogr. Sci.*, **1972**, *10*, 649.
11. H.M. McNair and G.L. Reed, *J. Microcol. Sep.*, **2000**, *12*, 351.
12. S.J. MacDonald and D. Wheeler, *Int. Lab.*, **1998**, 13C.
13. S.J. MacDonald and D. Wheeler, *Am. Lab.*, **1998**, *22*, 27.
14. C. Leonard, A. Grall and R. Sacks, *Anal. Chem.*, **1999**, *71*, 2123.
15. A.C. Lewis, K.D. Bartle and L. Rattner, *Environ. Sci. Technol.*, **1997**, *31*, 3209.
16. A. Peters and R. Sacks, *J. Chromatogr. Sci.*, **1992**, *30*, 187.
17. L.M. Blumberg, *J. High Resolut. Chromatogr.*, **1999**, *22*, 501.
18. F. David and P. Sandra, *Am. Lab.*, **1999**, *31*, 18.
19. R.E. Kaiser, R.I. Rieder, L.M. Lin, L. Blomberg and P. Kusz, *J. High Resolut. Chromatogr.*, *Chromatogr. Commun.*, **1985**, *8*, 580.
20. D. Repka, J. Krupcik, E. Benicka, T. Maurer and W. Engewald, *J. High Resolut. Chromatogr.*, **1990**, *13*, 333.
21. P.A. Leclercq, G.J. Scherpenzeel, E.A.A. Vermeer and C.A.Cramers, *J. Chromatogr.*, **1982**, *41*, 61.
22. L. Roach and M. Guilhaus, *Org. Mass Spectrom.*, **1992**, *27*, 1071.
23. G. Schomburg, *J. Chromatogr. A*, **1995**, *703*, 309.
24. A. Van Es, J. Rijks and C. Cramers, *J. Chromatogr.*, **1989**, *477*, 39.
25. A.J.J. Van Es, C.A. Cramers and J.A. Rijks, *J. High Resolut. Chromatogr.*, *Chromatogr. Commun.*, **1989**, *12*, 303.
26. J. Dalluge, R. Ou-Aissa, J.J. Vreuls, U.A.Th. Brinkman and J.R. Veraart, *J. High Resolut. Chromatogr.*, **1999**, *22*, 459.
27. R. Sacks, C. Coutant, T. Veriotti and A. Grall, *J. High Resolut. Chromatogr.*, **2000**, *23*, 225.
28. J.B. Phillips and J. Beens, *J. Chromatogr. A*, **1999**, *856*, 331.
29. A.J. Van Es, J.A. Rijks, C.A. Cramers and M.J.E. Golay, *J. Chromatogr.*, **1990**, *517*, 143.
30. R.T. Ghijssen, H. Poppe, J.C. Kraak and P.P.E. Duysters, The mass loadability of various stationary phases in gas chromatography, *Chromatographia*, **1989**, *27*(1/2), 60.
31. P.A. Leclercq and C. Cramers, *Mass Spectrom. Rev.*, **1998**, *17*, 37.
32. A. Van Es, J. Janssen, C. Cramers and J. Rijks, *J. High Resolut. Chromatogr.*, *Chromatogr. Commun.*, **1988**, *11*, 852.
33. L.M. Blumberg and M.S.Klee, *Anal. Chem.*, **1998**, *70*, 3828.
34. N. Dyson, *J. Chromatogr.*, **1999**, *842*, 321.
35. G. Gaspar, P. Arpino and G. Guiochon, *J. Chromatogr. Sci.*, **1977**, *15*, 256.
36. G. Gaspar, R. Annino, C. Vidal-Madjar and G. Guiochon, *Anal. Chem.*, **1978**, *50*, 1512.
37. P.G. Van Ysacker, H.-G. Janssen, H.M.J. Snijders, P.A. Leclercq, C.A. Cramers and H.J.M. van Cruchten, *J. Microcol. Sep.*, **1993**, *5*, 413.
38. C.C. Grimm, S.W. Lloyd and L. Munchausen, *Am. Lab.*, **1996**, Sept., S18.
39. F. David, D.R. Gere, F. Scanlan and P. Sandra, *J. Chromatogr.*, **1999**, *842*(1–2), 309.
40. M.M. Van Deursen, H.-G. Janssen, J. Beens, P. Lipman, R. Reinierkens, P.A. Leclercq and C.A. Cramers, *J. Microcol. Sep.*, **2000**, *12*(12), 613.
41. M.M. Van Deursen, J. Beens, H.-G. Janssen, P.A. Leclercq and C.A. Cramers, *J. Chromatogr.*, **2000**, *878*, 205.

42. P.G. Van Ysacker, H.M. Snijders, H.-G. Janssen and C.A. Cramers, *J. High Resolut. Chromatogr.*, **1998**, *21*(9), 491.

43. C.P.M. Schutjes, High speed, high resolution capillary gas chromatography, columns with a reduced inner diameter, PhD thesis, Eindhoven University of Technology, The Netherlands, 1983.

44. P.G. Van Ysacker, J.G.M. Janssen, H.M.J. Snijders and C.A. Cramers, *J. High Resolut. Chromatogr.*, **1995**, *18*, 397.

45. J.N. Driscoll, in *Detectors in Capillary Chromatography* (eds. H.H. Hill and D.G. McMinn), Chemical Analysis Series **121**, Wiley, New York, 1992, p. 69.

46. M. Van Deursen, J. Beens, H.-G. Janssen and C. Cramers, *J. High Resolut. Chromatogr.*, **1999**, *22*(9), 509.

47. M. Van Deursen, M. Van Lieshout, R. Derks, H.-G. Janssen and C. Cramers, *J. High Resolut. Chromatogr.*, **1999**, *22*(2), 119.

48. V. Jain and J.B. Phillips, *J. Chromatogr. Sci.*, **1995**, *33*, 55.

49. E.U. Ehrmann, H.P. Dharmesana, K. Carney and E.B. Overton, *J. Chromatogr. Sci.*, **1996**, *34*, 533.

50. J. De Zeeuw, J. Peene, H.-G. Janssen and X. Lou, in *Proc. 21st Int. Symp. Capillary Chromatography and Electrophoresis*, Park City, Utah, 1999, p. 16.

51. E.J. Guthrie and H.E. Schwarz, *J. Chromatogr. Sci.*, **1986**, *24*, 236.

52. C.A. Cramers, G.J. Scherpenzeel and P.A. Leclercq, *J. Chromatogr.*, **1981**, *203*, 207.

53. A.J.J. Van Es, High speed narrow bore capillary gas chromatography, PhD thesis, Eindhoven University of Technology, The Netherlands, 1990.

54. C.A. Cramers, F.A. Wijnheymer and J.A. Rijks, *J. High Resolut. Chromatogr., Chromatogr. Commun.*, **1979**, *2*, 329.

55. P.A. Leclercq, *J. High Resolut. Chromatogr.*, **1992**, *15*, 531.

56. J.C. Giddings, *J. Chromatogr. A*, **1995**, *703*, 3.

57. J.C. Giddings, *Multidimensional Chromatography—Techniques and Applications* (ed. H.J. Cortes), Marcel Dekker, New York, 1990, pp.1–23.

58. J.B. Phillips, R.B. Gaines, J. Blomberg, F.W.M. Van der Wielen, J.M.D. Dimandja, V. Green, J. Granger, D.G. Patterson, L. Racovalis, H.J. de Geus, J. de Boer, P. Haglund, J. Lipsky, V. Sinha and E.B. Ledford, *J. High Resolut. Chromatogr.*, **1999**, *22*, 3.

59. R.M. Kinghorn and P.J. Marriott, *J. High Resolut. Chromatogr.*, **1998**, *21*, 620.

60. E.B. Ledford, in *Proc. 23rd Symp. Capillary Gas Chromatography*, Riva del Garda, 2000, CD-ROM.

61. J.M.D. Dimandja and D.G. Patterson, in *Proc. 23rd Symp. Capillary Gas Chromatography*, Riva del Garda, 2000, CD-ROM.

62. J. Blomberg, P.J. Schoenmakers, J. Beens and R. Tijssen, *J. High Resolut. Chromatogr.*, **1997**, *20*, 539.

63. J. Beens, H. Boelens, R. Tijssen and J. Blomberg, *J. High Resolut. Chromatogr.*, **1998**, *21*, 47.

64. J. Beens, R. Tijssen and J. Blomberg, *J. High Resolut. Chromatogr.*, **1998**, *21*, 63.

65. J. Beens, R. Tijssen and J. Blomberg, *J. Chromatogr. A*, **1998**, *882*, 233.

66. G.S. Frysinger and R.B. Gaines, *J. High Resolut. Chromatogr.*, **1999**, *22*, 251.

67. R.B. Gaines, G.S. Frysinger, M.S. Hendrick-Smith and J.D. Stuart, *Environ. Sci. Technol.*, **1999**, *33*, 2106.

68. Z.Y. Liu, S.R. Sirimanne, D.G. Patterson, L.L. Needham and J.B. Phillips, *Anal. Chem.*, **1994**, *66*, 3086.

69. R.B. Gaines, E.B. Ledford and J.D. Stuart, *J. Microcol. Sep.*, **1998**, *10*, 597.

70. G.S. Frysinger, R.B. Gaines and E.B. Ledford, *J. High Resolut. Chromatogr.*, **1999**, *22*, 195.

71. P.A. Leclercq and C.A. Cramers, *J. High Resolut. Chromatogr., Chromatogr. Commun.*, **1985**, *8*, 764.

72. J.B. Phillips and V. Jain, *J. Chromatogr. Sci.*, **1995**, *33*, 541.

73. N. Amirav, S.B. Tzanani, S.B. Wainhaus and S. Dagan, *Eur. Mass Spectrom.*, **1998**, *4*, 7.

74. A. Grall, C. Leonard and R. Sacks, *Anal. Chem.*, **2000**, *72*, 591.

75. C. Leonard and R. Sacks, *Anal. Chem.*, **1999**, *71*, 5177.

76. A. Van Es, J. Janssen, R. Bally, C. Cramers and J. Rijks, *J. High Resolut. Chromatogr., Chromatogr. Commun.*, **1987**, *10*, 273.

77. M. Van Lieshout, M. Van Deursen, R. Derks, J.G.M. Janssen and C.A. Cramers, *J. Microcol. Sep.*, **1999**, *11*, 155.

78. P.G. Van Ysacker, High-speed narrow-bore capillary gas chromatography: theory, instrumentation and applications, PhD thesis, Eindhoven University of Technology, The Netherlands, 1996, pp. 90–91.

79. P.G. Van Ysacker, J.G.M. Janssen, H.M.J. Snijders, P.A. Leclercq, H. Wollnik and C.A. Cramers, in *Proc. 16th Int. Symp. Capillary Chromatography*, Riva del Garda, 1994, Hüthig Verlag, Heidelberg, 1994, pp. 785–796.

8 Multidimensional and comprehensive multidimensional gas chromatography

Philip J. Marriott and Russell M. Kinghorn

8.1 Introduction

Certain landmark developments have altered the course of gas chromatography: invention of the flame ionisation detector, introduction of fused-silica columns, the benchtop mass spectrometer, to name a few. Can multidimensional gas chromatography (MDGC) be considered to be one of these? In terms of introduction of MDGC, the Deans switch was an important enabling technical innovation. However, to be of major impact on the total method, the innovative step must have a significant impact on the practice of GC. In this respect, MDGC may be considered to lack the test of acceptance. It is not widely employed in research or industry, but those who have the expertise to use it routinely will acknowledge that it does hold the promise that many of the pioneers who led the way in MDGC predicted. One may be led to the perception that there are more reviews of MDGC than original papers in the recent past! But there is real reason to believe that this is about to be changed dramatically.

The MDGC method has recently undergone a 'renaissance'. The theoretical work of Giddings, who published reviews claiming that a much greater separation power should be possible by multidimensional GC, is now being realised. It took a major technical step forward to show that the ideas of Giddings could be implemented. This has now been embodied in the comprehensive gas chromatography method (GC × GC), and its elegance, power and sensitivity are now at the vanguard of gas chromatography research. Indeed, it has led one of the present authors to entitle one talk "The Phoenix of Multidimensional Gas Chromatography Rises". It is to be hoped that the GC × GC approach finally gains broad acceptance as a universal ultrahigh-resolution method. In this chapter we shall try to capture that aspiration.

In the pages that follow, some methods that are not necessarily multidimensional in the conventional understanding of MDGC are briefly described, primarily to ensure that the coverage is reasonably comprehensive, and hopefully to make a more logical progression in discussion.

8.2 Gas chromatography as one dimension in multidimensional analytical methods

Multidimensional analysis (MDA) in its broadest sense involves, in general, any analysis in which two different modes are used, such as TGA-EGA (thermo-gravimetric analysis with evolved gas analysis, using a method such as mass spectrometry), or any other hyphenated (or 'joined') system one might imagine. It need not be just instrumental methods that are coupled, but for the purposes of the present review we shall limit ourselves to instrumental approaches.

A limited view of multidimensional analysis in chromatography—where the two instrumental analysis methods (the sequential operation of first and second dimensions) constitute different methods of analysis but with only one of the methods being a chromatographic or separation dimension—is typified by the following instrumental arrangements: GC-MS; GC-FTIR; HPLC-diode array; HPLC-MS; HPLC-ICPMS; GC-AED; TLC-FID; TLC-HPLC and so forth. Note that some of these incorporate spectroscopic detection methods (e.g. ICPMS, MS, OES [optical emission spectroscopy]) and strictly speaking we should include GC-FID. The flame ionisation detector is not a particularly informative detection system, giving only an indication of the presence/absence of carbon-containing species and an indication of the relative abundances thereof. But this is as valid a second dimension as using, say, AES (atomic emission spectroscopy), where we might be monitoring only a specific emission line from a certain selected element. A more complete discussion of the above is beyond the scope of the present chapter, and the reader is advised to look at other chapters or in books such as that by Brinkman [1].

8.3 Multidimensional separation methods

Multidimension separation may be simply defined as a method of analysis in which different separation methods are employed sequentially. Cortes reviewed a variety of these coupled separation techniques in 1992 [2].

Typical examples of two-dimensional separation techniques are GC-GC, HPLC-GC, HPLC-CE, HPLC-HPLC, TLC-HPLC, and so forth. The first of these is the primary concern of the present chapter. It is easy, then, to extend the discussion to three-dimensional analysis, again with multidimensional separation, where a spectroscopic detection method is employed post-separation, or even where a bulk sample separation step is used prior to the other separation dimensions. These can be exemplified by such arrangements as GC-GC-MS, or LC-GC-GC. While the former of these is known and not uncommon in the literature, the latter is currently rare.

It is useful to mention briefly the role of a reasonably popular hyphenation mode, that of HPLC-GC [3]. Note that GC will invariably be the second of the separation steps, since this allows the technically easier approach of introduction of the liquid effluent from the HPLC separation into the GC. This experiment is usually conducted to permit prior class separation of a sample into molecular species of simpler composition. Thus a hydrocarbon oil sample can readily be separated into aliphatic, naphthene, aromatic and bitumen classes, with each separately introduced into the GC finish. Given the complexity of oil analysis, this is a straightforward simplification tool that gives fewer over-lapping components, where otherwise there would be complicated overlap of, for example, the aromatic components within the aliphatic compounds. Beens and Tijssen described such an approach to separate oils into cuts of a middle distillate sample [4]. It is only required to choose a suitable eluent that is acceptable as a GC injection solvent, and to decide what injection mode is required for the GC introduction, such as large-volume injection to permit the greatest amount of sample delivery into the GC. Note that as an automated approach this has benefits over an off-line method such as elution column chromatography–gas chromatography, which has been a mainstay tool in areas such as petroleum analysis and organic geochemistry for many years. Recently, testing of bitumen and its fumes using this approach has been reported for establishing the mutagenicity of the samples [5]. Applications in essential oil analysis and foods and flavours are likewise relevant [6, 7], and in this case selective isolation of desired components from the matrix may be one advantage of LC-GC. Thus, sample clean-up considerations may be a driving factor.

Having established the general definitions of MDA and multidimensional separations, we must now develop the logic that leads to a need for MDGC.

8.3.1   Scope of conventional capillary gas chromatography: rationale for higher-resolution methods

Jennings has given an account of general capillary gas chromatography tech-nology [8]. While the technology necessarily includes the important aspects of injection and detection modes that have been employed to provide the best possible analytical result for analysis of a sample, it is generally accepted that the method revolves around the capability of the column to provide separation of chemical components. The progression of gas chromatographic column inno-vation is based on three different, but necessarily related, areas; developments in phase technology; developments in improved column capacity; and develop-ments in the technical implementation of different modes of capillary GC. We shall discuss these briefly here, primarily to place MDGC into the context of trying to achieve better separation.

8.3.1.1   Phase chemistry
Many stationary phases are available for packed-column GC analysis. Through an understanding of the retention indices of sample components, and polarity of the available phases, the best stationary phase for a particular analysis may

be chosen or new phases may be introduced to give the required separation. Thus, for a sample of limited complexity, an optimum separation may be determined. However, within the low-resolution packed-column area, there is a realisation that merely shifting component relative retentions by the use of stationary phase polarity has limitations, since there will eventually be no 'vacancy' in the chromatographic scale to provide the extra capacity needed for component resolution. The next step is to improve column performance by reducing component dispersion or increasing efficiency during the chromatographic process.

8.3.1.2   Capillary column efficiency

Capillary GC was introduced not long after the gas chromatography method itself [9]. The technical requirements to produce stable phases on glass (fused-silica) columns have limited the range to phases used in capillary GC; however, the enhanced efficiency of capillary GC can be presumed to reduce the requirements to have such an extensive phase library available. Thus, low-resolution packed GC in which phase selectivity was a significant experimental variable gave way to capillary GC in which peak efficiency was now the major consideration, and this clearly played on the chief advantage of the narrow-bore column. Since maximum efficiency for a solute can be approximated by the formula for minimum plate height H (equation 8.1),

$$H_{min,theor} = r_c \left(\frac{1 + 6k + 11k^2}{3(1 + k)^2} \right)^{1/2} \tag{8.1}$$

then for a typical 25 m column of inner radius $r_c = 0.25$ mm, and a solute of retention factor $k = 10$, we have potentially 110 000 plates. If this solute eluted in 8 min, then its base width in time units would be about 6 s, and for a total 60 min analysis time for a sample, if each peak had this bandwidth in a temperature-programmed run, then the column capacity would be 500–600 peaks. Other ways to increase the column's capacity (the total number of peaks that can be separated by the column, if they were to all be just baseline resolved) would be to use a narrower-bore column, or one of greater length. This efficiency may be thought to be capable of providing a high degree of separation power; however, Davis and Giddings treated this problem statistically [10, 11] and showed that if the components of a separation behave independently (i.e. are randomly distributed throughout the available separation space) and if we require all peaks to be completely resolved, the number of components present in a sample should not be more than about 20% of the column's peak capacity. For instance, if the column capacity is 500 and if the sample has 100 peaks, then statistically we may expect almost all to be resolved. But as sample complexity increases, the probability of unresolved peaks will increase. There are two provisos to add to this discussion. First, samples will frequently contain many more components than this (even if many are at trace level) and, second, the concept of chromatographic 'independence' or randomness of the peaks in respect of their properties will be unlikely to be fulfilled, and so the probabilistic model is unlikely to apply to a real sample. Thus, if the analysis conditions are defined as a temperature program from 60°C to 280°C, it will be unlikely that, for any given sample, the components will be equally spread over this complete temperature range. Hence we will have the components compressed into a region of lower capacity and the need for even greater resolution arises.

8.3.1.3   Technical innovations

The limitations imposed by the use of a single column have been realised for many years. In 1978, Bertsch reviewed the area of what we generally refer to as two-dimensional gas chromatography in the first issue of *Journal of High Resolution Chromatography and Chromatography Communications* [12]. This was subsequently brought up to date with a two-part review by the same author in 1999 [13] and 2000 [14]. The history of multidimensional gas chromatography (MDGC) commences very soon after the advent of the GC method itself. Deans in 1968 recognised this as an effective way to increase resolution [15], while Schomburg considered the further resolution capabilities of capillary columns in this regard [16]. There are two reasons for the early interest in MDGC. The first is one that is a common characteristic of scientific inquiry—to explore what possibilities exist for extending the GC method. The second arose from a more practical consideration—there were deficiencies with the standard technique that limited its application, and hence the search for improved or enhanced opportunities that address these deficiencies was undertaken. The first reason need not be considered further, since we are more interested in the outcomes that are eventually adopted in the laboratory. The second is instructive, since it allows us to appreciate what advantages the MDGC method is purported to offer. It is to be hoped, then, the scientific enquiry direction and the practical problem-driven direction will converge. Unfortunately, even though some truly impressive separations have been described in the literature, a most telling quote from the recent review of Bertsch cuts to the heart of the status of MDGC as a tool for the routine analyst. He hoped that "the revolution in capillary column GC would be mirrored in the development of instrumentation for two-dimensional gas chromatography", but instead observed that "on the contrary, tentative steps taken by a few manufacturers and suppliers of chromatographic equipment fizzled out". He ascribed this largely to the introduction of gas chromatography–mass spectrometry. This will be further developed below.

8.3.2   Approaches to coupled column methods for improved separation in gas chromatography

Among the methods reported that have explored coupled columns for enhanced separation, we can discern a few types of approaches. One is the approach of altering analysis selectivity without enhancing peak capacity. Another is to increase capacity by employing physical separation/isolation techniques that can be incorporated into the chromatographic elution process. Figure 8.1 is a simplified generic diagram of any coupled column system. Traditional goals of MDGC may be broadly classified into three areas, as shown in Table 8.1, with the benefits and procedure indicated. The interface may be any of a range of devices

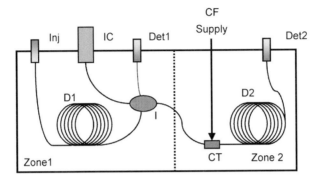

Figure 8.1 Schematic diagram of a typical generic multidimensional gas chromatography instrument: Inj, injector; I, interface device; IC, interface control; Det1, Det2, detectors 1 and 2; D1 and D2 are dimensions (columns) 1 and 2; zone 1 and zone 2 are locations of the two columns—they may be in the same or different ovens; CT, static cryotrap; CF supply, cryofluid supply.

Table 8.1 Traditional MDGC experimental methodologies

Method	Reason for implementation	Implementation procedure
Heartcutting	Increase the separation/resolution power for given zones of a sample	Valve or flow switching allows a small section of effluent from one column to be passed to a second column of greater selectivity. The zone is referred to as a heartcut
Backflushing	Exclude less volatile constituents from having to be completely passed through a gas chromatography column. This can speed up sample analysis	Once the desired sample components (higher volatility) have been passed out of the first column, the flow at a valve may be altered to allow carrier gas to purge through the first column in a reverse flow, and pass the heavier components out of the column inlet (usually to exit though a purge valve or split valve for a capillary GC injector)
Trace concentration	To increase the amount of analyte introduced to a capillary column (and possibly also exclude components present in larger quantities), and so increase sensitivity of analysis	Usually a column of lower phase ratio (i.e. comprising a larger loading of stationary phase) precedes the analytical column, such as a packed column–capillary column arrangement

or methods, and determines what experiment is conducted with the instrument. Table 8.2 lists a number of different approaches to employing coupled GC column systems. Included in this table is the technique of multiplex GC, which has historical interest in concepts of pulsing bands for introduction into GC [17, 18]. A selection of these approaches will be reviewed in the sections below.

8.3.2.1   Directly coupled columns

If a column containing one phase type is directly connected to a column of a second phase type, the elution of sample components is moderated by the passage through the two columns. We can simplify the understanding of this process by stating that the selectivity of the separation is a linear combination of the selectivities of the two separate columns. In this case, the interface in Figure 8.1 is a simple direct column connection, and interface control (IC) is not used. The benefits of this approach were summarised by Jennings [19], who showed that a new phase type could be synthesised on the basis of the success of this linear combination, to produce an intermediate phase chemistry and generate a predicted new phase type that gave the desired performance. In this case, the column does not necessarily have greater capacity, but has more desirable selectivity. It does not qualify to be termed a multidimensional separation, even though two separate columns (which we might call dimensions) are used. We shall strictly reserve the term multidimensional to the case where we isolate components from one column and pass them to a second column, and hence there is a need for valves and/or other instrumental devices to intercede in the chromatography process, and where the capacity of the system is increased through incorporation of this step. This classification of coupled columns will not be discussed further.

8.3.2.2   Selectivity tuning experiments

Sandra et al. proposed that two serially coupled columns could be used for the tuning of the overall polarity of the column pair [20]. Tuning can be achieved by variation in pressure setting between each column and/or by temperature variation of each column (where each column is in a separate heated zone); for this Figure 8.1 uses a variable-pressure inlet at the IC provided to the interface device (I). This technique is also termed the method of multichromatography, as proposed by Hinshaw and Ettre [21, 22]. The optimisation of the dual-oven technique was described by Repka et al., where each column's temperature was altered isothermally over a limited range in 10°C steps and simulated results were presented [23]. Subsequent work by Engewald's group based their interpretation of the tuning experiment upon retention index considerations [24]. This method has been revisited by Sacks and co-workers [25, 26], developing optimisation strategies to obtain the best resolution of a synthetic set of compounds, and recently time-of-flight mass spectrometry has been used to

Table 8.2 Coupled column GC methods (refer to Figure 8.1)

Method	Function of interface device I	Resultant benefits to chromatographic process
Directly coupled columns	No valving; direct connection	Similar to 'mixed-phase' column
Multichromatography (pressure tuning) process	Variable pressure provided at interface between the two columns	Provides some degree of selectivity variation through varied relative column contributions
Conventional multidimensional gas chromatography	Mechanical switching valve, or pressure switching	Diversion of components from column 1 to column 2, to achieve increased selectivity/separation
Targeted multidimensional gas chromatography	No valve; direct coupling with cryogenic modulation	Easy collection–transfer of whole peaks or peak group to column 2
Comprehensive gas chromatography	Rapid modulation of peaks from column 1 to column 2 using various modulation technologies	True multidimensional gas chromatography over total chromatographic space
Multiplex chromatography	Modulation of the input distribution to the analytical column	While not a combined column method, this technique paved the way for other modulation methods

record the MS of the components eluted in a fast GC experiment. Lourentzeas *et al.* [27] used experimental design methods to optimise the same experimental approach, but additionally studied temperature changes along with midpoint pressure to obtain the best conditions. While the pressure tuning experiment does use a valve between the two columns, the key to this method is that the pressure is varied at this so-called midpoint section. This has the effect of altering the overall analysis selectivity such that each components' retention parameter is varied simply from the change in pressure, which causes a change in relative time spent on each column. Thus, if the midpoint pressure is increased, the elution time on column 1 is increased (i.e. the pressure drop across column 1 is reduced, leading to an increase in retention time on that section) and so that column has a greater contribution to the apparent total system 'polarity'. If column 1 is the more polar, then the system will act as if it is more polar than when a lower midpoint pressure is used. Clearly, then, if two components co-elute at a certain midpoint pressure, and if they have different retention mechanisms on each column, simply altering the midpoint pressure will potentially separate the peaks. Again, it should be appreciated that this coupled column approach is still not strictly a multidimensional approach, but since it is relevant to the overall schematic diagram shown in Figure 8.1, it is appropriate to include this brief discussion here.

8.3.3   *Conventional multidimensional gas chromatography*

The realisation of multidimensional gas chromatography was mathematically emphasised by Giddings [28], who published a theoretical paper entitled "Two-Dimensional Separations: Concepts and Promise". He stated that

> Sequential displacements, as those occurring in column chromatography or in two-dimensional thin layer chromatography, are far more adaptable because optimum conditions can be applied separately to each step. One can carry out the first displacement in one medium under one set of conditions and transfer the linear array of zones to the edge of a 2D system for the second displacement.

This statement underpins and encapsulates the fundamental principle of two-dimensional chromatography and, coupled with modern chromatographic instrumentation and techniques, may be easily applied to a vast array of samples with varying degrees of analytical requirements. In Figure 8.1, interface I is a switching system, which may be a valve or a live T-switch. IC is the valve drive mechanism, or a pressure balance flow. A cold trap will often be provided to collect heartcut sections from D1. Either a single- or dual-oven system may be used. Heartcut timing is determined by initial analysis on D1 with the effluent going to Det1. The invention of the Deans switch in 1968 was an important early development for multidimensional gas chromatography [29–31]. This fluidic valve conveniently allows the transfer of the effluent from one column to another, second, dimension, column or equivalent process, while maintaining the chromatographic integrity of column pressure (and hence flow and linear velocity) and temperature. Over time, the Deans switch has been manufactured with a higher degree of consideration of chromatographic principles, therefore removing possible dead volumes inside fittings and active sites where analytes may be adsorbed in a reversible or irreversible fashion. This, incorporated in modern gas chromatographs, has given the chromatographer access to high-speed detectors, electronic pressure control, precisely timed external events and predefined pressure/flow algorithms, and allows the simpler implementation of multidimensional-based chromatography. Schomburg presented a comprehensive account of the technical application areas of MDGC [32], referring to this as a sampling technique, which upon reflection is an accurate description of the MDGC method. We effectively introduce the total sample into dimension 1 and then sub-sample the regions of interest into a second separation step. A detailed schematic diagram of the regulation control of the MDGC instrument accompanied Schomburg's review. The discussion was developed around examples of different types of coupled columns, and to what problems they could be applied. Column switching in GC was reviewed by Willis, including systems, packed–packed, packed–capillary and capillary–capillary column formats, and quantitative aspects [33]. A study by Blomberg *et al.* described a glass union based on fused-silica press-fit units designed to perform reliably as a Deans switching device prior to the detector [34] so that column effluent could be switched to either an FID or an MS detector. The same principle can operate as a regular live switch, and its construction is schematically presented in Figure 8.2.

The cryogenic component of the multidimensional GC system is usually located at or just downstream of the section joining the two coupled columns. In this arrangement a first column separation precedes a second column to which discrete segments or groups of peaks from the first column are passed. Each of these transfer steps constitutes a heartcut event.

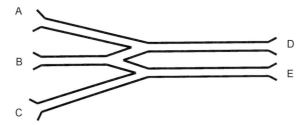

Figure 8.2 Sketch of a Deans switch-type interface. This device was used in a switching system to pass effluent to either of two detectors, but could also operate to switch flows to different columns. In the described system: A,C are switching gas supply; B is analytical column; D is to FID; E is to MS.

The first column effluent is normally passed through a section of nonretaining column leading to a monitor detector until the heartcutting is performed, at which time flow switching causes the first column flow to be diverted to the second column. The solutes that have been heartcut may be cryogenically trapped at the start of the second column, so that they are presented to the second column with minimal input band broadening, effectively representing a sharp injection profile. This arrangement can be used to advantage when only a single oven system is used to house both GC columns. In a dual-oven apparatus the need for a cryofocusing step may be less important, depending on the analysis requirements. This is because an effective trapping procedure can be implemented via the vapour pressure of the analytes, the phase ratios of the two separation columns and the dual-oven temperatures during the heartcutting operation. By reducing the phase ratio and oven temperature on the second column, and hence reducing the vapour pressure of the analytes, the heartcut fractions are effectively focused, therefore removing the requirement of a cold trap. This phenomenon was termed phase ratio partition focusing (PRPF) by the authors [35], which adequately describes the fundamental principles involved. Figure 8.3 shows two processes, where the heartcuts are collected together and then analysed together on dimension 2 (Figure 8.3a) or analysed as separate runs in the second dimension (Figure 8.3b). Figure 8.4 puts this process into a chromatographic context, where three components are collected together and analysed on column 2.

Many multidimensional designs and concepts have been described in detail in the book by Cortes [36] and the reviews by Schomburg [32, 37], de Geus [38], Krock [39] and Bertsch [13]. Operationally, the heartcut sections are normally all trapped together, the oven is cooled, the cryogenic trap fluid supply is turned off, and then the oven temperature is programmed up. In Krock and Wilkins'

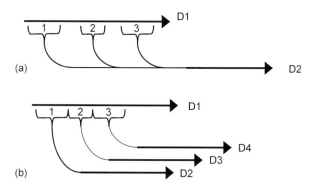

Figure 8.3 Schematic diagram showing different ways to employ a heartcut process. In (a), heartcuts are sent to the same second dimension, collected and analysed as one run. In (b), the separate heartcuts are individually analysed in the second dimensions, which could be separate columns or the same column.

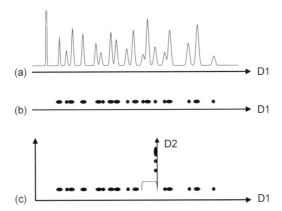

Figure 8.4 Relationship between the chromatographic analysis on the first column (D1) shown in (a), the same chromatogram represented as peak zones in (b), and selection of a heartcut zone, and its elution on the second dimension (D2), shown in (c).

system [39], a series of sequential heartcuts from the first-dimension column are delivered to parallel traps for storage prior to separate analysis of each trap's content on the second column. In this arrangement, both FTIR and MS detection could be selected. It is possible to operate the system without the cryogenic trap, but some measure of resolution may be lost in some situations if the peaks transferred to the second column undergo relative retention reversal [35]. The detector attached to the first column will therefore display a full chromatographic analysis, with the heartcut sections absent, while the second column detector shows the separation of just the heartcut compounds, which should now clearly exhibit different selectivities (and degree of separation) from those on the first column. In most cases the two columns have different selectivities by virtue of their different polarities; however, the second column could, for instance, consist of a chiral stationary phase, leading to enantiomer resolution. The exact requirement for choice of the two participating columns will be dictated by their individual separation mechanisms, how the solutes of interest interact with the phases, and other demands of the analysis.

A cryogenic modulation interface has been recently suggested as being able to provide a simplification to the normal multidimensional GC set-up [40, 41] and, while as yet few applications have been demonstrated, it has a number of benefits. It has been called targeted cryogenic multidimensional gas chromatography. Based on the technology described under cryogenic modulation in the section on comprehensive gas chromatography below, it offers the following opportunities.

For samples with not too much complexity, and where there are suitable differences in neighbouring/overlapping peaks to allow differentiation based on selective retention mechanisms on different phases, it should be possible to resolve all components of interest.

For samples of greater complexity, if only limited components are of interest, and provided they meet the requirements stated above, then the target components should be resolvable; complete peaks of interest are cryotrapped and so are concentrated in space, so that once they are pulsed to the second column they recommence their migration as a very high-concentration band.

Provided the second column is operated under fast GC conditions, response at the detector is very much increased. An additional attractive advantage is that the system is simple to set up and operate, and a regular data system can easily handle the data derived from this method. Figure 8.5 illustrates the concept, where broad peaks on the first capillary column are focused and pulsed to column 2, which operates on a fast time scale. Thus the peaks appear tall and narrow, giving better sensitivity. Figure 8.6 shows an example; in this case, normal and targeted analysis runs are presented, with expansion of selected pulsed zones showing that originally overlapping peaks are now very well resolved. There would appear to be a useful place in normal MDGC for this approach.

8.3.3.1 Selected applications of conventional multidimensional gas chromatography

MDGC as a standard test procedure. One of the few standard methods that specifies a MDGC approach is that of the test method for paraffins, naphthenes and aromatics in petroleum distillates [42]. Specific information obtained by the test include a range of components that are reported as single groups of compounds (e.g. hydrocarbons with boiling points > 200°C; aromatics boiling at C_9 and above), and the only individual components measured are benzene and toluene (for the C_6 and C_7 aromatics) and cyclopentane (the

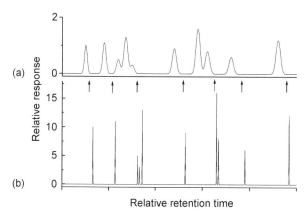

Figure 8.5 Simulated targeted multidimensional gas chromatography using the cryogenic modulation interface. (a) The normal capillary GC result is used to determine which solutes are to be co-trapped, as shown by the zones between the arrows. (b) The trapped peaks are rapidly pulsed to a short second dimension, where they elute on a fast time scale with greater response.

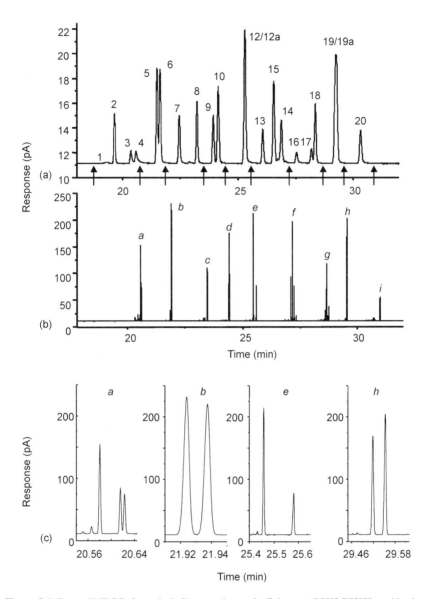

Figure 8.6 Targeted MDGC of a semivolatile aromatic sample. Column set BPX5-BPX50 combination, with 5.0 m BPX50 column. (a) Normal GC analysis, showing incompletely resolved peaks 5 and 6, and very poorly resolved peaks 12/12a, and 19/19a. (b) Each group of peaks shown by the arrows in (a) was cryogenically focused and pulsed into the BPX50 column at the time indicated by the arrow. Thus nine sets of pulsed peaks are seen. (c) Expansion of the groups a, b, e and h shows that now all peaks are resolved. Note that e (peaks 12/12a) and h (= peaks 19/19a) now show excellent resolution. Peaks 5 and 6 do increase in resolution but, because they are isomers, their increase in resolution is not so great.

only C_5 naphthene). The test method is applicable to hydrocarbon mixtures, including virgin, catalytically converted, thermally converted, alkylated and blended naphthas. In reading this method, the analyst may be led to the belief that MDGC is indeed a complex, technically demanding technique.

PCB analysis.   Specific congener analysis may be a required analytical target for PCB studies. The complexity of PCB mixtures (with a maximum possible 209 congeners) where many coelutions may occur is an ideal application for MDGC [43, 44]. The high degree of selectivity for PCB shape will permit unresolved components from column 1 to be separated on column 2. The function of the first column is also to exclude from the heartcut zone those components that may impair the column 2 analysis by causing co-elution on that column. Figure 8.7 is a typical series of chromatograms that illustrate how the data are acquired and employed for MDGC in this application. Results are presented for the selected congeners eluted on the analytical column (Figure 8.7a), Aroclor 1254 on the monitor detector (Figure 8.7b), Aroclor 1254 with seven heartcuts selected to pass the seven congeners of interest to the analytical column (Figure 8.7c) and results for the analysis of the seven heartcuts as a single chromatographic elution on the second column (Figure 8.7d). The process or sequence of analyses might be as follows:

1. Select appropriate columns; this will be done on the basis of the required component separations and data such as retention indices or relative retentions on the participating columns.
2. Determine suitable chromatographic conditions on each dimension.
3. Analyse each target congener (or small mixture standard) on each chromatographic column, acquiring retention data.
4. Set up the MDGC system, and obtain pre-column and analytical column retention data; determine heartcut timing.
5. Study technical PCB samples to investigate whether adequate resolution is obtained.
6. Analyse samples; conduct suitable quantitative standard runs.

Chiral MDGC has been used to determine enantiomeric ratios of chiral PCBs in shark samples [45] using an achiral/chiral column combination. A number of congeners were found to be essentially racemic, while one was found to have a small enantiomeric excess.

Toxaphene analysis.   Toxaphenes are a mixture of polychlorinated camphenes. As a class of compounds, these represent one of the most challenging separations, since technical toxaphene may contain several hundred compounds, and thousands of theoretically possible congeners if all are considered. MDGC has been reported to offer higher separation power, revealing many co-elutions [46]. A subsequent study focused on enantiomer ratios of bornane congeners

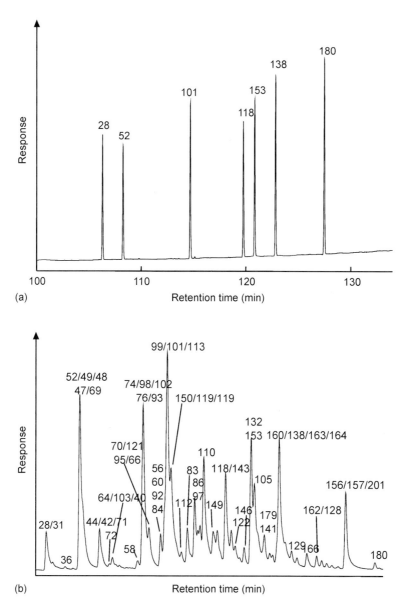

Figure 8.7 MDGC of Aroclor 1254. (a) Elution of a seven-congener standard of the PCBs on the analytical (second) column. (b) Monitor channel (first column) detector response for the Aroclor mixture. (c) As in (b), but with the heartcut regions now absent from the monitor detector response. (d) Analysis of the heartcut regions on the analytical detector. The target congeners should be directly comparable with the standard shown in (a).

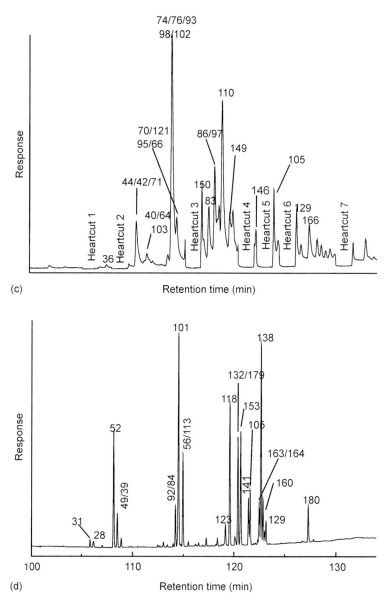

Figure 8.7 (continued).

in biological samples, with specific target analytes better characterised when using MDGC analysis [47]. Co-eluting impurities compromised single-column analysis when attempting to determine enantiomeric ratios; many components were found in the heartcut chromatograms, perhaps even more than would have been thought likely.

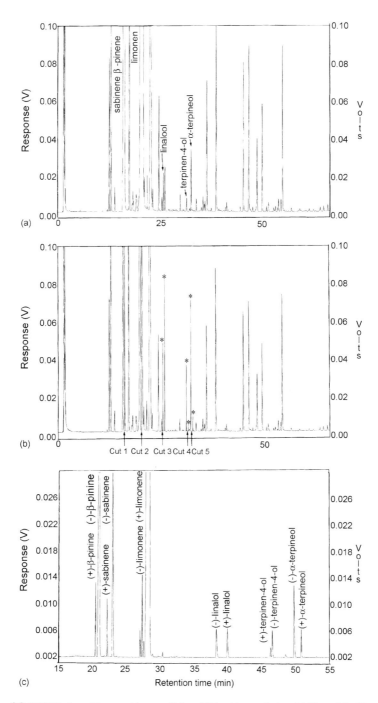

Figure 8.8 MDGC of a cold-pressed lemon oil from [48], with permission. (a) The original lemon oil GC result, with target peaks indicated. (b) The same analysis as in (a), but with the five heartcut regions implemented and the respective peaks absent from this chromatogram. (c) Chiral GC analysis of the five heartcut sections.

Essential oil analysis. This is one of the major application areas of MDGC, but even in this instance it appears to be limited to a few specialist laboratories. This interest stems from the potential complexity of the samples, variation in varieties of plants causing different ratios of components to be found, the importance of chiral analysis to characterisation of samples, and adulteration studies. Each of these can be aided by MDGC approaches. Since the oils are often dominated by a relatively few compounds in high concentration, overlap with minor components makes the analysis of the latter difficult. Additionally, chiral analysis can be fraught with uncertainty if accurate enantiomer ratios are required in circumstances where complete resolution from impurities cannot be guaranteed. Thus Mondello *et al.* applied MDGC to analysis of enantiomeric distributions of monoterpene hydrocarbons and alcohols of lemon oil using a dual-oven system [48]. Figure 8.8 illustrates the quality of results they were able to produce. The enantiomeric distributions of various components were assessed over several months for cold-pressed lemon oil, and further for distilled and commercial oils. The extraction procedure did not appear to influence the enantiomer distribution found. The procedure also allowed differentiation of the lemon oil from other citrus oils.

Other typical studies. Using a window diagram approach, Annino and Villalobos [49] were able to reduce the number of problem pairs of compounds before implementing the MDGC method. This was applied to target chlorinated compounds in a mixture of potential interferents. Methylcitric acid is a metabolite that acts as a diagnostic marker for certain diseases and may exist in four stereoisomeric forms. Their trimethyl ester derivatives may be analysed reliably by MDGC [50]. Flavour components have been studied by Mosandl and co-workers, who reported the separation of dill ether and its stereoisomers using MDGC with a two-oven system and a live T-switch [51]. The authentic dill oil was analysed with respect to stereoisomers that had been synthesised in the laboratory. Tobacco smoke is acknowledged as one of the most complex samples of environmental concern that is analysed by GC. By coupling a Carbowax column to an OV-1701 column, Gordon *et al.* [52] analysed contiguous sections of the first column's effluent, and presented second-dimension chromatograms that were still exceptionally complex, even though the sampling (heartcut) time was only of the order of 1.5–3 min in many cases. Some 306 compounds were identified. It is likely that this is still an underestimation of the true complexity, and this would be an interesting task for the comprehensive GC × GC technique (see below).

8.4   Comprehensive two-dimensional gas chromatography

The proposed comprehensive technique conceptually conceived by Giddings [53] (see Figure 8.9) is clearly the most powerful procedure for achieving the

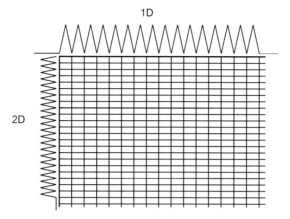

Figure 8.9 Diagram of Giddings' concept of peak capacity in a comprehensive chromatography experiment.

maximum degree of component separation, for all components, of a complex mixture. This diagram illustrates that there is no specific target region where the multidimensional process is applied, but rather it is achieved over the whole sample. In this way, there is no need to employ timed heartcuts for isolation of groups of compounds from a first-column elution. This is, therefore, a much simpler way to approach a sample analysis in 2D. However, it does not address many questions that might not have been apparent to Giddings, mainly because, without the technical implementation of his proposals, there would have been little value in considering the implications of data handling that the method has since revealed. Also, the figure might suggest that each individual peak is completely subjected to the second dimension separation in one step. It does correctly convey the multiplicative capacity that the system offers, so that total peak capacity is the product of the capacity of each dimension. A simple calculation suggests that system capacity is now an extraordinarily large number. If the first column has a total adjusted run time of 90 min, with average peak spacing for exactly baseline-resolved peaks of 10 s over this range, then the capacity of this dimension is 540 peaks. If the second dimension has the ability to resolve 15 peaks, (without regard to the actual mechanism that is required to achieve this), then system capacity is $540 \times 15 = 8100$ separable components. Given that GC $\times$ GC requires pulsing of peaks, we would have 40 500 (!) separate peaks in the chromatogram report if the average component gave five individual pulses. This suggests that even though no sample is likely to have the particular random chemical composition to take advantage of the polarity and volatility considerations that permit such a high component resolution on the two columns, there is still a tremendous opportunity to realise significant component separation in difficult regions of the sample where many components overlap on the first column. The second-dimension capacity is much smaller than that of the first dimension because we typically use a short column, with elution times of 3–5 s.

We can alter the diagram in Figure 8.9 to show the actual process of GC × GC a little better—at least in respect of the way it is conducted today. Figure 8.10 is the modified diagram, and while it does not alter the system capacity, it does indicate that the second dimension operates at a faster repetition rate than represented by the peak width duration on the first column. In actual experimentation, we will not have the uniformity of peak spacing, which again here is intended to show capacity considerations; rather we will have different degrees of neighbouring peak overlaps. We show here also that peaks in the second dimension are narrower than those in the first dimension, as expected. Figure 8.11 illustrates how we can view the instrumental set-up for GC × GC.

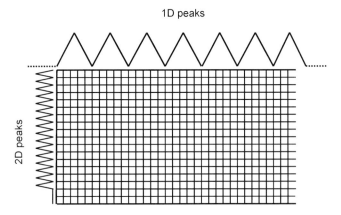

Figure 8.10 Modification of Giddings' concept to show the faster second-dimension analysis completed in a fraction of the time of a peak width in the first dimension.

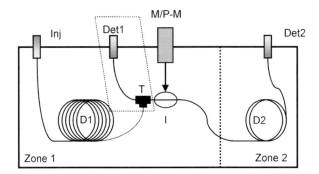

Figure 8.11 Instrumental arrangement for a comprehensive gas chromatography experiment. T, T-union allowing for splitting of effluent to a first detector; I, interface device; M/P-M, modulation/pulsing mechanism.

This diagram incorporates an optional T-union. Although normally not included, it is easily added and might be useful to provide a separate chromatogram of the progress of elution of compounds from the first column. In this way a comparison may be made of the separation at the end of D1 and that resulting from the GC × GC procedure. Ordinarily, however, the information provided by the final GC × GC separation is perfectly adequate, and does contain much more information.

Without the T-union, the GC × GC arrangement involves a directly coupled pair of columns, with no valving or flow switching. The interface or modulation/pulsing mechanism is the heart of the procedure, and is conceptually an easily appreciated process, even if its technical implementation has taken some years to realise.

8.4.1   General benefits of GC × GC

The advantages and characteristics offered by contemporary GC × GC may be summarised as follows:

1. Simplicity of the system set-up, because there are no valves and the columns need only be joined directly. The main consideration will be modulator performance and reliability.
2. Peak compression in most systems results in greater mass concentration of peak zones, and so greater detection sensitivity.
3. Fast analysis is employed on the second column, and this can be achieved by any or a combination of short columns ($\sim$1 m), narrow inside diameter capillaries (0.1 mm), thin-film columns ($d_f \sim 0.1$ μm), and greater carrier linear velocity (if column i.d. is reduced when going from the first to the second column); this allows rapid pulsing of the modulator without excessive overlap of successive pulsed bands.
4. The above fast analysis is of the order of 3–5s; this limits band broadening of the pulsed zones produced by the modulator and subsequently detected at the outlet of column 2, and maintains high peak sensitivity.
5. A tremendous increase in the number of resolved peaks.

In comparison with packed and normal capillary GC, we can offer a suggested improvement in the peak capacity achieved by using GC × GC, as shown in Table 8.3 based on a similar table elsewhere [54]. Note that each component gives multiple pulses in GC × GC, so the chromatographic report consists of many pulsed peaks.

8.4.2   Understanding comprehensive gas chromatography

According to the general principle of rapid elution on a second column with respect to peak widths on the first dimension, we can represent the two processes

Table 8.3 Suggested separable peaks using different gas chromatography methods. This table is only for comparison purposes, and will depend on the sample type and actual column performance in the experiment

Number of components	100 components	1000 components
Number of components separated and identifiable in packed GC	25 (1/4th of total)	100 (1/10th of total)
Number of components separated and identifiable in capillary GC	50 (1/2 of total)	250 (1/4th of total)
Number of peaks in the peak report if each component produces five modulated pulses in capillary GC	250 peaks	1250 peaks
Number of peaks in the peak report if each component produces five modulated pulses in capillary GC × GC *plus* the orthogonal system resolves ten times as many components	75 components resolved: 375 peaks	500 components resolved: 2500 peaks

as shown in Figure 8.12. The short, fast repetitive elutions correspond to the second dimension column and these are necessarily constrained by the time of pulsing of solute from column 1 to column 2. This does not mean that elution of peaks on column 2 has to occur within this time; it is possible for peaks to elute from D2 after the next pulse has been injected into D2. But it is preferable that peaks from one pulse do not physically intermingle in the second column, since this would decrease their separation. Figure 8.13 shows a comparison

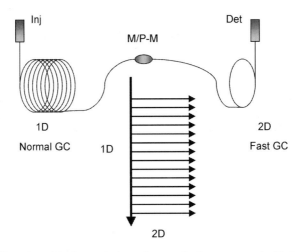

Figure 8.12 Illustration of the fast analysis on the second column compared with the slow (normal) analysis conducted on the first column.

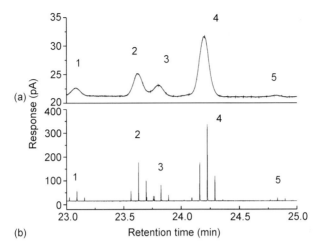

Figure 8.13 Comparison of a normal capillary GC analysis of components in tea tree essential oil (a) with the results obtained by using a modulation mechanism in a cryogenic system (b). Note the different response scales. Peaks numbered 1–5 are the same peaks in each chromatogram.

of normal capillary GC peaks for a section of a tea tree essential oil analysis (Figure 8.13a; peak half-widths ~6 s) with the pulsed GC result for the same region (Figure 8.13b; peak half-widths ~150 ms). The peak height for the largest peak of component 4 on the lower trace is about 320 pA, compared with the normal result of 11 pA—an increase of some 30-fold. Component 5 is hardly recognisable on the upper trace, but gives distinct peaks on the lower trace, and the overlapping components 2 and 3 show some degree of separation when pulsed together to the second column. By repeating the modulation process over the whole chromatographic region of interest, we build up a sequence of pulses for each compound, as shown in Figure 8.13. Each pulsed set constitutes that compound, and its total area is contained within these pulses. Hence mass is conserved and no quantitative data are lost. If we now line up the successive second-dimension chromatograms, we find that the pulses of a component will be essentially neighbouring, and they appear at about the same time after the pulsing event. By converting the data into a matrix, we can then use a suitable software package to show the peak as a contour plot. GC × GC only requires an appreciation of the role and efficacy of the pulsing process and the relative separation power/selectivity of the two participating columns. This will be highlighted below.

8.4.3   Methods for modulation of peak signals between two dimensions

Phillips, Venkatramani and Xu [55–57] introduced a metal-coated capillary at the junction between two columns that had a separate power supply to modulate the heat of the junction and so pulse migrating bands to the second column. Although good results could be obtained, the system was not sufficiently reliable to stand routine cycling. This system was replaced by a rotating thermal modulator or sweeper that has undergone much development since its introduction by Phillips and Ledford [58]. This system in its most advanced version offers the following components: primary dimension column, connected to an uncoated column section, connected to a thick film accumulator (modulator) column, connected to another uncoated column section, connected to the second dimension column. The system also has a separately heated oven zone for the second column to allow temperature tuning of this column compared to the first column.

The function of the uncoated sections is to ensure that the chromatographic band is essentially travelling unretained (i.e. with the carrier gas flow) in these sections. This is important when the band exits from the accumulator segment, since it must be presented as a sharp packet of solute to the second column.

The accumulator or modulator column performs the function of retaining the migrating band sufficiently to prevent it from breaking through the segment while the previous pulsed band or bands elute on the second column. Thus it is normally a thick-film column. The sweeper rotates over this column at a speed slower than the carrier flow velocity to thermally desorb the retained components, and then inject them into the gas phase to the second uncoated column and towards the second column. The thermal desorption must be carried out at up to 100°C higher than the prevailing oven temperature to allow sufficient heat to be transferred to the retained components and reduce their retention factors enough to have them reside essentially completely in the mobile phase. The features of this system and its development and problems have been reviewed [59].

Another system capable of producing very high 2D resolution data analogously to the above system is the diaphragm valve modulator developed by Synovec [60,61]. In this instance a very small slice of the effluent eluting from the end of the first column is diverted by operation of the valve so that it arrives at the head of the second column. This is capable of delivering pulses of 10 s of milliseconds width to the second column. In the same conceptual framework as the above system, overlapping peaks on the first column, when simultaneously passed to the second column, will be resolved according to different retention mechanisms provided these apply to the two dimensions. This system differs from that in 'regular' comprehensive GC × GC in that no zone focusing or band concentration occurs. It has been used in very fast analyses with chemometric methods providing interpretation of the results and, even though band concentration is not increased, some signal enhancement has been reported [62].

The third system for production of comprehensive two-dimensional GC × GC is the cryogenic modulator developed by Marriott and Kinghorn [63]. This system was originally demonstrated to be useful in a range of different modes, such as to enhance signal prior to detection, and offer improvement in injection performance [64]. Additionally, it is applicable to multidimensional gas chromatography by allowing material to be efficiently pulsed as a rapid band into a second column, whereas previous cryotraps were much slower to modulate between their low and high temperature range if a secondary electrical heater was employed; this approach can then be extended to GC × GC [65]. The construction of the system has been described in detail [66], with the pneumatic modulation drive mechanism shown. The essence of the system is that a hollow cryotrap is supplied with liquid coolant and the cryotrap region can be moved back-and-forth along the capillary column. Its low thermal mass permits the capillary to cool down very rapidly (when the cryotrap is located over the column) and to heat up (when the trap is moved and the circulating stirred oven heat comes into contact with the previously cold column section). The same system has been used for a range of semivolatile aromatics, and was comparable to the thermal modulation system [67]. The process of remobilisation of bands has been shown to be fast (low milliseconds), and modulation frequency at least as fast as 1 Hz is possible. The system has been demonstrated to allow flexible operation, with normal GC (when the CO_2 coolant is not applied), whole peak trapping (similar to the targeted mode described above) and GC × GC all conducted in the one analytical GC run [68]. The general performance of the system has been reviewed elsewhere [69, 70].

8.4.4   Orthogonality of separation

Maximum separation efficiency arises if the two dimensions of a separation obey some orthogonality criteria, which requires proper tuning of the GC × GC method [71]. Orthogonality means that they are decoupled in respect of the mechanism by which separation (or identification) is performed. Thus GC-MS can be considered an orthogonal two-dimensional technique because the GC separation and mass spectral identification possess no dependent relationship. In GC × GC we have two gas chromatographic dimensions. In order to achieve orthogonality, and decouple the two dimensions, we should ensure that the mechanisms of separation on each column are independent [72]. This is most readily achieved by using two columns of different polarities, and the most common approach is to use a first dimension of a nonpolar phase and the second of a more polar phase. Thus we have dispersion forces in the nonpolar column leading to a volatility separation; in the second column, to which compounds of isovolatility are delivered, specific molecular interactions between solute and phase now play a role to differentiate different chemical compounds in this dimension.

Phillips and Beens [72] reviewed the relevance of column orthogonality and discussed it in terms of two columns placed in one oven (hence the two columns are approximately at equal temperature when the solute passes through column 2). The retention factor on the second column was given by

$$k_{second} = \beta^{-1} \exp\left[\frac{\Delta\mu^0_{second}}{\Delta\mu^0_{first}}\right] \qquad (8.2)$$

where β is the phase ratio in the second column and $\Delta\mu^0$ is the chemical transport potential for a substance moving from mobile to stationary phase. Hence a particular component's retention on the second column is related to its respective interactions on both columns. Stronger interaction forces in column 1 will reduce the interaction in column 2, and lead to a lower retention k_{second} value. Conversely, stronger interaction in the second column will increase k_{second}. As an example, if two analyses are conducted sequentially, at different temperature program rates, then the lower program rate will elute solutes from the first column at a lower oven temperature. Since they enter the second column at lower temperature, they will have greater retention on this column than if they had been delivered to the column at higher temperature. Their chemical potential on column 2 will be higher at the lower temperature.

Note that the dispersive force on different stationary phases will be similar and so, if the second column is more polar, the dispersive force will have little effect on solute separation on this column, since this has effectively been cancelled out by their co-elution on the first column. The slight differences in polar interactions can then be enhanced to produce resolution. This would be an interesting concept to test in studies such as chiral separations, which have not yet been widely studied in GC × GC. Ledford *et al.* presented a discussion on the ordering of chromatograms [73] and illustrated the methodology in cartoon style, representing the ability of columns to separate on the basis of colour and shape and/or size. Depending on the characteristics of the two dimensions, different orders of compounds within the two-dimensional space are realised. This is also referred to as producing structured GC × GC chromatograms, and such results are not possible by any other means. This was then demonstrated by the patterns produced in the chromatogram of a jet fuel. The ability of GC × GC to allow classes of compounds to be grouped into distinct patterns of elution significantly aids understanding of the molecular make-up of complex samples, as shown by Phillips and Beens [72], Beens *et al.* [74] and Frysinger and Gaines [75] for petroleum-based samples. The particular clusters of compounds shown tend to be the alkylated aromatics within specific aromatic 'zones' in the two-dimensional space. It also suggests that the GC × GC separation has a novel role in characterisation of these samples.

Tuning of the separation has also been suggested to be aided by variability of temperature between the two dimensions, whether in a separate oven or with a small purpose-built heated zone in which column 2 is located. The elevated temperature should track that of the main oven but be advanced by a suitable temperature. de Geus *et al.* used independently temperature-programmable ovens to study the retention times on the second column [76].

Beens *et al.* [74] refer to the criteria for proper tuning of the GC × GC system as

1. choice of the correct dimensions of the second column in relation to the modulation time;
2. choice of an appropriate temperature program in relation to the dimensions of the first column and in relation to the substances to be separated;
3. determining a mass flow rate that is suitable for both columns.

In this study they used a range of different column sets, and in one case employed a different temperature for the second column from that for the first column.

8.4.5 *Presentation of GC × GC data*

We can summarise the process of GC × GC analysis in the following stepwise manner.

1. Perform normal capillary GC analysis.
2. Perform modulated analysis.
3. Expand segments of the modulated chromatogram to identify where overlapping peaks from column 1 are resolved on column 2. This allows a decision to be made whether the column 2 phase chosen is appropriate, and that elution conditions are adequate on column 2.
4. Ensure that sufficient pulses are generated for each first column peak.
5. Since each component eluted from column 1 now appears as a series of second-column pulsed peaks (e.g. four or more peaks) in the two-dimensional analysis, the time–response data can be transformed into a matrix format based on the sampling frequency and the modulation rate, and by presentation in a 2D space we get a contour plot for each peak. Thus the chromatogram appears as a rectangular space of (total analysis time × second dimension time), with contour peaks distributed throughout the space.

The above points can be considered further.

It is probably usual first to perform a 'normal' capillary GC run on a sample and then conduct the modulated analysis. In this case the analysis will be done sequentially on the coupled column. The normal GC analysis then serves to illustrate how the modulated run improves the analytical result, primarily in terms of sensitivity. Note, however, that in our work we often do not do the normal GC run—it is superfluous because the data or information contained in the GC × GC result exceeds that in the normal run, although we must take note of the present data processing constraints for GC × GC analysis.

By expanding the pulsed data it is easy to identify where resolution of compounds from the first column is obtained. This will indicate whether the resolution is adequate, and possibly allow determination of whether different conditions are required for the second column (stationary phase, or parameters affecting efficiency such as length, column inside diameter, flow rate, etc). The expanded trace will also show how many pulsed second-dimension peaks are obtained for each component. We would normally want at least four. This is because we then transform the data into a 2D space, and presumably the more pulsed peaks we generate the better will be the contour plot generation process. Clearly, trying to interpret a GC × GC chromatogram based on small expanded segments of data will be tedious, and the 2D plot summarises the total separation in an acceptable manner.

Finally, the analysis should be completed by a qualitative and/or quantitative report on the components identified in the sample, and the normal requisite data such as height, area, etc. At present such a reporting package is not available and, even though impressive results for complex sample analysis are available, with quantitative and qualitative data presented, this is far from an automated process. For instance, it is possible on some in-house systems to group peaks together and report their total areas, but this is still a hands-on process, and a suitable data processing package must be developed.

Applications of essential oil analysis serves to exemplify the above process. Figure 8.14 is a comparison of the normal and pulsed GC × GC results for tea tree oil. Note that all the features of the chromatogram are readily recognised in the GC × GC result. If we presented Figure 8.14b on the same scale as Figure 8.14a, considerably more peaks would be seen, now raised out of the baseline. Figure 8.15 is of a lavender sample, showing some different regions in expanded format. The three later peaks ($\sim$44 min) do not show much co-elution, but the tiny peaks at about 30–33 min show interleaving pulses of components, which are clearly seen in the topmost expanded trace, where the separate 4 s pulses are clearly displayed. As stated above, to perform such expansions and try to interpret the results is tedious, so in Figure 8.16 results for vetiver oil are given, with an approximate lining up of the pulsed GC × GC trace (lower) with the 2D separation space shown in contour plot format in the upper window. Peaks 1 and 2 are the two later prominent peaks, and peak 3 is the tailing peak. The largest peak is peak 4, and is located in the 2D space. Note that each dot in the 2D space is a real chromatographic peak. Figure 8.17 is a heating oil sample, analysed on a BPX5-BPX50 column set. Peaks eluting at later 2D retention time are aromatic components, and those between 2D = 0.5–1.0 s are alkanes. Many such 2D chromatograms are now available in the literature, but this is still a science in its infancy and there is much to anticipate in this area in the coming years.

8.4.6 Applications of GC × GC

The applications of GC × GC have tended to be directed towards difficult separations, and specifically petroleum fractions because of the complexity of the

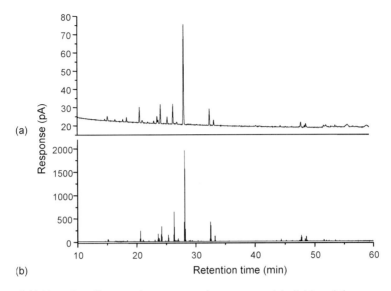

Figure 8.14 Normal capillary gas chromatogram of tea tree essential oil (a), and the same sample obtained with GC × GC cryogenic modulation conditions (b). A temperature program of 3°C min⁻¹ was used. The column set used comprised a BPX5-BP20 combination, with the second column about 1 m long. Note that the general chromatographic distribution appears essentially unaffected by the modulation process in terms of peak positions and overall presentation, except that the response is significantly different.

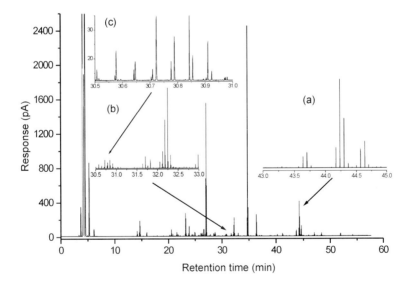

Figure 8.15 Lavender essential oil acquired under GC × GC cryogenic modulation conditions. A temperature program of 3°C min⁻¹ was used. The column set used comprised a BPX5-BP20 combination, with the second column about 1 m long. The insets show increasingly expanded time scales. Inset (c) is a 0.5 min display of selected very small peaks in the primary chromatogram, showing that the second-dimension resolution achieved for the pulsed zones is good.

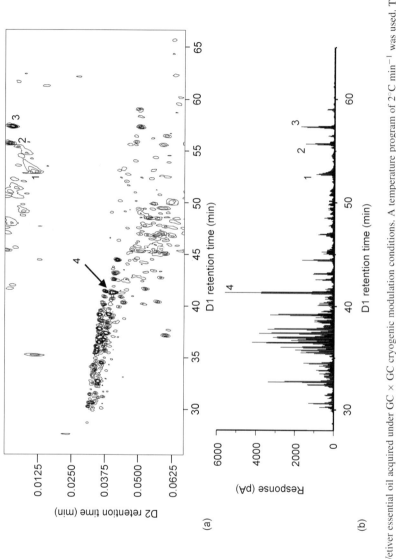

Figure 8.16 Vetiver essential oil acquired under GC × GC cryogenic modulation conditions. A temperature program of $2°C\ min^{-1}$ was used. The column set used comprised a BPX5-BP20 combination, with the second column about 1 m long. The lower trace is the pulsed chromatogram, and the upper trace is the two-dimensional separation space presentation for this sample. Note the apparent change in the sample composition (to more polar components) after about 42 min, when the components are much more spread out within the 2D space. Peaks labelled 1–4 are the same components located in each trace.

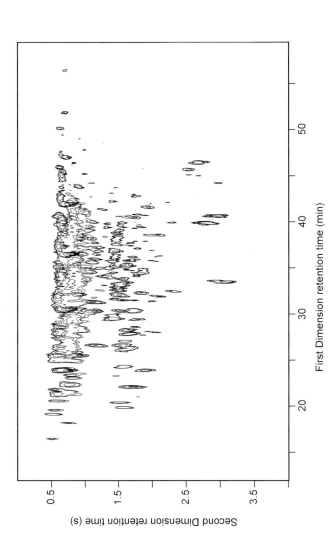

Figure 8.17 Analysis of a heating oil sample, acquired under GC × GC cryogenic modulation conditions. A slow temperature program of about 1°C min⁻¹ was used, from 50° C. The column set used comprised a BPX5-BP20 combination with the first about 2 m long and the second column about 1 m long.

challenge they pose. However, the general benefits of GC × GC are not lost on simpler samples, and more applications will soon demonstrate the universality of the approach. Table 8.4 lists a range of applications studies reported thus far.

It is apparent that there is a considerable diversity of applications, given the short period of the technique's existence and the few systems in use, although hydrocarbon separations predominate. As an increasing number of volatile and semivolatile samples are subjected to GC × GC study, the general applicability of the approach will be further strengthened.

Table 8.4 Application areas reported for comprehensive two-dimensional GC (GC × GC)

Application area	Brief details	Reference
Pesticides	Human blood serum sample	[77]
PCBs	Grouping of same Cl-number congeners on the same plane was found	[78]
Semivolatile organics	GC × GC and targeted operation compared	[40]
	Comparison of different modulation modes	[67]
Essential oil sample	Vetiver essential oil	[79]
Essential oil sample	Lavender and tea tree essential oil	[80]
Essential oil sample	Peppermint and spearmint essential oil	[81]
Atmospheric organics	Characterisation of organics implicated in ozone formation	[82]
Oxygenates in gasoline	Alcohols and ethers in a gasoline sample quantified	[83]
Chlorinated biphenyls and toxaphenes	Technical and biota samples studied	[76]
Petroleum and middle distillate oils	Tuning of separation to provide class separation	[74]
Sterols	Quantitative study of GC × GC and targeted MDGC operation	[41]
Kerosene	Jet-cooled cryogenic modulator performance studied	[84]
Kerosene	Longitudinally modulated cryogenic modulator performance studied	[85]
Oxygenates, saturates and aromatics	Fast analysis using two short columns	[86]
Benzene, toluene, ethylbenzene, xylenes (BTEX)	BTEX and other aromatics in gasoline	[87]
Weathered marine diesel oil	Samples compared for identification of oil spill sources	[88]
Marine diesel analysis	GC × GC-MS used in a very slow GC analysis step	[89]
White gas	BTEX and other components quantified using the diaphragm modulator	[61]
Petroleum samples	Qualitative and quantitative analysis of a variety of samples	[90–93]
Stabilised crude oil sample	Demonstration of cryogenic modulation	[65]
Alkylbenzenes	Chemometric study of detection limit enhancement	[62]
Fatty acids	Sample from marine organisms studied	[94]

A number of quantitative analyses are reported above, and there are several pertinent aspects that must be discussed in this respect.

1. The GC × GC technique presently suffers from lack of a reliable, effective and automated data analysis system for post-run data handling. Thus no complete peak report is available and considerable manual deconstruction and manipulation of the time–response data stream is necessary once it is acquired. The data stream consists of the pulsed peak data, and to assign relevant pulses to each individual compound is not a trivial exercise.

2. The peak compression effect in certain modes of GC × GC, combined with the fast second-column analysis time, means detection limits are improved, and peaks may be quantified to a much lower level than in normal capillary GC. The precision of peak measurement is also reported to be better in some studies.

3. In studies on quantification of oil components, the data quality is apparently improved over normal single-column analysis. The reduced level of peak coalescence or overlap means that individual peaks are now much purer, and so areas may be determined with more confidence. Also, it has been reported that it is almost always possible to locate a true detector baseline, free from chemical signal, and this allows more accurate chromatographic baseline detection.

4. The two-dimensional separation space produced by GC × GC allows for ready identification of individual compounds and classes of compounds, which may easily be grouped together and area-summed. Note that the data-handling procedures used for this are in-house, written specifically for this purpose. This now permits flame ionisation detection (FID) to be used quantitatively with little peak interference, whereas in single-column analysis mass spectral data are most likely required and this involves more convoluted quantitative response calculations or assumptions. Beens *et al.* commented favourably on the quality of their GC × GC data compared with single-column work for petroleum samples [90].

8.4.7   Standards and mass spectrometry studies with GC × GC

The very complexity of samples that are now resolvable in GC × GC poses a number of questions about how analysis should be conducted in both qualitative and quantitative modes. For instance, does the very existence of many more peaks (e.g. 10-fold more resolved peaks) require an order of magnitude increase in number of calibration graphs and/or standards used for an analysis? As stated above, if underlying interference peaks are absent, then FID should be usable, and since its response factors are well established, this should aid quantification. In this case it is only necessary to identify the component and apply the relevant response factor. This may still suggest that more standards are needed, and if so the analytical procedure must take cognisance of this. This is not an argument against the GC × GC technique, but rather points out the problems of relying on single-column analysis where fewer compounds are resolvable and many probably missed completely. If the required standards are not available, they will also not be available for single-column analysis. Identification through the use of mass spectrometry will soon be the major challenge in routine and research laboratories where GC × GC is studied. Just as GC-MS has become a primary tool in many laboratories, and was acknowledged earlier to be a reason for less effort being expended in improving the separation part of the analysis, there will likely be a need for similar identification power to support the separation achieved in GC × GC studies. Fast scanning time-of-flight mass spectrometry is the only current technology capable of handling the fast peaks produced [95,96], since spectral acquisition of at least 50 Hz will normally be required for peaks that are of the order of 200–400 ms wide at base, giving at least 10 spectra per peak and enabling sufficiently accurate peak shape and area to be obtained.

The mass sensitivity is again improved through the peak compression effect, and the spectral purity is increased where peak resolution gives less peak overlap, which may be significant in single-column analysis. However, one fundamental difference exists between the reliance on GC-MS analysis in single-column GC and its use in GC × GC. The 'low-resolution' single-capillary GC experiment may demand MS detection if there is no other way to obtain reliable quantitative data (e.g. in the case of severe peak overlap). However, the high-resolution GC × GC experiment may very well permit FID quantification of these same peaks. This simplifies the analysis. The GC-MS then serves to provide identification in a separate analysis, and once the peak positions are established, the samples can then be analysed by GC × GC-FID, provided the peak positions are not shifted in the different experiments. At present there are few data available on GC × GC-MS, although this is likely to change in the short term. Recent presentations at the 23rd International Symposium on Capillary Chromatography suggest that there is active research commencing in this area [97].

Acknowledgements

The authors acknowledge the support provided by SGE International, which enabled their ideas for the cryogenic modulation system to be realised, and support with consumables. Hewlett Packard are thanked for placing some model 6890 gas chromatographs in our laboratory, which has allowed further studies to be undertaken. P.J.M. thanks his current research students for providing some of the results reported in this chapter.

References

1. U.A.Th. Brinkman (ed.), *Hyphenation in Chromatography, Hype and Fascination*. Elsevier, Amsterdam, 1999.
2. H.J. Cortes, *J. Chromatogr.*, **1992**, *626*, 3.
3. K. Grob, *J. Chromatogr.*, **1992**, *626*, 25.
4. J. Beens and R. Tijssen, *J. Microcolumn Sep.*, **1995**, *7*, 345.
5. J. Blomberg, P.C. de Groot, H.C.A. Braqndt, J.J.B. van der Does and P.J. Schoenmakers, *J. Chromatogr.*, **1999**, *849*, 483.
6. L. Mondello, G. Dugo and K.D. Bartle, *LC-GC Int.*, **1998**, *11*, 26.
7. L. Mondello, G. Dugo and K.D. Bartle, *J. Microcolumn Sep.*, **1996**, *8*, 275.
8. W. Jennings, *Analytical Gas Chromatography*, Academic Press, London, 1987.
9. M.J.E. Golay, in *Gas Chromatography 1957* (eds. V.J. Coates, H.J. Noebels and I.S. Fagerson), Academic Press, New York, 1958.
10. J.M. Davis and J.C. Giddings, *Anal. Chem.*, **1983**, *55*, 418.
11. J.M. Davis, in Advances in Chromatography, vol. 34 (eds. P.R. Brown and E. Grushka), Marcel Dekker, New York, 1994, pp. 109–176.
12. W. Bertsch, *J. High Resolut. Chromatogr., Chromatogr. Commun.*, **1978**, *1*, 1, 85, 289.
13. W. Bertsch, *J. High Resolut. Chromatogr.*, **1999**, *22*, 647.
14. W. Bertsch, *J. High Resolut. Chromatogr.*, **2000**, *23*, 167.
15. D.R. Deans, *Chromatographia*, **1968**, *1*, 18.
16. G. Schomburg and F. Weeke, in *Gas Chromatography 1972*, (eds. S.G. Perry and E.R. Adlard), The Institute of Petroleum, UK, 1972, p. 285.
17. J.B. Phillips, D. Luu, J.B. Pawliszyn and G.C. Carle, *Anal. Chem.*, **1985**, *57*, 2779.
18. M. Zhang and J.B. Phillips, *J. Chromatogr.*, **1995**, *689*, 275.
19. W. Jennings, *Analytical Gas Chromatography*, Academic Press, London, 1987, pp. 171–174.
20. P. Sandra, F. David, M. Proot, G. Diricks, M. Verstappe and M. Verzele, *J. High Resolut. Chromatogr., Chromatogr. Commun.*, **1985**, *8*, 782.
21. J.V. Hinshaw and L.S. Ettre, *Chromatographia*, **1986**, *21*, 561.
22. J.V. Hinshaw and L.S. Ettre, *Chromatographia*, **1986**, *21*, 669.
23. D. Repka, J. Krupcik, E. Benicka, T. Maurer and W. Engewald, *J. High Resolut. Chromatogr.*, **1990**, *13*, 333.
24. (a) W. Engewald and T. Maurer, *J. Chromatogr.*, **1990**, *520*, 3. (b) T. Maurer W. Engewald and A. Steinborn, *J. Chromatogr.*, **1990**, *517*, 77.
25. R. Sacks, H. Smith and M. Nowak, *Anal. Chem.*, **1998**, *70*, 29A.
26. R. Sacks, C. Coutant, T. Veriotti and A. Grall, *J. High Resolut. Chromatogr.*, **2000**, *23*, 225 and references therein.
27. B. Lourentzeas, P.J. Marriott and J. Hughes, unpublished results.
28. J.C. Giddings, *Anal. Chem.*, **1984**, *56*, 1258A.
29. D.R. Deans, *Chromatographia*, **1968**, *1*, 18.
30. D.R. Deans, *J. Chromatogr.*, **1965**, *18*, 477.
31. D.R. Deans, *J. Chromatogr.*, **1981**, *203*, 19.
32. G. Schomburg, in *Sample Introduction in Capillary Gas Chromatography* (ed. P. Sandra), vol. 1, Alfred Heuthig Verlag, Heidelberg, 1985, pp. 235–261.
33. D.E. Willis, in *Advances in Chromatography*, vol. 28 (eds. J.C. Giddings, E. Grushka and P.R. Brown), Dekker, New York, 1989.
34. J. Blomberg and U.A.Th. Brinkman, *J. Chromatogr.*, **1999**, *831*, 257.
35. R.M. Kinghorn, P.J. Marriott and M. Cumbers, in *18th International Symposium on Capillary Chromatography*, vol. 2 (eds. P. Sandra and G. Devos), Hüthig, Heidelberg, 1996.
36. H. Cortes (ed.), *Multidimensional Chromatography: Techniques and Applications*, Dekker, New York, 1990.

37. G. Schomburg, *J. Chromatogr.*, **1995**, *703*, 309.
38. H.-J. de Geus, J. de Boer and U.A.Th. Brinkman, *Trends in Analytical Chemistry*, **1996**, *15*, 168.
39. K.A. Krock and C.L. Wilkins, *Trends Anal. Chem.*, **1994**, *13*, 13.
40. P.J. Marriott, R.C.Y. Ong, R.M. Kinghorn and P.D. Morrison, *J. Chromatogr. A*, **2000**, *892*, 15.
41. T. Truong, P.J. Marriott and N.A.P. Porter, *J. AOAC Int.*, **2001**, *84*, 323.
42. Standard test method for paraffin, naphthene, and aromatic hydrocarbon type analysis in petroleum distillates through 200°C by multidimensional gas chromatography, in *Annual Book of ASTM Standards 2000*, Section 5, *Petroleum products, lubricants, and fossil fuels*, Volume 05.01, *Petroleum products and lubricants* (I): D56-D2596, pp. 411–420.
43. R.M. Kinghorn, P.J. Marriott and M. Cumbers, *J. High Resolut. Chromatogr.*, **1996**, *19*, 622.
44. T. Anastasopoulos, P.J. Marriott and R.M. Kinghorn, *LC-GC Int.*, **1998**, *11*, 106.
45. G.P. Blanch, A. Glausch, V. Schurig, R. Serrano and M.J. Gonzalez, *J. High Resolut. Chromatogr.*, **1996**, *19*, 392.
46. J. de Boer, H.-J. de Geus, U.A.Th. Brinkman, *Environ. Sci. Technol.*, **1997**, *31*, 873.
47. H.-J. de Geus, R. Baycan-Keller, M. Oehme, J. de Boer and U.A.Th. Brinkman, *J. High Resolut. Chromatogr.*, **1998**, *21*, 39.
48. L. Mondello, M. Catalfamo, A. Cotroneo, G. Dugo, G. Dugo and H. McNair, *J. High Resolut. Chromatogr.*, **1999**, *22*, 350.
49. R. Annino and R. Villalobos, *J. High Resolut. Chromatogr.*, **1999**, *22*, 589.
50. F. Podebrad, M. Heil, A. Scharrer, S. Feldmer, O. Schulte-Mater, A. Mosandl, A.C. Sewell and H.J. Bohles, *J. High Resolut. Chromatogr.*, **1999**, *22*, 604.
51. S. Reichert and A. Mosandl, *J. High Resolut. Chromatogr.*, **1999**, *22*, 631.
52. B.M. Gordon, M.S. Uhrig, M.F. Borgerding, H.L. Chung, W.M. Coleman, J.F. Elder, J.A. Giles, D.S. Moore, C.E. Rix and E.L. White, *J. Chromatogr. Sci.*, **1988**, *26*, 174.
53. J.C. Giddings, *J. High Resolut. Chromatogr., Chromatogr. Commun.*, **1987**, *10*, 319.
54. P.J. Marriott, *Proceedings, 23rd International Symposium on Capillary Chromatography* (eds. P. Sandra and A.J. Rackstraw).
55. J.B. Phillips and C.J. Venkatramani, *J. Microcolumn Sep.*, **1993**, *5*, 511.
56. Z. Liu and J.B Phillips, *J. Microcolumn Sep.*, **1994**, *6*, 229.
57. J.B. Phillips and Z. Liu, *J. Chromatogr. A*, **1995**, *703*, 327.
58. J.B. Phillips and E.B. Ledford, *Field Anal. Chem. Tech.*, **1996**, *1*, 23.
59. J.B. Phillips, R.B. Gaines, J. Blomberg, F.W.M. van der Wielen, J.M. Dimandja, V. Green, J. Grainger, D. Patterson, L. Racovalis, H.J. de Geus, J. de Boer, P. Haglund, J. Lipsky, V. Sinha and E.B. Ledford, *J. High Resolut. Chromatogr.*, **1999**, *22*, 3.
60. B.J. Prazen, C.A. Bruckner and R.E. Synovec, *J. Microcolumn Sep.*, **1999**, *11*, 97.
61. C.A. Bruckner, B.J. Prazen and R.E. Synovec, *Anal. Chem.*, **1998**, *70*, 2796.
62. C.G. Fraga, B.J. Prazen and R.E. Synovec, *J. High Resolut. Chromatogr.*, **2000**, *23*, 215.
63. P.J. Marriott and R.M. Kinghorn, *Anal. Chem.*, **1997**, *69*, 2582.
64. R.M. Kinghorn, P.J. Marriott and P.A. Dawes, *J. Microcolumn Sep.*, **1998**, *10*, 611.
65. R.M. Kinghorn and P.J. Marriott, *J. High Resolut. Chromatogr.*, **1999**, *22*, 235.
66. R.M. Kinghorn, P.J. Marriott and P.A. Dawes, *J. High Resolut. Chromatogr.*, **2000**, *23*, 245.
67. P.J. Marriott, R.M. Kinghorn, R.C.Y. Ong, P.D. Morrison, P. Haglund and M. Harju, *J. High Resolut. Chromatogr.*, **2000**, *23*, 253.
68. P.J. Marriott and R.M. Kinghorn, *J. Chromatogr. A*, **2000**, *866*, 203.
69. P.J. Marriott and R.M. Kinghorn, *Anal. Sci.*, **1998**, *14*, 651.
70. P.J. Marriott and R.M. Kinghorn, *Trends Anal. Chem.*, **1999**, *18*, 114.
71. C.J. Venkatramani, J. Xu and J.B. Phillips, *Anal. Chem.*, **1996**, *68*, 1486.
72. J.B. Phillips and J. Beens, *J. Chromatogr.*, **1999**, *856*, 331.
73. E.B. Ledford, J.B. Phillips, J. Xu, R.B. Gaines and J. Blomberg, *Am. Lab.*, **1996**, 22.
74. J. Beens, J. Blomberg and P.J. Schoenmakers, *J. High Resolut. Chromatogr.*, **2000**, *23*, 182.
75. G.S. Frysinger and R.B. Gaines, *J. High Resolut. Chromatogr.*, **2000**, *23*, 197.

76. H.-J. de Geus, A. Schelvis, J. de Boer and U.A.Th. Brinkman, *J. High Resolut. Chromatogr.*, **2000**, *23*, 189.
77. Z. Liu, S.R. Sirimanne, D.G. Patterson, L.L. Needham and J.B. Phillips, *Anal. Chem.*, **1994**, *66*, 3086.
78. J.B. Phillips and J. Xu, *Organohalogen Compounds*, **1997**, *31*, 199.
79. P. Marriott, R. Shellie, J. Fergeus, R. Ong and P.D. Morrison, *Flavour Fragr. J.*, **2000**, *15*, 225.
80. R. Shellie, P.J. Marriott and C. Cornwell, *J. High Resolut. Chromatogr.*, **2000**, *23*, 554.
81. J.M.D. Dimandja, S.B. Stanfill, J. Grainger and D.G. Patterson, *J. High Resolut. Chromatogr.*, **2000**, *23*, 208.
82. A.C. Lewis, N. Carslaw, P.J. Marriott, R.M. Kinghorn, P. Morrison, A.L. Lee, K.D. Bartle and M.J. Pilling, *Nature*, **2000**, *405*, 778.
83. G.S. Frysinger and R.B. Gaines, *J. High Resolut. Chromatogr.*, **2000**, *23*, 197.
84. E.B. Ledford and C. Billesbach, *J. High Resolut. Chromatogr.*, **2000**, *23*, 202.
85. R.M. Kinghorn and P.J. Marriott, *J. High Resolut. Chromatogr.*, **1998**, *21*, 620.
86. R.B. Gaines, E.B. Ledford and J.D. Stuart, *J. Microcolumn Sep.*, **1998**, *10*, 597.
87. G.S. Frysinger, R.B. Gaines and E.B. Ledford, *J. High Resolut. Chromatogr.*, **1999**, *22*, 195.
88. R.B. Gaines, G.S. Frysinger, M.S. Hendrick-Smith and J.D. Stuart, *Environ. Sci. Technol.*, **1999**, *33*, 2106.
89. G.S. Frysinger and R.B. Gaines, *J. High Resolut. Chromatogr.*, **1999**, *22*, 251.
90. J. Beens, H. Boelens, R. Tijssen and J. Blomberg, *J. High Resolut. Chromatogr.*, **1998**, *21*, 47.
91. J. Beens, J. Blomberg and R. Tijssen, *J. Chromatogr. A*, **1998**, *882*, 233.
92. J. Blomberg, P.J. Schoenmakers, J. Beens and R. Tijssen, *J. High Resolut. Chromatogr.*, **1997**, *20*, 539.
93. J. Beens, R. Tijssen and J. Blomberg, *J. High Resolut. Chromatogr.*, **1998**, *21*, 63.
94. T. Truong, P.J. Marriott, P.D. Nichols and N.A.P. Porter, unpublished results.
95. R.J.J. Vreuls, J. Dalluge and U.A.Th. Brinkman, *J. Microcolumn Sep.*, **1999**, *11*, 663.
96. M.M. van Deursen, J. Beens, H.-G. Janssen, P.A. Leclercq and C.A. Cramers, *J. Chromatogr. A*, **2000**, *878*, 205.
97. Presentations by R.J.J. Vreuls and M.M. van Deursen, 23rd International Symposium on Capillary Chromatography, Riva del Garda, Italy, 2000.

9 On-line and at-line gas chromatography

Tom Lynch

9.1 Introduction

The petroleum and chemical industries are major users of gas chromatography [1] and have had a major influence on its development since the birth of the technique almost 50 years ago (e.g. [2]). In these industries the technique has historically been applied in two distinct modes of operation, viz. in a central laboratory or as an on-line method. Until recently, the practitioners and suppliers of these techniques have tended to be firmly positioned in one camp or the other, with minimal interaction between the two. This has resulted mainly from the hardware needs of the instruments being very different in these areas. For on-line analysis in the petroleum and chemical industry, the instruments are located in the process plant and are required to meet stringent regulations to ensure they are safe for operation in these hazardous areas. They are also required to be reliable and to provide continuous, unattended operation while exposed to a harsh environment. These requirements have traditionally limited the applicability of technology that is routinely used in laboratory instruments, as the cost of making these technologies robust enough to survive the process environment has been high.

However, in recent years the requirements of the industry and the development of new technologies has resulted in a greater coming together of the two techniques that has seen the technologies used in both types of instrument converge and has resulted in the launch of commercial at-line gas chromatographs that bridge the gap between the true on-line and laboratory instruments.

This chapter will review and discuss the recent developments and applications in on-line and at-line gas chromatography for chemical process analysis that have occurred in recent years and will attempt to speculate on what the future may hold for the technique. A literature search reveals that there is very little published on this subject and therefore most of the information published here has been obtained primarily from a combination of personal experience and opinion, together with consultation with experienced practitioners and many of the major instrument companies in the field. As a result, many of the references quoted will differ from normal in that they refer to manufacturers' publications and in many cases to publications and information freely available on the internet.

9.2   Why do we need process analysis?

In 1998, a Laboratory of the Government Chemist survey [3] of industry in the UK showed that analysis is an activity with a spend of £7.4 billion per annum that is carried out in about 16 000 laboratories (5000 of which are industrial) and reveals that 57% of analytical measurements are related directly to the control of a process.

Better process control leads to increased profitability, which can be gained by a reduction in operating costs (including a reduction in laboratory analytical costs) and increased production. The increased control of the process may also result in a reduction in the amount of off-specification product and/or a reduction in the utilities (e.g. water, steam, and power) required to run the plant. All of these can show a direct cost benefit through optimisation of process parameters. Furthermore, having a plant under total control can give a significant reduction in the emissions generated by the plant and therefore have a positive environmental impact.

In the petroleum/bulk chemical industry, product volumes are generally large and margins are small and therefore even a small improvement in process efficiency or reliability can make a dramatic change in profitability. As a result, process control is becoming increasingly important and process measurements linked to automated control systems are now considered essential, which in turn has led to them becoming key areas for development.

Process control may be achieved simply by measuring temperature, pressure or flow rate of the stream of interest, but it is often necessary to determine the detailed chemical composition of a multicomponent stream. In this case analytical chemistry provides a major data input into the control systems.

Many analytical techniques measure physical properties, but when the analysis requires multicomponent compositional data, generally spectroscopic and chromatographic methods dominate. In BP Amoco significant developments have been made in recent years in the application of both spectroscopy and chromatography in process analysis and in many plants both techniques operate side by side. The plants on the site where the author is based comprise both continuous and batch operations and range in production capacities from ∼ 5000 to 300 000 tonnes per annum. They employ a wide range of chemistries and many are now controlled using advanced computer-based distributive control systems (DCS). These systems require data to act upon and DCS implementation has been a major factor in promoting the development of on-line and at-line analytical techniques.

9.2.1   On-line, at-line or central laboratory?

The decision whether to employ on-line, at-line or laboratory-based gas chromatography depends on many factors, including cost, sampling issues, flexibility

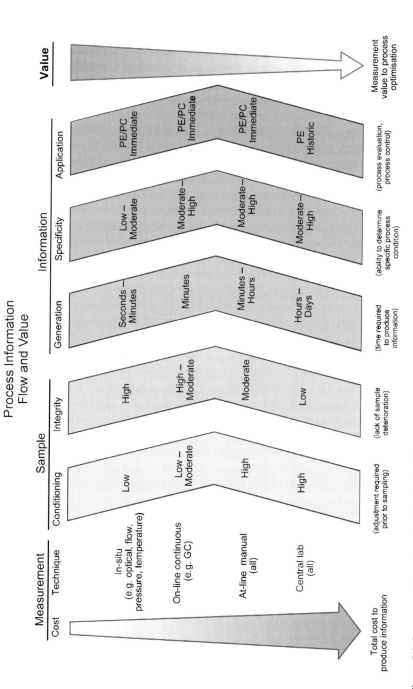

Figure 9.1 Summary guide to the principles involved in assessing the cost and complexity of a process measurement and correlating it with the value of the information received in relation to process optimisation. (Reproduced with permission of Douglas Jansen, Agilent Technologies, Inc., Little Falls, USA.)

and the control response time of the process plant. The type of analysis required, for example process streams or product quality, can also be a major factor in the decision. A useful summary guide to the principles involved in assessing which technology to use has been prepared by Douglas Jansen and is shown in Figure 9.1. This gives a simple graphical representation of how the cost and complexity of the process measurement can be correlated with the value of the information received in relation to process optimisation. The example shown here is a generalisation with numerous potential exceptions but it serves to demonstrate the principles involved and provides a useful tool to focus on the key issues to consider.

In the petroleum/bulk chemical industries, compositional data have tradition-ally been supplied by a central laboratory running scheduled analysis on a 24-hour, 365-days-a-year basis. In a modern production facility this is considered to be too costly in manpower and also inflexible to the demands of modern process plants. The long time delay in receiving the information can limit the ability of the plants to respond rapidly to optimise performance and to respond to market requirements.

On-line gas chromatography can increase the frequency of the analysis and release the plants from the restrictions of the laboratory schedule, but it has often been limited (usually owing to cost restraints) to the analysis of a single stream or component and has traditionally been inflexible in terms of chro-matographic performance. On-line chromatographic data processing systems have also traditionally been less flexible and user-friendly than their laboratory counterparts. In addition, the sampling systems and lines associated with on-line gas chromatography can frequently exceed the cost of the actual instrument by a significant margin. These limitations led to the search for a compromise solution, which has resulted in the development of not only the traditional on-line gas chromatographs but also the at-line approach, as well as a renewed interest in modifying what were considered to be laboratory-based instruments for application in process environments.

9.3   Trends and recent developments in process gas chromatography

Until recently the design and features available on on-line process gas chro-matographs had showed little change for a number of years and the instruments produced by the major manufacturers differed mainly in cosmetic appearance rather than in the technology applied. This resulted from the operational require-ments and limitations imposed by the environments in which they had to operate. These chromatographs have been specifically designed with the single aim of producing reliable compositional data almost on a continuous basis to allow the control of the process.

These systems have the following key features.

1. They are located on the process plant and are therefore designed for operation in harsh and hazardous environments.
2. They employ the highest safety standards with fail-safe design.
3. They are generally designed to measure only the key components required to control the process.
4. They use predominantly packed columns and isothermal operation, with extensive use of valves to obtain chromatographic performance with fast cycle times.
5. Fully integrated hardware and software which communicates directly with the plant control regime.

A comprehensive treatment of traditional on-line process gas chromatography is given by Annino and Villalobos [4], which is recommended for those desiring a more detailed insight into the subject.

In recent years the increasing demands for higher performance from process analysis instrumentation and the drive to reduce costs for chemical manufacturing processes have led to a re-evaluation of the features required for a process control measurement. For example, when considering the cost of the analysis, the cost of the analyser was traditionally seen as a major factor. This was particularly true when the capital cost of building the manufacturing plant was the major cost considered. However, when considering the overall profitability of a chemical process plant over its lifetime, the running costs of the plant on a day-to-day basis are often of prime importance. It is interesting, therefore, to look at the lifetime costs of running these instruments, and a typical cost breakdown for an on-line process gas chromatograph through a 15-year lifetime is shown in Figure 9.2.

The results are surprising because they show that the actual cost of the analyser accounts for only 9% of the total cost, with a staggering 43% going on maintenance and 19% on utilities (i.e. gases and power). Also, during its lifetime over half of the original analyser cost will have to be spent on spares to keep it running. This type of cost analysis has led not only to a refocusing of the design and performance of on-line gas chromatograph instruments but also to the consideration of alternative routes for obtaining process analysis data by gas chromatography. This has led to a range of technical developments that can be broadly categorised as follows:

- the development of conventional on-line gas chromatography instrumentation
- the development of gas chromatography-based transmitter instruments
- the application of laboratory-style gas chromatographs as on-line instruments
- the development of at-line gas chromatography as a process analysis tool

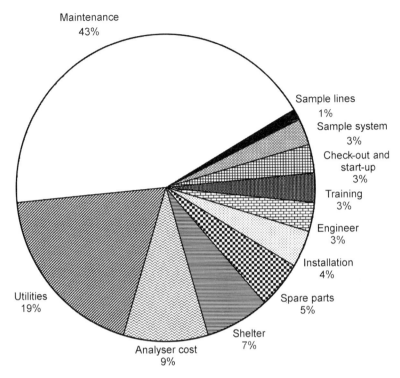

Figure 9.2 Typical cost breakdown for an on-line process gas chromatograph through a 15-year lifetime. (Reproduced with permission of Richard Elsenheimer, BP, USA.)

The remainder of this chapter will consider each of these themes and discuss some examples to illustrate the key issues and developments.

9.4 Recent developments in conventional on-line process gas chromatography instrumentation

The conventional on-line process gas chromatograph showed little change for many years as the quest for reliability and consistency tended to overcome the desire for performance. There was also a considerable body of opinion through the late 1980s and early 1990s that gas chromatography was a mature technique that was past its peak in terms of developments. However, as with many other areas of technology the microchip revolution over the latter part of the twentieth century has offered many new opportunities for improvement over a range of key performance areas for gas chromatography. These have ranged from the development of accurate and precise pressure and flow control technologies, and the routine implementation of wall-coated open tubular (capillary) columns, through to the advanced microprocessor monitoring and control of instrument hardware and the development of integrated software products for data capture and processing. This section will consider some of these recent technological advances and how they have influenced the development of on-line process gas chromatography.

9.4.1   The application of live switching techniques

Conventional process gas chromatography makes extensive use of multidimensional systems to achieve the required speed and chromatographic efficiencies. Although there are many ways in which multidimensional chromatography can be employed to change flows through columns, only a few are used extensively in process chromatography:

- distribution: e.g. the transfer of a group of components from one column to other columns with differing characteristics to facilitate component resolution and/or speed of analysis
- backflushing: e.g. to reduce analysis time by reversing the flow in the column such that highly retained components are eluted from the column either to vent or to a detector to provide a total backflush result
- heartcutting: e.g. when a specific component of interest is 'cut' from the first column onto a second column; often used in trace analysis to prevent bulk components from reaching the detector

Traditionally these have been achieved by the use of valve switching techniques based on mechanical valves that have been employed to switch the flows through the various columns and detectors in the system for over 40 years [5]. Rotary, slide and diaphragm valve designs have been employed, but they can have major disadvantages: they are subject to wear and leakage and therefore often require extensive maintenance; sample components come into contact with materials that are not ideal surfaces for chromatography and can therefore be affected through adsorption and/or memory effects; they may have temperature limitations; and the hardware is generally not suited for capillary columns and certainly not for the narrow-bore columns now being employed for high-resolution fast chromatography.

In 1968 Deans [6] published the technique of valveless switching that now bears his name, and the Deans principle has been employed extensively for many applications over the years. Deans employed in-line restrictors and two pressure controllers (one for the head pressure and one for the mid-point pressure) to balance flows while redirecting them through the two-column system by opening and closing external valves. The Deans method was limited to only a few switching modes and was mainly applicable to packed columns. The act of switching causes high pressure differences in the column coupling piece and this causes large flow fluctuations that can result in varying retention times and cut windows.

However, as the application of open tubular capillary columns became more and more widespread there was a desire to employ them in on-line process analysis systems; the major breakthrough that allowed this was the development of live capillary column switching by Siemens scientists, which was patented in 1976 [7]. The live column switching is based on the Deans approach except that it employs very small pressure differences across a special coupling T-piece, which results in very small flow differences. This gives a constant flow through all the columns and restrictors, resulting in more stable detector baselines and also allowing the use of intermediate detectors to monitor the cut vent. The live switch T-piece is part of a pneumatic bridge that operates in a very similar manner to an electrical Wheatstone bridge and contains four flow restrictors, two of which are variable, such that the differential pressure across the bridge is adjustable and can be instantly switched by operation of a solenoid valve. The two capillary columns are connected by the live T-piece (shown schematically in Figure 9.3) which is in fact two T-pieces coupled together by a precisely dimensioned piece of platinum–iridium capillary. It is designed with minimum dead volume to be fully compatible with capillary columns and the desired sample components only come in contact with the platinum capillary thus minimising adsorption effects.

A typical live switching manifold is shown schematically in Figure 9.4.

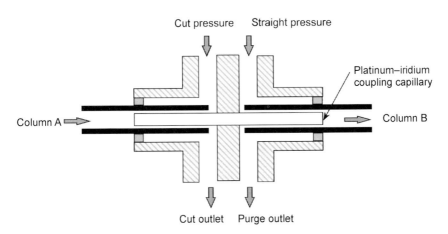

Figure 9.3 Schematic diagram of the Siemens live switching T-piece. (Reproduced with permission of Friedhelm Mueller, Siemens Applied Automation, Karlsruhe, Germany.)

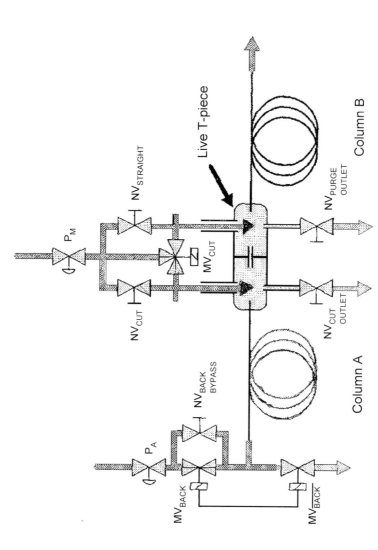

Figure 9.4 Schematic diagram of the Siemens live switching T-piece incorporated in a dual-column manifold. (Reproduced with permission of Friedhelm Mueller, Siemens Applied Automation, Karlsruhe, Germany.)

Different live T-pieces are required for different column diameters as the coupling capillary has to be inserted inside the separating columns. Ideally, the internal diameter of the separating columns should be 20% greater than the diameter of the coupling capillary. A single T-piece can be employed to perform all the common techniques employed in process chromatography. For example, a simple two-column system with a single T-piece between the columns can be operated in cut mode to prevent early- or late-eluting components such as the major peak passing onto the second column. It can then be switched into the straight mode to allow the components of interest to pass onto the second column and then switched back to the cut mode to vent. The T-piece can also be operated in backflush mode on the first column to remove heavier components while the components of interest are eluted from the second column to the main detector.

The combination of capillary columns with live switching offers the following advantages over conventional mechanical valve systems:

- highly inert as the sample only contacts inert material
- minimises diffusion effects
- very low dead volume maintains high resolution achieved by capillary columns
- allows use of capillary columns down to 0.2 mm i.d.
- fast switching permits selection of slices as small as one second in duration with complete transfer to the second column
- backflush, heartcut, and distribution capabilities in a single valve
- constant flow after the first column allows the use of a monitor/intercolumn detector
- can be used for temperature ranges between -100 and $400°C$
- no moving parts therefore minimal maintenance

A comprehensive description of the theory and practice of live switching is given by Meuller *et al.* [8] and is recommended for those desiring a more complete description of the technique.

This technique has been available in Siemens process gas chromatographs for a number of years and has been extremely popular in Germany and mainland Europe. Unfortunately, it has not had much exposure in the United States and the United Kingdom and the application of this technique has been limited in these countries. However, as will be discussed later, the acquisition of Applied Automation Inc. by the Siemens Corporation in 1999 should see a dramatic increase in the application of this extremely powerful technique.

9.4.2   The development of plug-and-play gas chromatography hardware

The high maintenance cost of on-line process gas chromatographs has already been discussed and this has led to many manufacturers looking at how they could improve their instruments performance in this area. The rapid developments in the personal computer world led to the introduction of 'plug-and-play' technology whereby any component of a computer system could simply be plugged in to the appropriate connector and would then automatically be recognised by the host computer and configured accordingly.

This development saved the operators many hours of frustration in trying to get the components to work together. The concept has been adopted by both laboratory and process gas chromatograph manufacturers to aid maintenance. In the traditional process gas chromatograph, a breakdown situation would be investigated by a technician who would examine the instrument to source the cause of the failure. The failed components could then be isolated and repaired or replaced. This could be very time-consuming and often meant the analyser was unavailable for long periods. In the most extreme case of a critical safety measurement, this could mean that the plant was shut down, with significant financial consequences. The Applied Automation Company [9] in the United States (now part of Siemens) adopted the 'plug-and-play' concept for their Advance Maxum™ instrument, which was launched in 1998 and employed totally modular plug-and-play electronics. Each module in the instrument includes a built-in microprocessor to make it a self contained 'smart' device and all the devices are connected with a snap-together serial bus. Detector modules are multifunctional, with a single thermal conductivity detector (TCD) module containing six independent measurement cells and two reference cells, whereas the flame ionisation detector (FID) module also includes two filament-type thermal conductivity detectors. Each detector module is equipped with a high-speed 32-bit microprocessor to provide direct signal processing with up to one thousand points per second digitisation, which allows fast chromatography techniques to be employed. The processor also provides all application hardware control and each unit connects directly to the electronics module through a plug-in feed-through assembly. Therefore, in the event of a failure, the instrument engineer simply installs a complete detector unit that is automatically recognised by the system controller. This significantly simplifies installation and maintenance, and reduces instrument downtime.

The instrument also makes extensive use of electronic pressure control, which allows automatic control of the carrier and other gases throughout the system and can give improved retention stability and also minimise pressure fluctuations during column switching.

This increased flexibility also allows the analyser to be reconfigured easily if the original application does not perform as required or if changes to the chemical process require new analytical methods.

The Advance Maxum instrument has applied state-of-the-art technology to produce a process gas chromatograph that in many ways eclipses even the most modern laboratory gas chromatographs in terms of functionality, flexibility and ease of maintenance.

9.4.3   Parallel chromatography

A previous section described one approach to improving the performance of conventional valve-switched multidimensional chromatography by utilising 'live switching' techniques. This section describes another approach to improvement by employing a technique called parallel chromatography [10] that was developed for application in the Advance Maxum Instrument by Applied Automation Inc. (prior to their acquisition in 1999 by Siemens.)

Conventional column switching applications using rotary or slide valves were complex and employed a large amount of hardware that could be difficult to maintain and set up. A typical traditional configuration for the analysis of a petroleum natural gas type that analyses for nitrogen through to C_{6+} is shown in Figure 9.5. This system employs four rotary valves and five columns to separate 11 components and divert them to a single TCD detector sequentially and produces a typical output chromatogram as shown in Figure 9.6.

In addition to being complex and difficult to maintain, the serial nature of this approach also gives a relatively long cycle time of 8.5 min. Parallel chromatography breaks this complex set of hardware into a set of simple standardised column trains called applets, each of which performs one specific part of the overall process. These applets operate in parallel, thus giving a simultaneous output from a number of detectors, which can significantly reduce the analysis time.

A typical applet configuration is shown in Figure 9.7 and comprises a single 10-port valve that combines the functions of injection and backflush and employs only two micro-packed separation columns. Each applet requires a dedicated carrier gas supply and, depending on the application, may employ one or two dedicated detector systems. The use of parallel chromatography employing four applets for the petroleum natural gas application is illustrated by the chromatograms shown in Figure 9.8, where a reduction in cycle time from 8.5 to 3 min has been achieved.

However, in order to successfully apply this approach, Applied Automation had to make significant improvements to the hardware employed for these systems as indicated below.

9.4.3.1   Switching valves

Traditional switching valves that employ rotors, sliders or diaphragms are one of the highest-maintenance items in a process gas chromatograph and if these were employed in parallel systems the higher maintenance costs would probably outweigh the benefits accrued by faster cycle times. It was therefore necessary to develop an improved valve; the Applied Automation 10-port valve uses air or carrier gas for activation and can be used as a combined vapour injection and backflush valve. It is a diaphragm valve with no sliding or rotating movement that has very low internal dead volume, and the diaphragm actuation movement is measured in tenths of a millimetre and therefore it is capable of carrying out

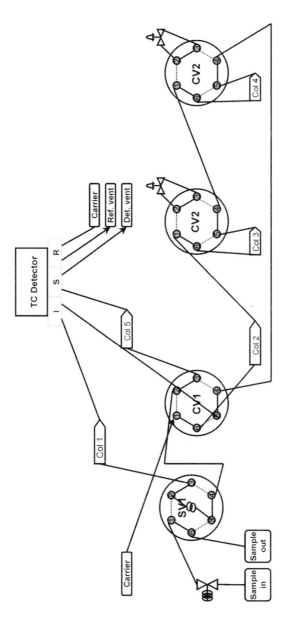

Figure 9.5 Manifold for a typical process gas chromatograph for natural gas analysis employing the reverse column step with double trap and bypass set up. Schematic diagram of the Siemens live switching T-piece. (Reproduced with permission of Bob Farmer, Siemens Applied Automation, Bartlesville, USA.)

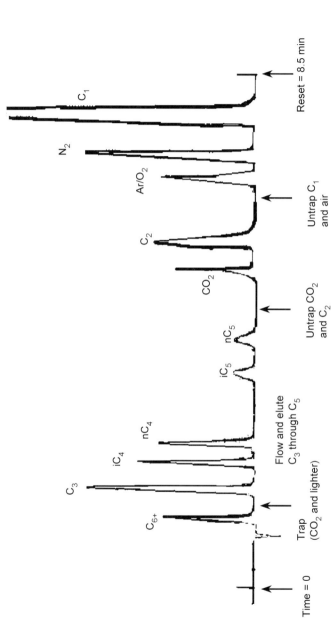

Figure 9.6 Typical chromatogram set for the separation of 11 components in natural gas by conventional sequential analysis utilising injection, backflush, two stop and go valves and one TC detector in an analysis time of 8.5 min. (Reproduced with permission of Bob Farmer, Siemens Applied Automation, Bartlesville, USA.)

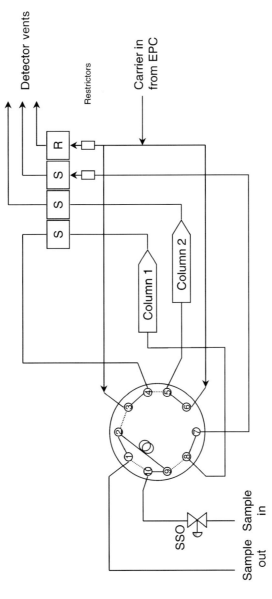

Figure 9.7 Manifold for a typical applet for natural gas analysis. (Reproduced with permission of Bob Farmer, Siemens Applied Automation, Bartlesville, USA.)

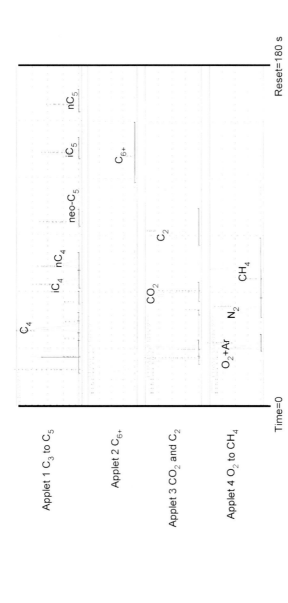

Figure 9.8 Typical chromatogram set for the separation of 12 components in natural gas using four independent applets with a total cycle time of 3 min. (Reproduced with permission of Bob Farmer, Siemens Applied Automation, Bartlesville, USA.)

in excess of 10^7 cycles for a clean gas application. This gives a valve lifetime in excess of 10 years for a cycle time as short as 1 min.

9.4.3.2 Detectors

The multiple application nature of parallel chromatography requires a large number of interim and final detectors in order that each applet can be independent and truly flexible. This required the development of multichannel detector modules and Applied Automation approached this in two ways. For thermal conductivity detection they developed a module that contained two banks of four thermistor bead-based detectors. Each bank contains four matched beads, thus giving three measurement and one reference cell per bank. The flame ionisation detector module also includes a filament-based thermal conductivity detector that can be employed as an intercolumn detector. The Advance Maxum oven can accommodate three independent detector modules and therefore can have up to three flame ionisation–conductivity combinations, or 18 measurement thermal conductivity detectors, or any combination of the two detector modules. This allows up to nine applets to be installed per oven, with simultaneous and continuous monitoring of all column vent, intercolumn separation and final output measurement responses. The detector electronics modules have scan frequency of up to 1 kHz and can therefore be employed for high-speed chromatography.

9.4.3.3 Separation columns

To make maximum use of the low dead volume of the 10-port valve and the separation power of each applet, the application of micro-packed columns was also developed. One-eighth-inch internal diameter packed columns are probably the most widely applied packed columns in process gas chromatograph instruments and a reduction in this diameter to 1/16 inch combined with the optimisation of the packing particle size, stationary phase coating and the reproducibility of the column packing procedure can give improved separation in a shorter time. The component design of the Advance Maxum can allow the use of capillary columns, but the 10-port valve and associated components for parallel chromatography were specifically designed to operate with 1/16-inch micro-packed columns. Furthermore, the improved separation efficiency and column-to-column reproducibility permits the definition of standard separation protocols while the improved signal-to-noise ratio permits lower detection limits despite the lower sample capacity.

9.4.3.4 The chromatography oven

In order to be able to cope with the large range of applications offered by parallel chromatography, the chromatography oven has to be big enough to be able to accommodate all this extra application hardware while remaining easily accessible for maintenance. The Advance Maxum oven can be operated as either a single oven, capable of housing three detector units with a maximum of eight applets, or as a dual-oven version capable of housing two detector modules with two banks of three applets in each temperature zone.

9.4.3.5 *Integrated control and chromatography data handling software*

Laboratory chromatographers have become accustomed over the last twenty years to employing advanced and flexible data collection and processing software that has evolved through the 1970s and early 1980s from basic computing integrators to sophisticated multichannel data systems running on mainframe-style computers. These systems were ideal for large installed bases of laboratory instruments as they provided access to any gas chromatograph data channel from desktop terminals that could be sited in the office away from the laboratory. However, they were very costly to purchase, install and maintain, and if the mainframe computer or the server network went down then all the instruments' data were lost.

However, as with many other areas of technology the advent of personal computer (PC) workstations saw the development of data-handling systems to run on these platforms that could also provide control for gas chromatograph instruments systems. These were typically capable of handling up to four instruments and many of them could be accessed by modem link, allowing analysts to operate them from remote locations.

In facilities with a large number of instruments this resulted in a large number of stand-alone PCs and inflexibility, which caused problems with data storage and maintenance. The development of networked PC-based systems and client server networks has overcome many of these deficiencies and allowed interfacing with the networks that are now commonplace in almost every office and laboratory environment. With these systems it is possible for the analyst in his office to control and examine data from an instrument situated in a laboratory on the other side of the world.

Scientific Software Inc. has been a leading developer of PC-based software under its own EZChrom banner and has also produced customised packages for many other suppliers including instrument manufacturers. The EZChrom Elite package provides a comprehensive data-handling and instrument control capability for a number of laboratory GC instruments from different manufacturers. It has now also been applied in the Advance Maxum and has provided an extremely powerful system whereby any GC on the network can be accessed from any location. In terms of parallel chromatography, the system allows for the simultaneous collection of data from the multidetector arrays used and also for control of the multiple electronic pressure control and valve modules that are required to achieve this simultaneous multi-method technology.

The technique of parallel chromatography is not new but its application in process gas chromatography has been limited by the lack of hardware specifically designed for the job. This has been addressed by the hardware advances that have been incorporated into the Advance Maxum platform, which has provided the option of a true multi-application instrument that can also allow in-built redundancy by duplicating method manifolds. This can be useful both for reliability issues and to employ each manifold as a calibration/performance check.

9.5 The development of gas chromatograph-based transmitter instrumentation

As already discussed, the lifetime cost of a process analytical instrument has many components of which the actual cost of the gas chromatograph instrument can be a minor contribution, with installation, services and maintenance costs often predominating. Most process measurements such as pressure, temperature and flow, employ instruments of the sensor/transmitter type, specifically developed to be low-cost, low-maintenance and robust, which provide a simple output signal that can readily be interfaced with process control systems. The Process Analytic Division of Rosemount™ Analytical Inc. [11] has developed the transmitter concept for process gas chromatography to produce an instrument that is specifically designed with the key value drivers of lower costs of installation, operation and maintenance.

The key components of the GCX™ Gas Chromatograph Transmitter are shown in Figure 9.9. The instrument is modular, with the electronics isolated

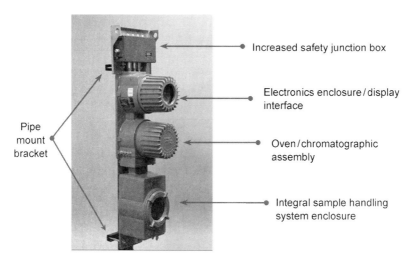

Figure 9.9 The Rosemount Analytical GCX transmitter showing the modular components. (Reproduced with permission of Rosemount Analytical, Inc.)

from the physics of the chromatography by housing each unit in purpose-designed explosion-proof circular enclosures. The modular construction of the GCX™ gives maximum uptime, with most repairs limited to pull out and replacement of the modules. The electronics modules are constructed of 'euro-card' slide rail-mounted circuit boards that can be removed and replaced without the use of tools. The units also contain sophisticated self-diagnostics that allow the service engineer to identify the cause of a problem before arriving on site. Each electronic module is equipped with a viewing window that shows a range of LED status displays giving a local indication of the instruments status.

The gas chromatograph oven module employs a directly heated mandrel with the columns wound around it and supports multiple detectors, valves and column configurations, which allows a wide application range while minimizing the service requirements. The design is also claimed to have lower power consumption and does not require oven purge air, which can lead to a significant reduction in the utilities cost through the life of the analyser. The supply and sample gas connections are mounted on an annular ring with needle valve adjustments accessible by removing the outer cover. An exploded view of a typical GCX transmitter system is shown in Figure 9.10.

Gas pressure and flow control are provided by a patented device called a fluidic transistor or fluistor. These novel thermopneumatic devices provide precise control of the carrier and detector gases while occupying only ~20% of the space of conventional flow controllers. They have an operating pressure range of 0–100 psig (0–700 kPa), a flow range of 0.5–50 SCCM (standard cubic centimetres per minute), an average response time of 500 ms and an accuracy of ±1%.

The oven/chromatography and electronic modules are constructed as complete units and therefore allow easy changing with replacement of a new oven unit possible in ~60 min.

There is also an explosion-proof optional sample system enclosure that can be mounted below the chromatography module. The modules have been designed to be mounted out in the field, often directly onto the pipe carrying the components to be analysed. This can significantly reduce the cost of installation as it minimises the requirements for analyser shelters and extensive sampling systems to transport the sample to the analyser house.

The first production instruments for gas analysis based on TCD detectors were released in early 1998, with the FID instrument following before the end of the same year and a liquid sample injection valve system following in early 1999.

The features and benefits described do have some limitations, in that the systems can currently only deal with a maximum of 20 components and two methods in an instrument. They also have limited data-handling and data-processing capability and are currently primarily based on packed-column technology.

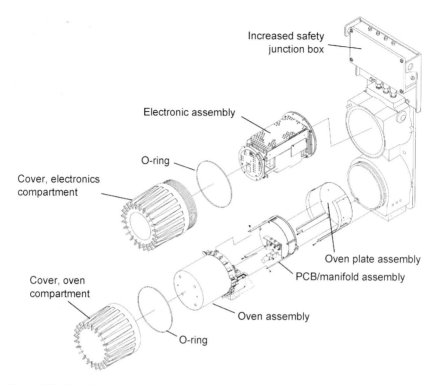

Figure 9.10 Exploded view of a typical GCX transmitter showing the modular construction of the instrument. (Reproduced with permission of Rosemount Analytical, Inc.)

However, many applications can be achieved using the GCX transmitter system and the concept looks set to be a success, particularly for analysis of streams such as natural gas, where it can provide BTU values to either ISO-6976 or AGA-21 with a typical cycle time of 6 min.

9.6 The application of laboratory gas chromatograph instruments as on-line process analysis instruments

There are many process environments in which the intrinsic safety of instruments is not an issue. For example, in processes where there is no risk of fire or explosion or in many pilot plant environments these restrictions might not apply and conventional laboratory-style gas chromatograph instruments have a role to play as on-line analysers. This section considers some of the areas where these instruments have made an impact. The instruments range from conventional off-the-shelf laboratory gas chromatographs equipped with valve-based sampling systems to specialised instrumentation that has evolved from a laboratory instrument into a fully fledged process instrument. There is also a growing interest in locating laboratory gas chromatographs in specialised enclosures to allow them to be employed in hazardous environments.

Almost all of the major laboratory gas chromatograph manufacturers have options for gas or liquid sampling valves that allow their instruments to be plumbed into a process sample stream. These have found widespread use in nonhazardous areas and pilot plants areas. For example, Perkin Elmer [12] has a long-standing cooperative agreement with Arnel to produce customised solutions and specialised valve installations, and Agilent Technologies [13] (formerly Hewlett Packard) and their partner AC Analytical [14] have produced a range of customised analysers for the petroleum industry based on standard methods such as those of the ASTM. However these systems are still more laboratory or at-line based rather than true on-line systems.

Unicam Chromatography [15] has taken this concept further and has effectively moved from being a company that supplied primarily laboratory instrumentation to one that has developed a niche area in process analysis GC systems. The company produce a wide range of analyser systems based on its Pro GC model and each system is guaranteed to perform the specified analysis upon installation. This has been particularly true in the use of the Valco pulsed discharge helium ionisation detector (PDHID) which it has employed to develop special systems for the analysis of trace components in industrial and high-purity gas streams. The PDHID, which is illustrated schematically in Figure 9.11, is a

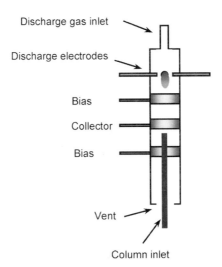

Figure 9.11 Schematic diagram of a pulsed helium ionisation detector. (Reproduced with permission of Martin Smith, Unicam Chromatography, UK.)

nonradioactive detector that uses a low-power pulsed DC discharge in helium as the ionisation source.

The eluent from the column flows counter to the helium from the discharge zone and eluting components are ionised by photons emitted from the discharge zone. The resulting ions are focused towards the collector electrode by two bias electrodes. The detector is essentially nondestructive and has a linear response for organic compounds of over five orders of magnitude. Detector response is universal (except for neon) for components that can be eluted from a gas chromatograph and it is also very sensitive, with detection limits typically in the low picogram range.

However, the detector does have some drawbacks in that it requires very high-purity helium because impurities such as oxygen, nitrogen and water will give a high background response. This is also important when changing supply cylinders, as any air that is introduced into the carrier and detector feed lines will produce a response on the detector and this may take a long time to purge from the system. The detector is very sensitive to any leak in the plumbing system and also the type of components used in the gas stream; Valco recommends avoiding any components other than stainless steel in the gas manifolds and the use of gold-plated ferrules for critical connections on the detector.

To ensure maximum performance and stability of the PDHID detector, the Unicam analyser systems employ high-quality VCR fittings and have a helium-purged valve oven with all valves tested for integrity by mass spectrometry. There is also a high-performance helium clean-up system fitted to each analyser to ensure ultrapure helium for the detector. Application areas for this type of system include the analysis of trace impurities in products from the high-purity gas industry and the semiconductor industry, and in nuclear reactor coolant gases. An example chromatogram for the determination of ppb levels of methane and carbon monoxide in helium is shown in Figure 9.12 and a summary list of applications is given in Table 9.1.

The PDHID is an excellent detector that could have great potential in process analysis systems if the problems associated with helium purity can be resolved. It offers significant benefits in terms of sensitivity and universal application that make it worth development effort.

Unicam has also developed a specialised software product called PRO LINK that provides a seamless connection of the analyser to the process control system. It uses fourth-generation programming tools to provide custom software solutions and operates in conjunction with Microsoft Excel™ to give tabular presentation of results, statistical calculations and process control signals.

An example of this type of system is the application to filling of liquid carbon dioxide cylinders, which has become very topical in recent years in the food and drink industry. The application tests the bulk chemicals as they arrive at the bottling plant prior to bulk storage. If extraneous materials are found, alarms are set off that stop the delivery from taking place. On the cylinder filling line each

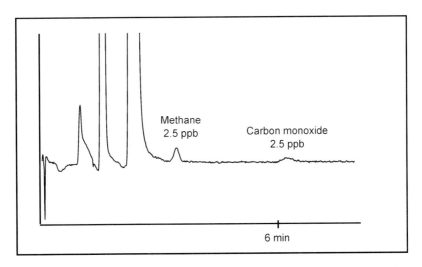

Figure 9.12 Typical Pro-GC chromatogram using a pulsed helium ionisation detector for the determination of ppb level impurities in helium. (Reproduced with permission of Martin Smith, Unicam Chromatography, UK.)

Table 9.1 Summary of trace impurity analysis systems using the Unicam PDHID specialised applications

Pure gas	Trace impurities measured
Hydrogen	$Ar + O_2, N_2, CH_4, CO, CO_2$
Helium	$H_2, Ar + O_2, N_2, CH_4, CO, CO_2$
Oxygen	Ar, N_2
Carbon dioxide	$H_2, Ar + O_2, N_2, CH_4, CO$
Hydrogen chloride/chlorine	$H_2, Ar + O_2, N_2, CH_4, CO, CO_2$
Silane/boron trichloride	$H_2, Ar + O_2, N_2, CH_4, CO, CO_2$
Nitrous oxide	$H_2, Ar + O_2, N_2, CH_4, CO, CO_2$
Nuclear reactor coolant gas	H_2, Ar, N_2, CH_4, CO

cylinder is sampled for analysis and, if it falls outside specification, an alarm is set off and the filling process is stopped. If the cylinder is within specification, a report is generated that acts as the quality control report that is shipped with the cylinder.

Effective as the Unicam systems are, they still have the disadvantage of being based on instruments that were fundamentally designed as laboratory instruments and therefore they cannot be used in hazardous areas. In the United States, Wasson Inc., [16] has sought to rectify this by developing purpose-built enclosures and sampling systems to house Agilent Technologies (formerly Hewlett Packard) instruments. The company designs, configures, and tests turn-key analyser systems specifically to customer requirements. Its process gas chromatograph can handle up to 16 different streams, which can be high-pressure liquids as well as gases. Dual-train injection allows the instrument to simultaneously inject samples from two different streams onto parallel but separate analyser subsystems. A smart instrument calibration system enables single or multiblend calibrations to be carried out automatically for all programmed methods, with inbuilt quality control tests and recalibration when results exceed specified tolerances.

Controlled clean-up procedures and double block and bleed sampling system construction are used to prevent cross-contamination of streams and other sample-related problems. The systems are equipped with wide-ranging communication protocols for communication with Distributed Control System (DCS) and other equipment. The process instruments are contained in a purged cabinet that can be employed in the field without an extra analyser shelter. They are available with heating and air-conditioning options and can meet Class 1, Divisions 1 and 2, and Groups C and D regulations.

These examples show that there is a place for this type of technology in process analysis and for certain niche applications there are clear benefits, but it remains to be seen whether this approach will seriously challenge the conventional process analysis instruments in the future.

9.7   The micro gas chromatograph

The current interest in miniaturisation of analytical systems is not new; initially interest was in producing small compact units that could be employed as portable instruments in the field and that could also be employed in space exploration. In both cases compactness and lightness were key parameters in the development. There are many variants of the portable GC concept available, with some targeted particularly at the environmental monitoring market.

The MTI corporation in the United States developed a micro gas chromatograph based on silicon micromachined components that has been applied very successfully, with over 2500 units installed wordwide by 1998. These instruments were compact and light enough to be employed on the space shuttle.

Major gas chromatograph manufacturers such as Varian Chrompack [17] and Agilent Technologies [18] have micro GCs that are based on the original MTI instrument. The MTI corporation was acquired in 1998 by Agilent technologies (then Hewlett Packard) and the MTI instruments are now being developed and marketed under the Agilent name. These micro GCs are increasingly being employed in process applications for the analysis of gas streams.

In the Agilent range of instruments there are a number of options available to choose from depending on the application. In all systems the chromatography is performed by modules that include injector, column, detector and, if required, a heater. The unheated versions can analyse gas samples containing compounds with boiling points up to 150°C and the heated version up to 220°C. Portable versions are available with an in-built power supplies and carrier gas bottle.

The micromachined injector has been proved reliable for over 2.5 million cycles and can be employed in either manual or automatic mode. It is available either as a fixed-volume model of 4 µl or in a variable-volume model that is user-selectable between 0.5 and 15 µl with the gaseous sample introduced by an internal sample pump. The micromachined thermal conductivity detector has an internal volume of only 240 nl, a linear dynamic range of greater than 10^5, and a precision of $\pm 2\%$ RSD at constant temperature and pressure. Typical detection limits are ~ 10 ppm for many components.

The Agilent instruments are available in a number of configurations with models containing from one to four separate chromatography modules, and they have been used for a wide range of applications including vent gas, water and soil headspace, landfill gas, furnace gas, scrubber bed efficiency, leak detection, environmental monitoring and natural gas analysis.

The Agilent Website has a list of over 45 reference chromatograms and conditions, which are summarised under the following headings:

- permanent gases
- hydrocarbons
- halogenated hydrocarbons
- aromatics, alcohols and ketones
- amines and phosphines
- sulfur compounds

In addition to the general-application instruments, specialised units for refinery gas and natural gas analysis are available.

The miniaturisation of the components not only improves the portability of the system but also allows a significant increase in analysis speed, which is illustrated by the natural gas analysers that allow for an extended natural gas analysis up to C_8 and beyond in a cycle time of less than 100 s. A chromatogram showing a typical natural gas analysis to C_8 is shown in Figure 9.13.

The portable versions can be run from a laptop computer, with results produced in seconds. It is also possible to store hundreds of files and chromatograms for re-analysis at base or to download results automatically to commonly used spreadsheet packages.

At the time of writing, these micro gas chromatographs are not suitable for the analysis of liquid samples, nor are they certified for use in hazardous environments, which currently limits their application for on-line analysis. However, Agilent has recently introduced a heated sample vaporiser that extends the capabilities of the micro gas chromatographs to sample and measure liquefied gases such as liquefied petroleum gas (LPG) streams. This system can easily accommodate the equivalent throughput of several complex and large conventional laboratory gas chromatographs while occupying a quarter of the

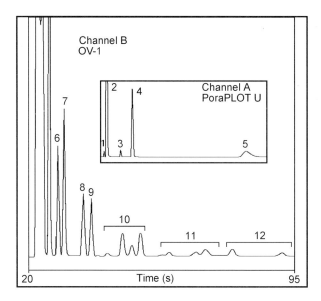

Figure 9.13 Natural gas separation obtained from a microGC with a cycle time of 95 s. Peaks: 1, nitrogen; 2, methane; 3, carbon dioxide; 4, ethane; 5, propane; 6, isobutane; 7, n-butane; 8, isopentane; 9, n-pentane; 10, hexanes; 11, heptanes; 12, octanes. (Reproduced with permission of Agilent Technologies, Inc., Little Falls, USA.)

bench space. The compact design and operational simplicity make the system ideal for use outside the laboratory. The system uses a specially designed heated pressure-reducing regulator to convert LPG stored in sample vessels to a vapour phase for analysis.

This is a step in the right direction and, if similar developments can be made for liquid samples in the near future, the application of this type of technology will certainly increase significantly in the process analysis arena.

9.8   Recent developments in at-line gas chromatography

9.8.1   Why do we need at-line gas chromatography?

At-line gas chromatography was introduced in the BP Amoco plant at Hull owing to the failure of laboratory and on-line gas chromatography to meet fully the needs of the manufacturing plants. Laboratory analysis was based on a schedule of samples that was infrequent and inflexible, with the time from sampling to results often being several hours. On-line instruments overcome this by performing dedicated analysis on a fixed cycle time, which can have many advantages but can also limit the flexibility of the analyser and dramatically reduce the number of streams a single instrument can handle. In addition, the costs of stringent safety features required for operation in process environments, the analyser housing and the sample conditioning system can often significantly exceed the cost of the gas chromatograph instrument, as illustrated in Figure 9.2.

9.8.2 What are at-line gas chromatographs?

On-line gas chromatographs are located on the manufacturing plant and are physically connected to the process they are monitoring, with the samples piped from the process to the instrument. As a result of their location and this physical connection with the process chemicals, they have to be designed to be safe, particularly where there is a risk of explosive atmospheres.

At-line gas chromatographs are also located on the manufacturing plant, usually in the plant control room building, which is a safe area. They are based on highly automated laboratory gas chromatographs but are not plumbed directly into the plant sample streams. Samples are collected in the conventional manner as for laboratory analysis and are then sub-sampled into autosampler vials and loaded into the at-line instrument. Since the instruments are operated by manufacturing process technicians who have no real experience in analytical measurement, the equipment must be easy to use and require minimal operator intervention.

9.8.3 At-line gas chromatography development in BP Amoco

On the BP Amoco Hull site, the first at-line systems were based on Packard gas chromatographs equipped with 'one shot' sampling systems. With these the process operator could present a sample bottle to the instrument and a single analysis could be performed. These instruments were very successful but still had significant drawbacks in terms of limited applicability and the fact that an operator was required to visit the instrument to introduce each sample. The development of reliable gas chromatograph automation and PC-based data systems led to instruments that allowed greater flexibility of analysis, and in the early 1990s two Perkin Elmer Autosystem GCs were successfully implemented as at-line instruments. The key improvement for these systems was the inclusion of the autosampler; this could be loaded with a number of samples and the instrument would run them automatically.

As the gas chromatograph hardware and software developed, the flexibility of the systems increased and around 1995 the current generation of at-line instrumentation was first introduced as a result of collaboration with Chrompack. These are highly automated and are operated by process operators by simple push button control. The key to their success is their flexibility and high degree of automation.

Analysis of the key information needed for process control is used to define the requirements for the configuration of the analyser; after this, consideration is given to the range of other streams that may be analysed for the plant. The aim is to analyse as many of the streams as possible without jeopardising the critical analyses. The systems have to be capable of multistream analysis over wide concentration ranges (from bulk component to trace) and this required the development of automated multiwash and injector cleaning programs to reduce sample carry-over between crude and pure sample streams and maintain data quality.

The key development needed to achieve a successful implementation was a practical user interface that included software and hardware to allow process operators to operate the equipment simply and effectively. This was developed together with Chrompack using their Maestro data system, which was a derivative of Ezchrom.

Different analytical methods are used for different sample streams depending on the location of the sample vial in the colour-coded autosampler tray, with each colour representing a different sample stream. The colour-coded regions have been pre-specified in the software to link the sample stream to the correct method. The operator interface comprises a VDU and a small keyboard. The operator controls the instrument by selecting a single keystroke from the keyboard; each key corresponds to a particular set of samples and analyses. Once the operator has made the selection, the VDU display shows a representation of the autosampler tray with the streams to be run highlighted. Figure 9.14 shows the selection of eight different sample streams to be analysed and also the positions of the relevant clean and calibration check vials required for the run. The operator then places the sample vials containing the streams shown in the colour-coded autosampler tray and the instrument does the rest. The system is therefore very simple for the operators to use, which is critical for successful application.

The development of an at-line gas chromatography system is only one aspect of a successful application since, once the instrument is in place, it has to operate reliably and efficiently and with adequate technical support. The instruments are maintained and serviced regularly by laboratory staff who, in addition, can also access any of the systems remotely to check on performance. The remote access facility can also be made available to the equipment supplier, enabling a faster response in the event of problems that cannot be readily rectified by in-house staff. This remote access facility can be particularly useful in sorting out software problems rapidly and effectively.

9.8.4   The benefits of at-plant gas chromatography

In terms of the day-to-day operation of the process, faster results are obtained by removing the need for transportation of the samples from the point of manufacture to a centralised laboratory, resulting in reduced costs and better process control.

In addition, the frequency and choice of the analysis of specific sample streams may be varied as appropriate, enabling the plant to focus on any particular

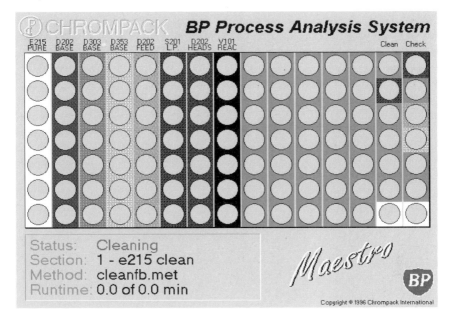

Figure 9.14 Example of the BP/Chrompack at-line GC user interface showing a schematic representation of the autosampler tray with expected sample streams highlighted. (Reproduced with permission of Kevin Lythe, BP Chemicals, Hull, UK.)

part of the process. In some plants a ~ 4-fold increase in the number of samples run on the at-plant gas chromatograph has been observed compared to when only the laboratory was employed. This is particularly beneficial during start-up and troubleshooting, when individual process components (e.g. a distillation column) can be brought up to specification or throughput can be optimised in a much shorter time than was possible when relying on laboratory analysis. This feature has been successfully employed after maintenance shutdowns, when the cost benefit has justified the at-line installation almost immediately. Because the configuration of the at-line analyser is extremely flexible, the instrument can be rapidly reconfigured to analyse different sample streams or even utilised on a different process if future requirements change. This can be particularly useful for batch processes where a plant may be used to produce many different products. In this case, the at-line gas chromatograph, based on a laboratory gas chromatograph, can easily be converted to suit the process requirements. All these features have led to extremely fast payback times in justifying the capital expenditure on the systems.

9.8.5 The current status of at-line gas chromatography

Although laboratory gas chromatography can be considered a mature technique, it has been given a new impetus in recent years by significant developments in software and hardware control, particularly in the field of at-line process analysis. Developments in software user interfaces, automation, miniaturisation and networking capabilities have led to the introduction of new instruments specifically developed and marketed for the on-line and at-line market.

This is illustrated by the change in the views of the instrument manufacturers in recent years. The major instrument manufacturers have now realised that this is an area of high interest for industry and have developed new products to meet the growing requirements of the market. Recently, Unicam described new software for at-plant gas chromatograph analysers and Hewlett Packard (now Agilent Technologies) launched a new gas chromatograph specifically aimed at the at-line market. This instrument evolved from market surveys and discussions with major manufacturers based on a 'gas chromatograph use model' that defined main use areas as R&D, production, and distribution. The production area was further categorised into QA and QC laboratories and at-line and on-line application.

Analysis of this survey revealed that the production sector had the largest requirement for gas chromatograph instruments, as illustrated in Figure 9.15, which when combined with the relative projected growth in instrument requirements (Figure 9.16) led to a re-evaluation of the strategy.

In 1997 Hewlett Packard saw its flagship 6890 instrument as meeting the needs of all these sectors. However, as a result of these studies, it was clear that the prime application sector for the 6890 was as an R&D instrument and that the other application sectors required instruments specifically designed for their needs. In March 1998 the company acquired the MTI micro gas chromatograph, which had applications particularly in the distribution/field portable sector, but even this in combination with the HP6890 could not fully meet the requirements

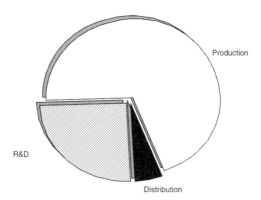

Figure 9.15 GC instruments employed by use sector in the chemical/petrochemical/petroleum industries. (Reproduced with permission of Douglas Jansen, Agilent Technologies, Inc., Little Falls, USA.)

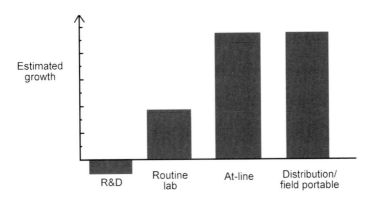

Figure 9.16 Relative estimated GC market growth by use sector in the chemical/petroleum industries. (Reproduced with permission of Douglas Jansen, Agilent Technologies, Inc., Little Falls, USA.)

of the production sector. This led to the development of the HP 6850 gas chromatograph specifically to meet the needs of the QA/QC laboratory and at-line portions of the production sector. This is a major step that will dramatically raise the profile of at-line gas chromatography analysis and should lead to many more companies implementing the technology.

Chrompack has now merged with Varian [19] and the new merged company is still actively pursuing the development of at-line GC analysers based on the systems already developed, and the other major instrument manufacturers are also keen on developing the at-line capabilities of their instruments.

9.9   Future developments in process gas chromatography

9.9.1   What will drive new developments?

The future development of process gas chromatography will be driven by a number of factors including the continued desire to reduce costs and improve accuracy, precision and cycle time. However, the encroachment of other measurement technologies such as spectroscopy into areas that were traditionally the domain of gas chromatography will also have a major impact.

Spectroscopic techniques such as near infrared and Raman spectroscopy are becoming more and more widely employed in process analysis. They have the advantage of very short cycle times in that the measurements are nonintrusive and do not require extraction of a sample from the process. The use of fibre optics means that the sample cell can be placed out in the process but the spectrometer and computer data handling system can be housed in the control room or a safe area, and the use of multiplexers has meant that a single spectrometer can service many measurement cells. This can mean significant cost savings in terms of installation and maintenance over gas chromatography. For bulk component analysis, spectroscopic techniques such as near infrared will become the techniques of choice. They have very high precision, which is important in process control applications where the trends in concentration are often more important than the accuracy of the measurement. Mass spectroscopy may also pose a significant threat to process gas chromatography if it can be made more reliable and cost-effective and capable of handling multicomponent liquid sample streams.

However, there is a new area of interest in on-line analysis in which it is required to measure product quality and provide trace analysis and where it is the accuracy of the analysis that is important. This is required to allow direct export of product without the necessity for intermediate storage in tanks. This is not currently an area of strength for spectroscopic techniques such as near infrared, which often rely on gas chromatography of complex mixtures in order to build their calibration models. The requirement for trace analysis will open a new window of opportunity for the process gas chromatography industry but will require the development of existing equipment in order to fully meet the needs of the customers. Accuracy instead of precision will be a major requirement and more sophisticated calibration routines will be required to produce the necessary confidence in the results to allow them to replace certificates of analysis as currently produced by central laboratories.

The requirement to reduce the total cost of ownership will be a major driver in the development process, with the reduction in maintenance costs and utility usage being key areas where significant impact could be made. Cycle times will also become a major area of focus as more sophisticated plant control systems require faster, more frequent data to keep the process plants operating at peak performance.

9.9.2   The future for on-line gas chromatography

This chapter has discussed some of the developments have occurred in recent years in on-line gas chromatography and the future will see further development along the same lines, with instrument companies following their chosen strategies and improving and refining their existing technologies.

The drive for cost reduction will continue, with maintenance and utility usage being key areas for attention. The development of plug-and-play and transmitter-type instruments will continue with the implementation of new and improved technology in the respective platforms. The wider implementation of electronic pressure and flow control for all gas streams will improve the chromatographic stability of on-line systems and also allow the chromatography parameters to be checked and set remotely over the network.

Development of current models of micro gas chromatographs into systems suitable for use in hazardous environments is likely in the near future as is the development of liquid injection systems, and this could have a major impact on their application in on-line process analysis regimes.

However, there will also be a coming together of some of these developments to generate new, more sophisticated instruments that could result in step changes in performance. Previous sections have discussed the development of live switching by Siemens for their process gas chromatographs and the development of the plug-and-play Advance Maxum platform by Applied Automation. These developments occurred prior to the merging of these two companies and at their first combined user meeting [20] they announced the intention to combine the best technologies from both into a new instrument to be available by the end of 2001. This is an exciting development, as much of the best technology available from both companies is complementary. For example, the multidetector and advanced data-handling capability of the Maxum and the live switching technology from the Siemens models will allow the development of parallel chromatography with fast narrow-bore capillary columns. They have also developed the live switching concept into a method of live sampling in which a liquid sample is vaporised and a measured sample is injected using developments of their valveless switching technology. This also opens the possibility of time-based variable-volume sample injection, which could be employed for streams of widely varying concentration ranges, such as the top and tail of a distillation column, or as a means of creating a calibration by variable-volume injection of a standard. Siemens has also presented [21] the results of recent research employing microfabrication techniques to produce chromatography components in silicon. These include detectors and live switching valves that have been designed specifically for use with narrow-bore capillary columns with process analysis in mind. Such systems will be equally at home in a larger multiapplication conventional process gas chromatograph or in a transmitter-type package, and if they can be mass produced cheaply enough they may even become disposable items for which maintenance means replacing a complete chromatography module.

The development of advanced software for monitoring and control will be required to allow self-diagnostics to implement predictive maintenance regimes. In these systems the operation of all components will be continually monitored and any deviation from normal operation flagged as a warning. This will be employed to ensure developing faults are identified, which will allow them to be rectified in routine maintenance before they cause system failure and thereby ensure maximum instrument uptime. The chromatographic performance of the system could also be monitored, for example peak resolution, widths and theoretical plate calculations, and warnings posted if the chromatography deteriorated. One certain thing is that the demand for on-line analysis is increasing, but if gas chromatography is to continue as a key process analytical technique in the petroleum and bulk chemical industry it will have to develop to become more reliable, accurate, precise and cost effective.

9.9.3   The future for at-line gas chromatography

The development of at-line gas chromatography to date has been possible mainly as a result of advances in instrument automation and control and also the increased computing power offered by the PC. However, to date these developments have resulted primarily from features introduced for laboratory-based gas chromatography instruments, with the introduction of electronic pressure control and advanced instrument control combined with data handling through Windows ™-based PC software packages as prime examples. This has now led to the introduction of commercial instruments aimed directly at this market, and as more industries begin to apply this technology the customer demands for new features specifically for at-line applications will increase.

Users will now be looking for other features that have particular benefits for their operations, which will inevitably lead to future developments specifically for at-line analysers. Cycle times will become more important, as will ease of maintenance. Plug-and-play instrumentation will be required, with quick release instrument panels to facilitate repair and minimise downtime.

The next generation of instruments will employ fast gas chromatography with improved control and automation and totally transferable methods. The provision of gas supplies also requires consideration. Most at-line sites will have a source of compressed air, but the provision of hydrogen may require hydrogen generators for carrier and fuel gas supply as an integral part of the gas chromatograph instrument. Whatever technology is employed, it must be easy to use and extremely reliable and require minimal maintenance. To achieve this, new modular hardware and software with plug-and-play capabilities will be required together with full self-diagnostics that will detect the onset of problems before they cause instrument failure.

To date, laboratory gas chromatographs have been employed primarily for at-line systems and have operated in a serial mode, with samples analysed in a sequence or at best in pairs on a dual-channel chromatograph. However, there is increasing interest in investigating the potential for applying more conventional process chromatographs in an at-line mode. For example, the Applied Maxum instrument with its multidetectors and parallel chromatography capability could be developed to perform multistream analysis simultaneously. The only requirement to realise this would be a suitable sampling system. The micro gas chromatograph technology discussed earlier could also have significant benefit for at-line applications, particularly where space is limited. It also has potential for parallel operation as even current models can contain up to four discrete chromatography modules in a briefcase-sized box. However, as with on-line application, this will require the development of suitable liquid sampling systems.

If the at-line market develops as predicted in the Agilent market surveys discussed earlier, these developments could be seen sooner rather than later.

Acknowledgements

The author acknowledges the contributions of all his BP Amoco colleagues and also the former Chrompack UK and International who were leading figures in the development and implementation of at-line gas chromatograph at the Hull site. He also acknowledges the discussions held with many of the instrument companies, especially Siemens Applied Automation, Fisher-Rosemount, Agilent Technologies and Unicam, who have been particularly helpful through their contributions and many discussions.

References

1. E.R. Adlard (ed.), *Chromatography in the Petroleum Industry*, Journal of Chromatography Library **56**, Elsevier, 1995, ISBN 0-444-89776-3.
2. D.H. Desty *et al.*, *Anal Chem.*, **1960**, *32*, 302.
3. The Analytical Market, *Annual Review of the Government Chemist*, LGC, Queens Road Teddington, Middlesex TW11 0LY, UK, 1998. www.lgc.co.uk
4. R. Annino and R. Villalobos, *Process Gas Chromatography, Fundamentals and Application*, Instrumental Society of America, 1992, ISBN 1-55617-272-9.
5. R.F. Wall, W.J. Baker, T.L. Zinn and J.F. Combs, *Ann. NY Acad. Sci*, **1959**, *72*, 379.
6. D.R. Deans, *J. Chromatogr.*, **1968**, *1*, 194.
7. Deutsche Patentschrift DE-PS 2955-387,07.12.1976.
8. H. Mahler, T. Maurer and F. Meuller, in *Chromatography in the Petroleum Industry* (ed. E. Adlard), Journal of Chromatography Library **56**, Elsevier, 1995, ISBN 0-444-89776-3.
9. www.aai-us.com
10. R.K. Bade, R.W. Dahlgren, S.B. Hunt and S.H. Trimble, in *Proc. Annual ISA Analytical Division Symposium*, **1999**, *32*, 85.
11. www.processanalytic.com
12. www.perkinelmer.com
13. www.agilent.com
14. www.analytical-controls.com
15. www.unicam.co.uk
16. www.wasson-ece.com
17. www.chrompack.com
18. www.chem.agilent.com
19. www.varianinc.com
20. Siemens Applied Automation European Users Meeting, Karlsruhe, Germany, 10–11 May 2000.
21. F. Mueller and U. Gellert, Process Gas Chromatography 2000+: Vision, Technology, Conception, *Book of Abstracts—Sixth Int. Symp. Hyphenated Techniques in Chromatography and Hyphenated Chromatographic Analyzers*, ISBN 90-74870-04-X.

Index

accelerated solvent extraction (ASE) 19, 23
accuracy and precision 200
activated carbon sorbents 3
activity coefficient 2, 250
adsorptive extraction of samples 3
atomic emission detector (AED) 175-181
    compound independent response 178
    isotope measurement 178
    sensitivity and selectivity 177
    simultaneous element detection 177
air analysis 111-112
aldehydes and ketones in air 5
alkyl mercury compounds with AED 180
ambient temperature, effect on GCs 123
amphetamines 119, 121
analytical reference materials 198
aqueous samples 8
Arochlor™ 1260
    by multidimensional GC 274-276
    in foodstuffs 47
arylene phases 94
ASTM Method 5134 125
ASTM Method 594 128
ASTM Method D-2887 96-97
at-line GC, recent developments of 324-329
atomic emission detectors (AED) 175-181
at-plant GC, benefits of 326-327

back flushing 218, 265
Ballschmiter PCBs 16
Ballschmitter-Zell numbering system for
    PCBs, 105
band broadening, extra-column 229
barbiturates 120-121
    by GC-FTIR 184
barometric pressure, effect on GCs 123
base peak in MS spectra 147
Belgian beers, headspace sampling of 39
Belgian dioxin crisis 41
boiling range of petroleum samples 96

calibration of methods 205
capillary column
    efficiency 263

materials 92
carrier gas for fast GC 210
chemical ionisation mass spectrometry
    (CI) 148-152
chemical warfare agents
    by GC-MS 173
    with AED 180
chiral compounds 118-119
chlorinated insecticides with AED 180
chlorinated pesticides 109
Chromosorb™ 3
Class 1, Divisions 1 & 2 instruments 322
classification of sample types 219-227
close-loop stripping 9
column dimensions 93
column efficiency 93
columns for plant analysis 314
comprehensive two-dimensional GC (GC ×
    GC) 208, 218, 249-256, 278-294
concentration range 202
cool on-column injection 27, 74-75
coupled column methods 264-268
coupled GC methods, summary of 267
cyanopropyl phases 94

data collection rate for FID 129
data handling software for plant analysis 315-
    316
DB-WAX™ 95
Deans switch 260, 268, 304-305
derivatisation 31
design qualification 194-196
detector time constant for fast GC 229-230
detectors for plant analysis 314
diazepam and chlorpromazine with AED 180
diesel fuel, fast GC of 234
dioxins in sewage effluent by GC-MS 173
direct injection 72
directly coupled columns 266
distributive control systems (DCS) 299
dynamic range of FID 129
dynamic sorptive extraction 7

electron capture detector (ECD) 122

electron impact ionisation (EI) for mass spectrometry 146-148
electronic pneumatic controls 122
electrospray ionization 7
Empore discs™ 10
EPA Method 502.2 97, 103
EPA Method 624 97, 103
EPA Method TO-14 111
EPA priority pollutants, list of 101-102
epichlorohydrin in air 7
equipment calibration 192
equipment qualification 193-194
essential oils
    by GC × GC 288
    by MDGC 278
ethion and chlorpyriphos 29
ethylene oxide in air 5
EZ-Flash™ system for rapid programming 240

fast GC
    with TCD 233
    practical consequences of 227-231
fast GC-MS 230, 236
fast temperature programming 208, 211, 240-242
fatty acid analysis 113, , 115-117
FFAP phase 95
field ionisation 152
flame ionization detector (FID) 122, 125-131
flame photometric detector (FPD) 122, 132-134
flow rate for FID 129
flow/pressure programming 217
fluorocarbons in plasma with AED 180
freeze concentration of aqueous samples 8
Freons™ 103

gas and liquid sampling valves 83-85
gas chromatography-mass spectrometry (GC-MS) 140-174
gas purifiers for detectors 135
gasoline analysis 95
comprehensive two dimensional GC (GC × GC) 278-294
    applications of 292
    benefits of 281
    cryogenic modulator 284-285
    data presentation 287-288
    diaphragm valve modulator 284
    MS 293-294
    orthogonality of separation 285-287

peak modulation methods 283
    thermal modulator 284
GC-Fourier transform IR (GC-FTIR) 181-186, 217
    by direct deposition 182
    detection limits 184
GC-MS-MS 161, 168-170
glass and quartz wool in injector systems 61-62
groups C and D regulations 322
guard columns 109
guidelines for retention time reduction 212-213
gum phase extraction (GPE) 7, 12

halogenated hydrocarbons, fast GC of 232
headspace sampling 2, 85-90
headspace sorptive extraction (HSSE) 5, 20-21
heartcutting 265
heating oil by GC × GC 288, 291
helium ionization detector (HID) 122
hyperthermal surface ionisation 152-154

injection band width 229
injection discrimination 60
injector parameters 80
inlet liners 59, 69, 73
installation qualification 196
internal and external standardisation 204
ion trap mass spectrometer 155-163
ionisation methods for GC-MS 145-154
isotope effects on FID response 125

jet separator for GC-MS 143-144
juvenile hormone by GC-MS 174

Kuderna-Danish evaporator 31

laboratory GCs for plant analysis 318-322
large volume injection 2
lavender oil by GC × GC 288-289
LC-GC 27
lemon oil by MDGC 277
light hydrocarbon gases 97
light pipe for GC-IR 182
limit of detection (LOD) 201, 228
limit of quantification (LOQ) 201-202
linearity of FID 130
linearity 202
live switching techniques 304-307
lorazepam in plasma by GC-MS-MS 174

magnetic sector mass spectrometers 165-167
mass resolution, definition of 154
mass spectrometers
    for plant analysis 330
    types of 154-167
matrix isolation GC-FTIR 182
method documentation 197
method transfer criteria 194
method validation 197-205
methyl citric acid by MDGC 278
micro GC for plant analysis 322-324
micro liquid-liquid extraction (LLE) 10
microwave plasma 175
microwave-assisted solvent
        extraction(MASE) 19, 23-24
milk, cheese and yoghurt by SBSE 14
minimum detection limit (MLD) for the FID
        128
minimum plate height equation 263
monoterpene enantiomers by MDGC 278
multicapillary columns 207, 237-240
multidimensional analysis 261-297
multidimensional GC (MDGC) 261
    applications of 272-278
    instrument 265
multi-hyphenated techniques, miscellaneous
        applications 186-188
muramic acid in bacteria by GC-MS-MS 174

narrow bore columns 207, 210
near IR and Raman spectroscopy for plant
        analysis 329-330
negative chemical ionisation 150
nitroaromatics in wine 65
nitrogen compounds in diesel with AED 181
nitrogen-phosphorus detector (NPD) 29, 122,
        131-132
nitro-PAHs in air particulates 35

on-line and at-line GC 298-333
on-line GC, cost analysis of 302-303
open split interface for GC-MS 144-145
operational qualification 196
Optic™ PTV injector 79
orange oil 29
organometallics in petroleum samples 181
organotin and lead compounds with AED 180
ovens for plant analysis 314-315

packed column inlets 70
packed vs capillary columns for fast GC 211
PAHs in soil 24

paraffins naphthenes and aromatics 272
parallel GC for plant analysis 309
particulate collection 3
patulin in apple juice by GC-MS 173
PCB analysis 43, 105-108, 151, 274
    by GC-MS 173
    in animal feed 44
    in chicken and pork fat 44
    in cod liver oil 46
    in human sperm 14
    in mackerel oil 46
    in sediments 24
peak deconvolution 218, 249
pesticide and herbicide residues 105-108
pesticides
    in water by GC-MS 172
    in wine 14-16
phenols
    in urine 32
    in water 31
phenylmethyl phases 94
phthalates in indoor air 5
PLOT columns 97
plug-and-play hardware 307-308
polyethylene glycol phases 95
polyimide coating 92
polysiloxanes 93
porous polymer sorbents 3
positive chemical ionisation 7, 149
process analysis, requirements of 299
process GC
    future developments 329-332
    recent developments 301-316
programmed temp. vaporization injection
        (PVT) 27-78
pulsed discharge helium ionisation
        detector 319-321
pulsed flame photometric detector 135
purge and trap sampling (P & T) 9
pyrolysis GC 17

quadrupole mass spectrometer 155-156
quenching of FPD signal 132

racemic amino acids 33
rapid GC with short wide bore columns 243
reduced parameters at optimum
        conditions 209
reflectron in TOF MS 164
relative response of FID for heteroatoms 125
repeatability and reproducibility 201
retention gap 77

robotic sampling systems 37
robustness 202

salting-out 2
sampling 1
sample
  backflash 64
  capacity 211, 228
  capacity *vs* film thickness 244
  enrichment methods 2
  fractionation 26
  preparation for capillary GC 1
  size 63
  storage 1
selected ion monitoring 167-168
selected ion storage in an ion trap MS 162-163
selective ion monitoring 7
selectivity tuning 266
D- and L-selenomethionine 35
septa 53
septum bleed 54-55
speeding up separations 213-219
solid phase extraction (SPE) 10
solid phase microextraction (SPME) 5, 12, 20
solid samples 17
  enrichment methods for 18
solvent evaporation 30
solvent focusing 67
sonication extraction 19
sorptive extraction 12
soxhlet extraction 19
Soxtec™ 19
specificity and selectivity of GC
    methods 198-199
split flow 57
split injection 57
split/splitless injection 53, 57
splitless injection 64
  purge time 66
stability of samples 203-205
stable isotope standards for GC-MS 170-171
static sorptive extraction 5
stationary phases 93
stationary phase focusing 67
stir bar sorptive extraction (SBSE) 12-13
substituted anilines 111, 114
substituted polycyclic aromatics by GC-
    FTIR 184
sulfur chemiluminescence detector 134
supercritical fluid extraction (SFE) 19, 25
switching valves for plant analysis 309-314

tea tree oil by GC × GC 289
Tenax™ 3
thermal conductivity detector (TCD) 122, 137
thermal desorption 3, 19
thermal focusing 67
time-of-flight (TOF) mass spectrometers 164-165, 231
tin compounds in environmental samples 24
tomato and strawberry volatiles by GC-TOF
    MS 174
toxaphenes by MDGC 274
trace analysis 228
trace concentration with two columns 265
transmitter instrumentation for plant
    analysis 316-318
triazine herbicides by GC-MS 173
triazines, extraction from water 12
Turbo-Vap concentrator™ 31
Twister™ 14

UKAS 196
ultrasonic extraction (UE) 23

vacuum outlet GC 207, 211, 217, 242-244
validation parameters 197
van Deemter curves 91
vetiver oil by GC × GC 290
volatile compounds 2

yoghurt, flavour compounds in 14